T0367750

Dynamic
Systems and Control
with Applications

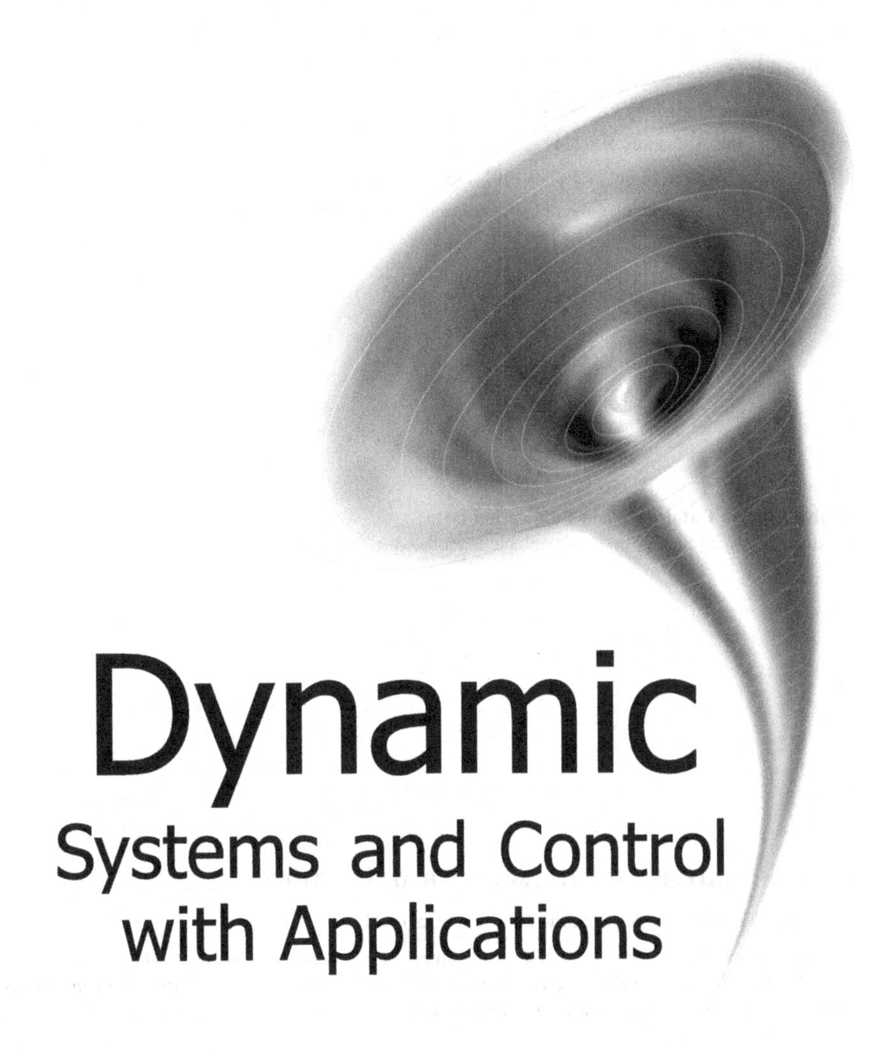

Dynamic Systems and Control with Applications

N. U. Ahmed
University of Ottawa, Canada

NEW JERSEY • LONDON • SINGAPORE • BEIJING • SHANGHAI • HONG KONG • TAIPEI • CHENNAI

Published by

World Scientific Publishing Co. Pte. Ltd.
5 Toh Tuck Link, Singapore 596224
USA office: 27 Warren Street, Suite 401-402, Hackensack, NJ 07601
UK office: 57 Shelton Street, Covent Garden, London WC2H 9HE

British Library Cataloguing-in-Publication Data
A catalogue record for this book is available from the British Library.

DYNAMIC SYSTEMS AND CONTROL WITH APPLICATIONS

ISBN-13 978-981-270-053-7
ISBN-10 981-270-053-6

Printed in Singapore

In memory of my Mother, Father,
Uncles and my Wife Feroza who gave so much.

Dedicated to my Sons Jordan, Schockley,
Daughters Pamela, Rebeka, Mona, Lisa and my Grandchildren,
Reynah-Sofia, Eliza, Pearl,
Kira, Jazzmine, Austin.

Contents

Preface

Over the last sixty years control theory has gone through enormous development beyond the recognition of the old masters of the subject. The phenomenal development of system and control sciences is due to the interaction of scientists, mathematicians, engineers, economists and experts from industry and military. The subject is one of the most remarkable confluence of applied and pure mathematics and the very real world applications. It is amazing that some of the scientific and engineering problems would require the very abstract mathematics that used to be considered beyond the reach of applied scientists and engineers.

There are many prominent areas of systems and control theory that include systems governed by linear and nonlinear ordinary differential equations, systems governed by finite dimensional stochastic differential equations, systems governed by partial differential equations including their stochastic counter parts and, above all, systems governed by abstract differential and functional differential equations and inclusions on Banach (even metric and topological) spaces including their stochastic counterparts. This remarkable advance of the field is due to the unprecedented interest and contribution of pure and applied mathematicians and engineering sciences. There is no question that this interaction would continue simply because there are many unsolved challenging problems and emerging new ones. Such problems are of interest to mathematicians, scientists and engineers.

In recent years significant applications of systems and control theory have been witnessed in diverse areas such as physical sciences, social sciences, engineering, management, finance etc. In particular the most interesting applications have taken place in areas such as aerospace (civilian, military), buildings and space structures, suspension bridges, artificial heart, chemotherapy, power system, hydrodynamics, plasma, magneto hydrodynamics, computer communication network, flow control of internet traffic, optical network. Importance of applications whereby theory is tested and new theories are developed is clearly recognized.

The subject is so vast in scope and applications, it is impossible to cover all these in one single book. There are several dozens of books written for engineers and mathematicians on systems governed by ordinary differential equations, half a dozen books on systems governed by partial differential equations, and similar number of books on stochastic differential equations, and a number of books on systems governed by deterministic and stochastic abstract differential equations on Banach spaces.

The objective of this book is to present a small segment of theory and applications of systems and control governed by ordinary differential equations and inclusions. It is expected that any reader who has absorbed the materials presented here would have no difficulty to reach the core of current research directions.

The book includes 11 chapters and an appendix. The first chapter is mainly for those who have very little mathematical background. The second and the third chapter deal with linear and nonlinear systems respectively, covering questions of existence, uniqueness and regularity properties of solutions. In addition to classical models, these chapters also include some aspects of impulsive systems. Chapter 4 is elementary; it introduces the reader to basic Lyapunov stability theory for ordinary and impulsive systems as they are required for later chapters. Chapters 5 and 6 cover observability, identification, controllability and stabilizability of systems governed by ordinary linear and nonlinear differential equations subject to regular (measurable) as well as impulsive inputs. Chapter 7 presents calculus of variations as the precursor to optimal control theory. Here both existence of solutions and necessary conditions of extremality for variational problems are presented. Chapter 8 deals with optimal control theory. Both existence of optimal controls and necessary conditions of optimality are presented in details covering some elements of uncertain systems. For non convex control problems relaxed controls are used to develop necessary conditions of optimality. From these general results Pontryagin's minimum principle follows as a special case. Chapter 9, presents linear quadratic regulator theory with and without control constraints covering some aspects of disturbance rejection. Chapter 10 deals with time optimal control both for linear and nonlinear systems including impulsive systems. Chapter 11 deals with stochastic systems driven by Brownian and fractional Brownian motions including counting processes. Stochastic dynamic models recently developed for fluid approximation of network traffic are presented and their ramifications discussed. The appendix contains Lebesgue integration and some prominent results from measure theory and abstract functional analysis often used in the text.

Target Audience: The book is targeted to first and second year graduate students of Engineering, Mathematics, Management, Finance and Researchers in the field of systems theory, control, optimization and their applications. Selected parts of the book have been taught to first year graduate students of Systems Science and a select group of senior undergraduate students of engineering at the university of Ottawa for over last 25 years. Students who have taken this course were better prepared for graduate studies and research.

Reading Guide: Materials presented in Chapters 1-6 can be taught at the graduate levels of engineering under the title "Linear and Nonlinear Systems" without requiring any special prerequisite. The materials presented in Chapter 1 and some of the materials from the appendix are sufficient. If required, Chapter 3 on nonlinear systems can be skipped and recalled only when needed in the study of stability, observability and controllability in Chapters 4, 5 and 6. Materials presented in Chapters 1, 2, 4 and parts of Chapter 9 can be chosen for a one semester course on "Linear Systems and Optimal Control". Materials presented in Chapters 7, 8, 10, 11 can be taught at the second year graduate level. A good prerequisite would be real analysis provided in Chapter 1, and the materials provided in the appendix. This course should lead the students to the current trends of research in the fascinating field of modern systems and optimal control theory. For Chapter 11, some knowledge of stochastic processes including Brownian motion and Poisson process is required.

The author would like to thank Professor K.L.Teo of Curtin University, Western Australia and Professor C.Charalambos of the University of Cyprus, for their encouragement throughout the period of writing this book. I would also like to thank my student Mr Cheng Li for his remarkable help during the preparation of the manuscript.

The author hopes that this book will inspire young mathematicians and engineers to discover the beauty of mathematics blended with and enriched by applications.

Chapter 1

Basic Mathematical Background

1.1 Introduction

In the study of systems, linear or nonlinear, mathematics plays a fundamental role. Without it, it is impossible to understand the complex behavior of any natural or artificial (man made) systems that we encounter as human species. Mathematics is perceived as a difficult subject, only meant for people born with special talents. The author believes that human mind is naturally logical and therefore once the underlying logic is captured or imparted by inspiring teachers, the mind can not stop. It continues to grow like wild fire and monuments of apparently complex thoughts are produced which are useful for the survival of mankind as intelligent beings.

Here we present only very basic elements of mathematics that we need to understand systems that can be modeled by linear or nonlinear ordinary differential equations and inclusions. When we watch any natural or manmade phenomenon as they evolve with time, we deal with mathematical models such as differential equations and we try to see and quantify the evolution of the process with time. To fully describe the state of the process at any point of time one may need to quantify a multiplicity of variables which we may call a vector. For example, if one wants to describe the weather at any moment of time, one must give the current temperature, barometric pressure, wind velocity, moisture, brightness etc. Clearly these quantities vary with time and this variation is governed by some suitable system of differential equations (stochastic possibly). This makes us interested in vectors and hence vector spaces.

1.2 Vector Space

We present here the most important axioms of a vector space. Let X denote an arbitrary set and let $F \equiv R/C$ denote the field of real or complex numbers.

The set X is said to be a vector space over the field F if it satisfies the following two basic properties:

$$(i): \text{if } x \in X \ \text{ and } \alpha \in F, \text{ then } \alpha x \in X \tag{1.2.1}$$

$$(ii): \text{if } x, y \in X \text{ then } \ x + y \in X. \tag{1.2.2}$$

For example,

$$X \equiv R^n \equiv \left\{ x = \begin{pmatrix} x_1 \\ \vdots \\ x_n \end{pmatrix} : x_i \in R, i = 1, 2, \cdots, n \right\}$$

is the set of all ordered n-tuples called a real vector space or a vector space of dimension n over the field of real numbers R. A simple example is the daily production of a farm that produces n distinct goods (products) which may be quantified by a vector $x \in R^n_+ \subset R^n$. This may vary from day to day. Here R^n_+ denotes the positive orthant of the real Euclidean space R^n. The total production of the farm over any number of days is given by the sum of vectors produced each day during the given period. Note that R^n_+ is not a linear vector space but it is a convex cone with vertex at the origin. Similarly if one wishes to describe the currents in m branches of a circuit and the corresponding voltages across the branches, one may use R^n with $n = 2m$.

Let $\mathcal{P}_n$ denote the set of all polynomials of degree equal or less than n. Clearly, for $p \in \mathcal{P}_n$ and $\alpha \in R$, we have $\alpha p \in \mathcal{P}_n$. For $p, q \in \mathcal{P}_n$, we have $p + q \in \mathcal{P}_n$. Hence $\mathcal{P}_n$ is a vector space. Let $M(n \times m)$ denote the class of all matrices having n rows and m columns. Clearly, for $A \in M(n \times m)$ and $\alpha \in R$, we have $\alpha A \in M(n \times m)$ and for $A, B \in M(n \times m)$, and $\alpha, \beta \in R$, we have $\alpha A + \beta B \in M(n \times m)$. Hence $M(n \times m)$ is also a vector space.

Often one has to deal with complex vector spaces. A complex vector space of dimension n may be denoted by E^n where each component x_i has a real and an imaginary part given by $x_i = r_i + j\sigma_i$ where $j \equiv \sqrt{-1}$. Thus $E^n = R^n + jR^n$. For example, if one wishes to describe the impedance of n branches of an electrical network one must specify n ordered n-tuples of complex numbers.

1.3 Normed Space

A normed space is a vector space X furnished with a norm $N(\cdot)$ and denoted by $(X, N(\cdot))$. The norm N is a real valued function defined on X which must satisfy the following properties:

$$(i) : N(x) \geq 0 \ \ \forall \ x \in X \tag{1.3.3}$$

$$(ii) : N(x) = 0 \ \text{ if and only if } \ x = 0, \tag{1.3.4}$$

$$(iii) : N(\alpha x) = |\alpha| N(x), \quad \forall \ x \in X, \alpha \in F, \tag{1.3.5}$$

$$(iv) : N(x+y) \leq N(x) + N(y), \quad \forall \ x, y \in X. \tag{1.3.6}$$

The standard notation for the function N is

$$N(x) \equiv \parallel x \parallel . \tag{1.3.7}$$

Here we present some examples of normed vector spaces.

(1): R^n with $N(x) \equiv \sqrt{\sum_{i=1}^n x_i^2}$,

(2): $\ell_p^n, 1 \leq p < \infty$, with norm given by $N(x) \equiv \left(\sum_{i=1}^n |x_i|^p \right)^{1/p}$; ℓ_p, with norm, $N(x) = \left(\sum_{i=1}^\infty |x_i|^p \right)^{1/p}$,

(3): $M(n \times m)$ with norm given by

$$N(A) \equiv \left(\sum_{1 \leq i \leq n, 1 \leq j \leq m} a_{i,j}^2 \right)^{1/2},$$

(4): Let I be any bounded interval and $C(I, R^n)$ denote the class of all continuous functions on I taking values in R^n. Furnished with the norm

$$N(f) \equiv \sup\{\parallel f(t) \parallel_{R^n}, t \in I\},$$

it is a normed vector space.

(5): $L_p(I), 1 \leq p < \infty$, with $N(f) \equiv \left(\int_I |f(t)|^p dt \right)^{1/p}$; and $L_p(\Omega)$ with norm

$$N(f) \equiv \left(\int_\Omega |f(\xi)|^p d\xi \right)^{1/p},$$

where $\Omega \subset R^n$.

(6): Weighted L_p spaces denoted by $L_p(\Omega, \rho)$, where ρ is a nonnegative function defined on Ω. The norm for this space is given by

$$N(f) \equiv \left(\int_\Omega |f(\xi)|^p \rho(\xi) d\xi \right)^{1/p}, \tag{1.3.8}$$

(7): In fact one can consider more general L_p spaces like $L_p(\Omega, \mu)$ where μ is any countably additive (see appendix for explanation) finite positive measure. In this case the norm is given by

$$N(f) \equiv \left(\int_\Omega |f(\xi)|^p \mu(d\xi) \right)^{1/p}, \tag{1.3.9}$$

Two functions $f, g \in L_p(\Omega, \mu)$ are said to be equivalent if the measure of the set in Ω on which they differ is a set of μ-measure zero. That is,

$$\mu\{x \in \Omega : f(x) \neq g(x)\} = 0.$$

Thus these spaces can be partitioned into equivalence classes. For each $f \in L_p(\Omega, \mu)$, one defines

$$[f] \equiv \{g \in L_p(\Omega, \mu) : \mu\{x \in \Omega : g(x) \neq f(x)\} = 0\}.$$

Since

$$\int_\Omega |f|^p \mu(dx) = \int_\Omega |g|^p \mu(dx) \text{ for all } g \in [f],$$

it is immaterial which one of the members of the equivalence class is used to determine the integral. The reader can easily appreciate this by reflecting on the equivalence class determined by the measure $\mu(\sigma) \equiv \int_\sigma \rho(\xi) d\xi$ (see 1.3.8). Let

$$\mathcal{N} \equiv \{f \in L_p(\Omega, \mu) : \mu\{x \in \Omega : f(x) \neq 0\} = 0\}$$

and define $\mathcal{L}_p(\Omega, \mu)$ as the quotient space $L_p(\Omega, \mu)/\mathcal{N}$. For $f, g \in \mathcal{L}_p(\Omega, \mu)$, $f = g$ if and only if $f - g \in \mathcal{N}$, that is, $\mu\{x \in \Omega : f(x) \neq g(x)\} = 0$. It is a standard practice to relabel the quotient space by the same symbol $L_p(\Omega, \mu)$. Throughout the book we adopt this standard practice.
(8): Three most commonly used function spaces from the classes described in (5)-(7) are those for $p = 1, 2, \infty$. For $p = \infty$, the norm is given by

$$N(f) \equiv ess - sup\{|f(x)|, x \in \Omega\}. \tag{1.3.10}$$

This is a very important vector space used extensively as controls. There is a significant difference between plain sup and $ess-$sup. In the case of plain sup, the smallest number α, that the amplitude (or norm) of the function does not exceed is called the sup. The ess-sup, on the other hand, is the smallest number α that the norm does not exceed except on a set of measure zero. That is, $\mu\{x : \parallel f(x) \parallel > \alpha\} = 0$. The L_2 space has special physical significance; it is related to physical energy and it has a special mathematical structure that makes it very useful in physical sciences. It is also a Hilbert space discussed later.

The physical significance of norm can be easily appreciated by thinking of the following example. Consider a cable of length L stretched between two hydro poles and suppose n balls of masses $\{m_1, m_2, \cdots, m_n\}$ are attached to the cable at positions $\{x_i, i = 1, 2, \cdots, n\}$. Wind sets the balls in motion. The state of the process at any moment of time is given by the positions and velocities of the balls at that instant of time. The intensity of the wind can be measured in terms of the intensity of vibration of the balls. Let $\{v_i, i = 1, 2, \cdots, n\}$ denote the velocities of the balls (displacement rate). Neglecting the mass of the cable, the total kinetic energy of the system is then given by

$$K.E \equiv \sum_{1 \le i \le n} (1/2) m_i v_i^2$$

This can be looked at as the weighted norm of the vector $v \equiv (v_1, v_2, v_3, \cdots, v_n)'$, so that

$$\| v \| \equiv \left(\sum_{1 \le i \le n} (1/2) m_i v_i^2 \right)^{1/2}. \tag{1.3.11}$$

Clearly this quantity is a very good indicator of the wind activities of the region where the cable is installed. The more intense the wind is the larger is the value of this quantity.

From the discrete we can easily pass into the continuum. Let $\{y(t, x), x \in [0, L]\}$ denote the displacement of the cable from the rest state with the corresponding displacement rate given by $\{y_t(t, x), x \in [0, L]\}$. Let $\rho(x)$ denote the mass density of the cable meaning that $\rho(x)\Delta x$ is the mass of the elementary section Δx. From the approximate expression we can now derive the exact kinetic energy of the cable by letting $n \to \infty$ yielding

$$\| y_t \|^2 \equiv \left((1/2) \int_0^L \rho(x) |y_t(t, x)|^2 dx \right). \tag{1.3.12}$$

Clearly this is a weighted L_2 space which is denoted by $L_2([0, L], \rho)$. In general we may define a positive measure $\mu(J) \equiv \int_J \rho(x) dx$ giving the mass of the section J. In that case we can rewrite the expression of the kinetic energy as

$$\| y_t \| \equiv \left((1/2) \int_0^L |y_t(t, x)|^2 \mu(dx) \right)^{1/2}. \tag{1.3.13}$$

This is an example of Lebesgue-Stieltjes integral as discussed in the appendix A. Note that this integral allows us to consider both distributed and point masses at the same time. If the cable is massless and there are n-point masses, we take $\mu(dx) = \sum_{1 \leq i \leq n} m_i \delta_{x_i}(dx)$ and we recover the expression (1.3.11). Here $\delta_{x_i}(dx)$ denotes the Dirac measure which equals one if $x_i \in dx$, and zero otherwise.

1.4 Banach Space

Definition 1.4.1. (Cauchy sequence) Let X be a normed space with norm $\| \cdot \|$. A sequence $\{x_n\} \in X$ is said to be a Cauchy sequence if

$$\lim_{n \to \infty} \| x_{n+p} - x_n \| = 0 \text{ for every } p \geq 1.$$

Definition 1.4.2. (Banach space) A normed space X is said to be complete if every Cauchy sequence has a limit. A complete normed space is called a Banach space .

We present here some examples of Banach spaces. The vector spaces, $R^n, E^n, \ell_p^n, 1 \leq p \leq \infty, M(n \times m)$ with the norms as defined in the previous section are finite dimensional Banach spaces. The normed spaces $\ell_p, C(I, R^n), C(\Omega, R^n), L_p(\Omega, \lambda), L_p(\Omega, \mu), 1 \leq p \leq \infty$, with norms as defined earlier are infinite dimensional Banach spaces.

Another class of spaces occasionally used is denoted by $L_p^{\ell oc}(\Omega, \Sigma, \mu)$ which consist of measurable functions which are only locally p-th power integrable. That is, for any set $\Gamma \subset \Omega$ with $\Gamma \in \Sigma$ and $\mu(\Gamma) < \infty$, we have

$$\int_\Gamma |f|^p d\mu < \infty.$$

The measure space (Ω, Σ, μ) is called σ-finite in the sense that $\Omega = \bigcup_n^\infty B_n$ for $B_n \in \Sigma$, and that $\mu(B_n) < \infty$ for each $n \in N$ but $\mu(\Omega) = \infty$. This is not a normed space and hence not a Banach space. In fact this can be furnished with an increasing family of seminorms turning it into a locally convex topological vector space. This contains all the $Lp(\Omega, \Sigma, \mu)$ spaces. Since we do not have any significant use of this space we do not mention this any further.

An important Banach space, denoted by $B(I, R^n)$, is the vector space of bounded measurable functions defined on the interval I taking values from R^n. The norm for this space is given by

$$\| f \| \equiv \sup\{\| f(t) \|_{R^n}, t \in I\}.$$

As we shall see later in Chapter 3, this Banach space is very useful in the study of impulsive systems.

1.4.1 Hilbert Space

Definition 1.4.3. (Hilbert space) A Hilbert space H, is a Banach space with spacial structure. It is furnished with an inner (or scalar) product $(,)$ as defined below.

For any two elements $f, g \in H$, the scalar product (f, g) is a real or a complex number. Let C denote the field of complex numbers. The map

$$\{f, g\} \longrightarrow (f, g)$$

has the following properties:

(H1) $(\alpha f, g) = \alpha(f, g)$, $(f, \alpha g) = \alpha^*(f, g), \alpha \in C, f, g \in H$ where α^* is the complex conjugate of α,
(H2) $(f, g_1 + g_2) = (f, g_1) + (f, g_2), \forall f, g_1, g_2 \in H$,
(H3) $(f, g) = (g, f)^*, \forall f, g \in H$,
(H4) $\| f \|^2 = (f, f) \quad \forall f \in H$.

Some examples of finite dimensional Hilbert spaces are R^n, E^n, ℓ_2^n, with the norms as defined in the previous section. The normed spaces $\ell_2, L_2(\Omega, \lambda)$, $L_2(\Omega, \mu)$, are infinite dimensional Hilbert spaces. We have seen in the physical example given in section 3, that the state of the cable at any point of time is described by its position and velocity at infinitely many points along its span.

For estimates of intensity of forces, energy or any other physically important variables often we may use inequalities. A very important inequality, called the Schwartz inequality, holds for Hilbert spaces.

Proposition 1.4.4. (Schwarz Inequality) For every $f, g \in H$,

$$|Re(f, g)| \leq \| f \| \| g \| . \tag{1.4.14}$$

Proof. Clearly if f and g are collinear, that is, $f = \beta g$ for some $\beta \in R$, the inequality is an identity. If not, it is clear that for every real λ

$$\| f + \lambda g \|^2 > 0.$$

Thus the polynomial,

$$P(\lambda) \equiv \| f + \lambda g \|^2 = \lambda^2 \| g \|^2 + 2\lambda Re(f, g) + \| f \|^2,$$

can not have a real root. However, being quadratic, $P(\lambda) = 0$, must have two roots and they must be complex. Since $P(\lambda)$ is real, these roots must be a pair of complex conjugate numbers. Solving the quadratic equation, the roots are found to be given by

$$\lambda_{1,2} = -\frac{Re(f,g)}{2 \parallel g \parallel^2}(+/-)\frac{1}{2 \parallel g \parallel^2}\sqrt{(Re(f,g))^2 - \parallel f \parallel^2 \parallel g \parallel^2} .$$

Since the roots must be complex, it is necessary that the discriminant be negative. Thus

$$(Re(f,g))^2 - \parallel f \parallel^2 \parallel g \parallel^2 < 0.$$

This means

$$|Re(f,g))| < \parallel f \parallel \parallel g \parallel,$$

in case f and g are not collinear. Hence, in general, including the collinear case, we have

$$|Re(f,g)| \leq \parallel f \parallel \parallel g \parallel,$$

proving the inequality. □

As a corollary of this result we have the triangle inequality.

Corollary 1.4.5. (Triangle inequality) For every $f, g \in H$

$$\parallel f + g \parallel \leq \parallel f \parallel + \parallel g \parallel . \tag{1.4.15}$$

Proof. Note that

$$\begin{aligned} \parallel f + g \parallel^2 = (f + g, f + g) = & \parallel f \parallel^2 + \parallel g \parallel^2 + 2Re(f,g) \\ & \leq \parallel f \parallel^2 + \parallel g \parallel^2 + 2|Re(f,g)| \end{aligned}$$

Using the previous proposition, it follows from this that

$$\parallel f + g \parallel^2 \leq \parallel f \parallel^2 + \parallel g \parallel^2 + 2 \parallel f \parallel \parallel g \parallel = (\parallel f \parallel + \parallel g \parallel)^2.$$

Hence we have

$$\parallel f + g \parallel \leq \parallel f \parallel + \parallel g \parallel .$$

This completes the proof. □

Note that for the real Hilbert space $L_2(\Omega, \mu)$, with μ a positive measure, the scalar product is given by

$$(f,g) \equiv \int_\Omega f(x)g(x)\mu(dx) \equiv \int_\Omega f(x)g(x)d\mu, \tag{1.4.16}$$

and the norm is given by

$$\| f \| \equiv \left(\int_{\Omega} |f|^2 \mu(dx) \right)^{1/2}. \tag{1.4.17}$$

An interesting example of a positive measure on Ω is

$$\mu(dx) = \lambda(dx) + \sum \gamma_i \delta_{x_i}(dx) = dx + \sum \gamma_i \delta_{x_i}(dx), \gamma_i \geq 0,$$

where $\delta_{x_i}(dx)$ denotes the Dirac measure concentrated at the point $x_i \in \Omega$. For any bounded set $\Gamma \subset \Omega$,

$$\int_{\Gamma} \mu(dx) = Vol(\Gamma) + \sum_{\{i: x_i \in \Gamma\}} \gamma_i.$$

Another example of a Hilbert space is given by $H^1(\Omega, \mu)$ which consists of functions (equivalence classes) which along with their first derivatives are square integrable.

$$H^1(\Omega, \mu) \equiv \{\phi : \phi \in L_2(\Omega, \mu), D\phi = (D_{x_i}\phi, i = 1, 2, \cdots, n) \in L_2(\Omega, \mu)\}.$$

The scalar product in this space is given by

$$(\phi, \psi) = \int_{\Omega} \phi(x)\psi(x)\mu(dx) + \sum_{1 \leq i \leq n} \int_{\Omega} (D_{x_i}\phi)(D_{x_i}\psi)\mu(dx),$$

with the corresponding norm,

$$\| \phi \|_{H^1} = \left(\int_{\Omega} |\phi(x)|^2 \mu(dx) + \sum_{1 \leq i \leq n} \int_{\Omega} |D_{x_i}\phi|^2 \mu(dx) \right)^{1/2}.$$

If $\Omega = I \equiv (0, T)$ and μ is the Lebesgue measure (length), $H^1(I)$ is the space of square integrable functions which are absolutely continuous with first derivatives being square integrable. The norm is given by

$$\| \phi \|_{H^1} = \left(\int_0^T |\phi|^2 dt + \int_0^T |\dot{\phi}|^2 dt \right)^{1/2}.$$

1.4.2 Special Banach Spaces

Now we consider a very important class of function (signal) spaces which are Banach spaces, not Hilbert spaces. For $f \in L_p(\Omega, \mu), 1 \leq p < \infty$, we denote the norm of f by

$$\| f \|_p \equiv \left(\int_{\Omega} |f|^p \mu(dx) \right)^{1/p},$$

where μ is a countably additive σ-finite positive measure on Ω (see appendix for the definition). For $p = 2$, this reduces to a Hilbert space. We shall prove that the $L_p(\Omega, \mu), 1 \leq p \leq \infty$ spaces are Banach spaces. There is no scalar product in these spaces in the sense defined for Hilbert spaces. For $f, g \in L_p$, (f, g) is not defined. In other words $\int_\Omega f(x)g(x)\mu(dx)$ does not exist or even makes sense. However there is a duality pairing as described below.

For special Banach spaces, like $L_p(\Omega, \mu)$ with μ any positive measure, we have similar results like those of Proposition 1.4.4 and Corollary 1.4.5. These are Hölder and Minkowski inequalities.

Proposition 1.4.6. (Hölder Inequality) For every $f \in L_p(\Omega, \mu)$ and $g \in L_q(\Omega, \mu)$ with $1 < p, q < \infty, (1/p) + (1/q) = 1$,

$$|(f, g)| \leq \parallel f \parallel_p \parallel g \parallel_q . \tag{1.4.18}$$

Proof. Define

$$a \equiv \left(\frac{|f(x)|^p}{\parallel f \parallel_p^p}\right), \quad b \equiv \left(\frac{|g(x)|^q}{\parallel g \parallel_q^q}\right) \tag{1.4.19}$$

and $\alpha \equiv (1/p)$, $\beta \equiv (1/q)$. Since the function $\log x, x \geq 0$, is a concave function, we have

$$\log(\alpha a + \beta b) \geq \alpha \log a + \beta \log b = \log(a^\alpha b^\beta).$$

Hence

$$a^\alpha b^\beta \leq \alpha a + \beta b. \tag{1.4.20}$$

Substituting the values of a, b, α, β from equation(1.4.19) into equation (1.4.20) we have

$$\left(\frac{|f(x)|}{\parallel f \parallel_p}\right)\left(\frac{|g(x)|}{\parallel g \parallel_q}\right) \leq (1/p)\left(\frac{|f(x)|^p}{\parallel f \parallel_p^p}\right) + (1/q)\left(\frac{|g(x)|^q}{\parallel g \parallel_q^q}\right). \tag{1.4.21}$$

Now integrating this with respect to the measure μ we obtain

$$\begin{aligned}
&(1/ \parallel f \parallel_p \parallel g \parallel_q) \int_\Omega \{|f(x)||g(x)|\}\mu(dx) \\
&\leq (1/p \parallel f \parallel_p^p) \int_\Omega |f(x)|^p \mu(dx) + (1/q \parallel g \parallel_q^q) \int_\Omega |g(x)|^q \mu(dx) \\
&\leq (1/p) + (1/q) = 1.
\end{aligned}$$

Hence we have

$$\int_\Omega \{|f(x)||g(x)|\}\mu(dx) \leq \parallel f \parallel_p \parallel g \parallel_q . \tag{1.4.22}$$

Clearly,

$$|(f,g)| \equiv \left| \int_\Omega f(x)g(x)\mu(dx) \right| \leq \int_\Omega \{|f(x)||g(x)|\}\mu(dx). \tag{1.4.23}$$

Hence, from the expressions (1.4.22) and (1.4.23) we obtain the inequality (1.4.18). This ends the proof. □

Remark. Note that the above result also holds for $p = 1$ and $q = \infty$, giving

$$|(f,g)| \leq \int_\Omega |f||g|d\mu \leq \| g \|_\infty \int_\Omega |f|d\mu = \| f \|_1 \| g \|_\infty$$

where $\| g \|_\infty = \mu - ess - sup\{|g(x)|, x \in \Omega\}$.

As a corollary of this result we have the following triangle inequality.

Corollary 1.4.7. (Minkowski inequality) For $f, g \in L_p(\Omega, \mu), 1 \leq p < \infty$,

$$\| f + g \|_p \quad \leq \quad \| f \|_p + \| g \|_p \,. \tag{1.4.24}$$

Proof. For $p = 1$, the result is obvious, since

$$\| f + g \|_1 \quad \equiv \quad \int_\Omega |f + g|d\mu \leq \int_\Omega |f|d\mu + \int_\Omega |g|d\mu \quad \equiv \quad \| f \|_1 + \| g \|_1 \,.$$

For $p > 1$, observe that

$$\begin{aligned} \int_\Omega |f + g|^p d\mu &= \int_\Omega |f + g|^{p-1}|f + g|d\mu \\ &\leq \int_\Omega |f + g|^{p-1}|f|d\mu + \int_\Omega |f + g|^{p-1}|g|d\mu. \end{aligned} \tag{1.4.25}$$

Applying the Hölder inequality (Proposition 1.4.6) to each of the terms on the right hand side of the above inequality, we obtain

$$\int_\Omega |f + g|^p d\mu \leq \left(\int_\Omega |f + g|^{(p-1)q} \right)^{1/q} \left\{ \| f \|_p + \| g \|_p \right\}, \tag{1.4.26}$$

for q satisfying $(1/q) + (1/p) = 1$. Since $p = (p-1)q$, it follows from this inequality that

$$\int_\Omega |f + g|^p d\mu \leq \left(\int_\Omega |f + g|^p \right)^{1/q} \left\{ \| f \|_p + \| g \|_p \right\}. \tag{1.4.27}$$

This, in turn, leads to the following expression,

$$\left(\int_\Omega |f + g|^p d\mu \right)^{1-1/q} \leq \left\{ \| f \|_p + \| g \|_p \right\}, \tag{1.4.28}$$

which means

$$\left(\int_{\Omega} |f + g|^p d\mu \right)^{1/p} \leq \left\{ \| f \|_p + \| g \|_p \right\}. \tag{1.4.29}$$

This is precisely the result sought as given by (1.4.24). This completes the proof. □

The following result states that all Lebesgue spaces are complete and hence are Banach spaces.

Theorem 1.4.8. For any positive measure space (Ω, B_Ω, μ) and for each $p \geq 1$, the vector space $L_p(\Omega, B_\Omega, \mu)$ is a Banach space.

Proof. We must show that every Cauchy sequence has a limit. Let $\{f_n\}$ be a Cauchy sequence in $L_p(\Omega, B_\Omega, \mu)$. Then $\{f_n\}$ is a Cauchy sequence measure and by a theorem of measure theory [69, 41], there exists a subsequence $\{f_{n_k}\} \subset \{f_n\}$ which is a Cauchy sequence almost every where and hence there exists an f such that

$$f_{n_k}(\omega) \longrightarrow f(\omega) \;\; \mu - a.e.$$

Since the limit of any almost everywhere convergent sequence of measurable functions is measurable the limit f is measurable, that is, $f \in M$. Clearly

$$|f_{n_k}(\omega)|^p \longrightarrow |f(\omega)|^p \quad \mu - a.e.$$

also. Then by Fatou's Lemma [see Appendix A.6.3], we have

$$\begin{aligned} \int_{\Omega} |f(\omega)|^p d\mu &= \int_{\Omega} \lim_{k \to \infty} |f_{n_k}(\omega)|^p d\mu \\ &= \int_{\Omega} \liminf_{k \to \infty} |f_{n_k}(\omega)|^p d\mu \\ &\leq \liminf_{k \to \infty} \int_{\Omega} |f_{n_k}(\omega)|^p d\mu. \end{aligned}$$

Since $\{f_n\}$ is a Cauchy sequence in L_p, it is norm bounded and hence it follows from the above inequality that $f \in L_p(\Omega, B_\Omega, \mu)$. Now we prove that

$$\lim_{k \to \infty} \| f_{n_k} - f \|_p = 0.$$

It is clear that for any fixed integer k

$$|f_{n_k}(\omega) - f(\omega)|^p = \lim_{s \to \infty} |f_{n_k}(\omega) - f_{n_s}(\omega)|^p \;\; \mu - a.e.$$

By use of Fatou's Lemma once again, we have

$$\int_\Omega |f_{n_k}(\omega) - f(\omega)|^p d\mu \leq \liminf_{s\to\infty} \int_\Omega |f_{n_k}(\omega) - f_{n_s}(\omega)|^p d\mu. \tag{1.4.30}$$

Now since $\{f_n\}$ is a Cauchy sequence in L_p, letting $k \to \infty$, it follows from the above inequality that

$$\lim_{k\to\infty} \| f_{n_k} - f \|_p = 0. \tag{1.4.31}$$

The fact that $\{f_n\}$ is a Cauchy sequence in L_p, also implies that, for every $\varepsilon > 0$, there exists an integer n_ε such that

$$\| f_n - f_{n_\varepsilon} \|_p < \varepsilon/3 \ \forall\ n > n_\varepsilon. \tag{1.4.32}$$

Using Minkowski inequality, the facts (1.4.31) and (1.4.32) and choosing k large enough such that $n_k > n_\varepsilon$, we obtain

$$\begin{aligned} \| f_n - f \|_p &\leq \| f_n - f_{n_\varepsilon} \|_p + \| f_{n_\varepsilon} - f_{n_k} \|_p + \| f_{n_k} - f \|_p \\ &< (\varepsilon/3) + (\varepsilon/3) + (\varepsilon/3) = \varepsilon \end{aligned}$$

for all $n > n_\varepsilon$. Since $\varepsilon > 0$ is otherwise arbitrary, this shows that every Cauchy sequence in $L_p(\Omega, B_\Omega, \mu)$, not just a subsequence of it, converges to a limit that belongs to the same space. This proves that the L_p spaces are complete and hence are Banach spaces. □

Proposition 1.4.9. Consider the Banach spaces $\{L_p(\Omega,\mu), 1 \leq p \leq \infty\}$ and suppose $\mu(\Omega) < \infty$. Then for $1 \leq p_1 \leq p_2 \leq \infty$, we have

$$L_{p_2}(\Omega,\mu) \subset L_{p_1}(\Omega,\mu). \tag{1.4.33}$$

Proof. We show that for every $f \in L_{p_2}(\Omega,\mu)$, we have $f \in L_{p_1}(\Omega,\mu)$. For $f \in L_{p_2}(\Omega,\mu)$ we compute its L_{p_1} norm,

$$\begin{aligned} \int_\Omega |f|^{p_1} d\mu = \int_\Omega |f|^{(p_1/p_2)p_2} d\mu &\leq \left(\int_\Omega |f|^{p_2} d\mu\right)^{p_1/p_2} \left(\int_\Omega 1 d\mu\right)^{(p_2-p_1)/p_2} \\ &\leq (\mu(\Omega))^{(p_2-p_1)/p_2} (\| f \|_{p_2})^{p_1}. \end{aligned}$$

Hence

$$\| f \|_{p_1} \leq c(p_1,p_2,\mu(\Omega)) \| f \|_{p_2}, \tag{1.4.34}$$

where the constant $c \equiv c(p_1,p_2,\mu(\Omega)) = (\mu(\Omega))^{(p_2-p_1)/p_2p_1}$. So for every f that belongs to $L_{p_2}(\Omega,\mu))$ also belongs to $L_{p_1}(\Omega,\mu)$ provided $p_1 < p_2$. Hence

$$L_{p_2}(\Omega,\mu) \subset L_{p_1}(\Omega,\mu), \ \forall\ 1 \leq p_1 \leq p_2 \leq \infty.$$

This completes the proof. □

Remark. In view of this result, whenever $\mu(\Omega) < \infty$, we have

$$L_\infty(\Omega, \mu) \subset L_p(\Omega, \mu) \subset L_1(\Omega, \mu), \forall\, p \in [1, \infty].$$

Note that the embedding constant c is dependent on the parameters indicated. One can verify that $c \in [1 \wedge \mu(\Omega), 1 \vee \mu(\Omega)]$. Some of the values are indicated below.

$$\begin{aligned}
&(1): \text{ for } p_1 = 1, c(1, p_2, \mu(\Omega)) = (\mu(\Omega))^{(p_2-1)/p_2},\\
&(2): \text{ for } p_2 = \infty, c = c(p_1, \infty, \mu(\Omega)) = (\mu(\Omega))^{1/p_1},\\
&(3): \text{ for } p_1 = 1, p_2 = \infty, c = \mu(\Omega),\\
&(4): \text{ for the trivial case } \ p_1 = p_2, c = 1.
\end{aligned}$$

Remark. It is clear from the inequality (1.4.34) that if $\mu(\Omega) = \infty$ the inclusion does not make much of a sense indicating that the inclusion may not even hold. Here is an example justifying the legitimacy of this suspicion. Let $\Omega = [0, \infty)$ and μ the Lebesgue measure. Consider the function

$$f(x) \equiv \frac{1}{x^{1/p}(1 + |\log x|)^{2/p}}, x \in [0, \infty), \infty > p \geq 1.$$

The reader can verify that $f \notin L_r([0, \infty), \mu)$ for $r \neq p$. For $r = p$ we have

$$\begin{aligned}
\int_0^\infty |f(x)|^p dx &= \int_0^\infty \frac{1}{x(1 + |\log x|)^2} dx\\
&= \int_{-\infty}^\infty \frac{1}{(1 + |y|)^2} dy = 2 \int_0^\infty \frac{1}{(1 + y)^2} dy = 2.
\end{aligned}$$

In summary, $f \in L_p$ and $f \notin \bigcup_{r \neq p} L_r$.

Remark. Let $\mu(dx) = dx$ denote the Lebesgue measure and $\nu(dx) \equiv (1/1 + x^2)dx$ be another measure both defined on R. Note that $\nu(R) = \pi$, so this is a finite positive measure. Then, one can easily verify that $L_p(R, \mu) \subset L_p(R, \nu)$ for all $p \geq 1$. For $f \in L_p(R, \nu), g \in L_q(R, \nu)$, with (p, q) being the conjugate pair, we have $fg \in L_1(R, \nu)$. Similar conclusions hold for the Gaussian measure $\nu_g(K) \equiv \int_K (1/\sqrt{2\pi})\{\exp -(1/2)x^2\}dx$.

Another interesting property of the Banach spaces, $\{L_p(\Omega, \mu), 1 \leq p \leq \infty\}$, is stated in the following result. First, we introduce some notations. Let

$$B_1(L_p) \equiv \{h : h \in L_p(\Omega, \mu), \| h \|_p \leq 1\}$$

denote the unit ball of the Banach space L_p. The unit sphere is denoted by ∂B_1, that is, these are the elements of B_1 which have norms exactly equal to one. This is the boundary of the unit ball.

Proposition 1.4.10. For every $f \in L_p(\Omega, \mu)$ there exists a $g \in B_1(L_q)$ where q is the conjugate of p in the sense that $(1/p) + (1/q) = 1$, such that

$$(f, g) = \| f \|_p .$$

Further if $1 < q < \infty$, there is only one such (unique) $g \in B_1(L_q)$.

Proof. Let $f(\neq 0) \in L_p(\Omega, \mu)$ be given. Define

$$g(x) \equiv (1/ \| f \|_p^{p-1}) |f(x)|^{p-1} sign f(x). \tag{1.4.35}$$

The signum function of a Borel measurable function is also Borel measurable. Product of measurable functions is a measurable function. Thus g as defined is a measurable function. By integration we show that this element belongs to the unit sphere $\partial B_1(L_q)$ where q is the number conjugate to the number p. Integrating $|g|^q$ and noting that $(p-1)q = p$, we find that

$$\int_\Omega |g|^q d\mu = (1/ \| f \|_p^{(p-1)q}) \int_\Omega |f|^{(p-1)q} d\mu = 1.$$

Thus $g \in \partial B_1(L_q)$. Now computing the duality product of f with the g given by (1.4.35), we obtain

$$(f, g) = \int_\Omega f(x) g(x) d\mu = (1/ \| f \|_p^{p-1}) \int_\Omega |f|^p d\mu = \| f \|_p . \tag{1.4.36}$$

This proves the first part. For the second part, we note that the ball $B_1(L_q)$ is strictly convex for all $q \in (1, \infty)$ in the sense that the line segment joining any two points f_1, f_2 of the ball B_1 given by

$$\{h \in B_1 : h = \alpha f_1 + \beta f_2, \alpha, \beta \geq 0, \alpha + \beta = 1\},$$

can not touch the boundary except possibly at the end points. We prove this by contradiction. Suppose there are two points $g_1, g_2 \in \partial B_1(L_q)$ such that $(f, g_1) = (f, g_2) = \| f \|_p$. Define $g = (1/2)(g_1 + g_2)$. Then clearly $(f, g) = \| f \|_p$. Since the ball is strictly convex g must be an interior point of the ball and so $\| g \|_q < 1$. But this implies that $|(f, g)| < \| f \|_p$. This is a contradiction and so g_1 must equal g_2. This ends the proof. □

Remark. Reversing the roles of $\{f, p\}$ and $\{g, q\}$ we obtain similar conclusions.

Let $L_p(\Omega,\mu)^*$ denote the class of continuous linear functionals on $L_p(\Omega,\mu)$. This is known as the dual space of the Banach space $L_p(\Omega,\mu)$. We state the following result without giving the proof. Readers interested in the proof may consult any standard book on functional analysis.

Proposition 1.4.11. The dual space $L_p(\Omega,\mu)^*$ is isometrically isomorphic to the linear space $L_q(\Omega,\mu)$ for all $\{p,q : 1 \le p < \infty, (1/p)+(1/q)=1\}$.

From now on, we will denote by E^* the dual (or the conjugate) of any Banach space E. For example, the dual of L_p is L_q, as seen above, and particularly the dual of L_2 is the L_2 space itself.

The dual of the space of continuous functions $C(\Omega)$, with Ω a compact subset of R^n, is the space of countably additive measures of bounded total variation denoted by $\mathcal{M}(\Omega)$. In fact this is also valid for any compact metric space Ω.

In addition to the norm topology, an L_p space can be furnished with the topology of weak convergence as defined below.

Definition 1.4.12. For any $p \in [1,\infty)$, a sequence $\{f_n\} \subset L_p$ is said to be weakly convergent to an element $f \in L_p$ if, and only if,

$$\int_\Omega f_n(x)g(x)\mu(dx) \longrightarrow \int_\Omega f(x)g(x)\mu(dx)$$

for every $g \in L_q$ where $q = (p/p-1)$ is the dual of L_p.

Definition 1.4.13. A Banach space E is said to be reflexive if it coincides with its second dual, that is, $E=(E^*)^* \equiv E^{**}$.

For example, all L_p spaces with $1<p<\infty$ have this property. This is clear from the fact that the dual of L_p is L_q for $q=(p/p-1)$ and that the dual of L_q is L_p. Thus $L_p^{**}=L_p$.

This is a very important class of Banach spaces having some properties which are similar to those of finite dimensional spaces as discussed in the following section.

1.5 Metric Space

A set M which is furnished with a measure of distance between any two elements of the set is called a metric space. A function $d : M \times M \longrightarrow R$ is called a metric if it satisfies the following properties

(M1): $d(x,y) \ge 0$, $\forall\, x,y \in M$, and $d(x,y)=0$ if and only if $x=y$
(M2): $d(x,y)=d(y,x)$ $\forall x,y \in M$
(M3): $d(x,z) \le d(x,y)+d(y,z)$ $\forall x,y,z \in M$.

Definition 1.5.1. A metric space $M \equiv (M, d)$ is said to be complete, if every Cauchy sequence has a limit.

Example 1. Every Banach space $B \equiv (B, \| \cdot \|)$ is a complete metric space with respect to the metric d given by $d(x, y) \equiv \| x - y \|$. Thus all $L_p(\Omega, \mu), 1 \leq p \leq \infty$, spaces are complete metric spaces.

Example 2. Let B be a Banach space with norm $\| \cdot \|$, and let Γ be any closed subset of B. Then (Γ, d) with $d(x, y) \equiv \| x - y \|$, is a complete metric space. But this is not a linear metric space, though, in this particular case, it can be embedded in the linear space B.

Example 3. Let D be a closed subset of R^n, then $C(I, D)$ is again a complete metric space with the metric inherited from the norm of the Banach space, $C(I, R^n)$, with supremum norm as defined earlier.

An example of special importance in control theory is the class of measurable functions with the metric topology induced by the topology of almost everywhere convergence.

Example 4. Let λ denote the Lebesgue measure on I and $M(I, R^n)$, the space of Lebesgue measurable functions defined on I and taking values in R^n. Define

$$d(x, y) \equiv \lambda\{t \in I : x(t) \neq y(t)\}. \tag{1.5.37}$$

This defines a metric on $M(I, R^n)$. Indeed, the function d, as defined here, clearly satisfies the axioms (M1) and (M2). For (M3), let $z \in M(I, R^n)$ and note that

$$\{t \in I : x(t) = y(t)\} \supset \{t \in I : x(t) = z(t)\} \bigcap \{t \in I : z(t) = y(t)\},$$

and hence

$$\{t \in I : x(t) \neq y(t)\} \subset \{t \in I : x(t) \neq z(t)\} \bigcup \{t \in I : z(t) \neq y(t)\}.$$

Thus

$$\lambda\{t \in I : x(t) \neq y(t)\} \leq \lambda\{t \in I : x(t) \neq z(t)\} + \lambda\{t \in I : z(t) \neq y(t)\},$$

which means $d(x, y) \leq d(x, z) + d(z, y)$. Hence d, as defined above, is a metric. One can prove that, the vector space $M(I, R^n)$ furnished with this metric is a complete metric space (M, d).

In relation to the example 4, one can easily check that $(M(I, D), d)$ is also a complete metric space whenever D is a closed subset of R^n.

Example 5. Binary codes are used in all modern communication systems today. A typical signal is an n-tuple, $x \equiv (x_1, x_2, \cdots, x_n)$, where for each i, $x_i \in \{0, 1\} \equiv F$. Let M denote the class of all such vectors. Here $F \equiv \{0, 1\}$ is a field with addition modulo 2, that is,

$$a \oplus b \equiv a + b \text{ (mod 2)}.$$

Thus $0 \oplus 1 = 1 \oplus 0 = 1$ and $0 \oplus 0 = 1 \oplus 1 = 0$. This modular addition can be easily performed by a logic circuit, called anti-coincidence circuit. Using this modular addition, one can then define a metric on the set M as follows:

$$\begin{aligned} d(x,y) &\equiv \text{No. } \{1's \text{ in } x_i \oplus y_i, 1 \le i \le n\} \\ &\equiv \text{No. } \{i : x_i \ne y_i, 1 \le i \le n\}. \end{aligned} \tag{1.5.38}$$

The reader can easily check that d, as defined here, satisfies all the axioms of metric spaces, (M1)-(M3). Note the amazing similarity of the metrics given by (1.5.37) and (1.5.38). In coding theory, this metric is known as the Hamming metric (after the name of its discoverer).

Definition 1.5.2. A subset K of a metric space (M, d) is said to be relatively compact (or precompact) if, for every $\varepsilon > 0$, K can be covered by a finite number of balls of radius ε. That is, there exists an integer n_ε and a finite set of points $\{x_i \in K, i = 1, 2, \cdots, n_\varepsilon\}$ such that

$$K \subset \bigcup_{1 \le i \le n_\varepsilon} B(x_i, \varepsilon)$$

where $B(x, \epsilon) \equiv \{z \in M : d(x, z) \le \varepsilon\}$. It is said to be compact, if it is closed and relatively compact.

Definition 1.5.3. A subset K of a metric space (M, d) is said to be sequentially compact if every sequence from K has a convergent subsequence with the limit being in K. A set $K \subset M$ is said to be conditionally compact, if its closure is compact.

Here are some examples of compact sets. A closed bounded set in R^n is compact. A bounded set in R^n is relatively compact. In fact any bounded set in a finite dimensional space is relatively compact. In infinite dimensional Banach spaces, this is not true. However, for reflexive Banach spaces, furnished with the weak convergence topology, we have the following result.

Theorem 1.5.4. A weakly closed bounded subset K of a reflexive Banach space is weakly compact, as well as, weakly sequentially compact.

One of the most frequently used Banach spaces is the space of continuous functions. Let (X, d) and (Y, ρ) be any two metric spaces. A function $f : X \longrightarrow Y$ is said to be continuous if the inverse image of any open set in Y is an open set in X, that is $O_X \equiv f^{-1}(O_Y)$ is an open set in X whenever O_Y is an open subset of Y. Denote by $C(X,Y)$ the space of continuous functions from X to Y. Furnished with the metric topology

$$\gamma(f, g) \equiv \sup_{x \in X} \rho(f(x), g(x)),$$

$C(X, Y)$ is a metric space. If (Y, ρ) is a complete metric space then so also is $(C(X, Y), \gamma)$. We are especially interested in the following Banach space. Let D be a compact subset of a metric space and Y a finite dimensional Banach space. Let $C(D, Y)$ denote the vector space of continuous functions furnished with the norm topology

$$\| f \| \equiv \sup\{|f(x)|_Y, x \in D\}.$$

Then it is a Banach space, that is, every Cauchy sequence $\{f_n\}$ in $C(D, Y)$ converges in the norm topology defined above to an element $f \in C(D, Y)$. In particular, let D be a closed bounded interval I of the real line R and $Y = R^n$. Then $C(I, R^n)$ is a Banach space. See appendix (Theorem A.5.4) for proof.

Definition 1.5.5. (Absolute Continuity) A function $f \in C(I, R^n)$ is said to absolutely continuous if, for every $\varepsilon > 0$, there is $\delta > 0$ such that, for any set of disjoint open intervals $\{I_i \equiv (t_i, t_{i+1})\} \subset I$,

$$\sum |f(t_{i+1}) - f(t_i)|_{R^n} < \varepsilon$$

whenever

$$\sum |t_{i+1} - t_i| < \delta.$$

We use $AC(I, R^n)$ to denote the class of absolutely continuous functions from I to R^n. It is evident that $AC(I, R^n) \subset C(I, R^n)$. The question of compactness of a set $G \subset C(I, R^n)$ is very crucial in the study of existence of solutions of differential equations and many optimization problems. This is closely associated with the notion of equicontinuity.

Definition 1.5.6. (Equicontinuity) A set $G \subset C(I, R^n)$ is said to be equicontinuous at a point $t \in I$ if, for every $\varepsilon > 0$, there exists a $\delta > 0$ such that

$$|f(t) - f(s)|_{R^n} < \varepsilon \;\; \forall \;\; f \in G$$

whenever $|t-s| < \delta$. The set G is said to be equicontinuous on I if this holds for all $t \in I$.

Following result is very important in the study of existence of solutions of differential equations.

Theorem 1.5.7. (Ascoli-Arzela) A set $G \subset C(I, R^n)$ is conditionally (relatively) compact if (i) it is bounded in the norm topology and (ii) it is equicontinuous. The set G is compact if it is relatively compact and the t section, $R(t) \equiv \{f(t), f \in G\}$ is closed for every $t \in I$.

Proof. Take any sequence $\{x_n\} \subset G$. We show that it has a convergent subsequence. Since I is a compact interval, for every $\delta > 0$, there exists a finite set $I_F \equiv \{t_i, i = 1, 2, \cdots, m = m(\delta)\} \subset I$ such that for every $t \in I$, there exists a $t_j \in I_F$ such that $|t - t_j| \leq \delta$. Since the set G is equicontinuous, the sequence $\{x_n\}$ is equicontinuous. Thus for any $\varepsilon > 0$, there exists a $\delta > 0$ such that

$$\sup_{n \in N} \| x_n(t) - x_n(s) \| \leq \varepsilon \text{ whenever } |t - s| \leq \delta.$$

Consider the sequence $\{x_n(t_j), j = 1, 2, \cdots, m\}$. Since G is a bounded set it is clear that each of the elements of the above sequence is contained in a bounded subset of R^n. Bolzano-Wierstrass theorem says that every bounded sequence of vectors in a finite dimensional space has a convergent subsequence. Thus for every sequence $\{x_n(t_j), n \in N\}$ there exists a convergent subsequence $\{x_{n(j)}(t_j)\}$. Let $\{n(1)\}$ be a subsequence of the sequence $\{n\}$ for which $x_{n(1)}(t_1)$ is convergent. Denote by $n(2)$ a subsequence of $n(1)$ for which $x_{n(2)}(t_2)$ is convergent. Carrying out this diagonal process we obtain a subsequence $\{n(m)\} \subset \{n\}$ such that along this subsequence every element of the set $\{x_n(t_j), j = 1, 2, \cdots, m\}$ is convergent. For any $t \in I$, there exists a $j \in \{1, 2, \cdots, m\}$ such that $|t - t_j| \leq \delta$. Clearly the required t_j depends on $t \in I$. We indicate this dependence by $t_j(t)$. Thus we have,

$$\begin{aligned} \| x_k(t) - x_r(t) \| \leq & \| x_k(t) - x_k(t_j(t)) \| + \| x_k(t_j(t)) \\ & - x_r(t_j(t)) \| + \| x_r(t_j(t)) - x_r(t) \| \\ \leq & 2\varepsilon + \| x_k(t_j(t)) - x_r(t_j(t)) \| \end{aligned}$$

for any $k, r \in \{n(m)\}$. Hence

$$\sup_{t \in I} \| x_k(t) - x_r(t) \| \leq 2\varepsilon + \sup_{t \in I} \| x_k(t_j(t)) - x_r(t_j(t)) \| .$$

Letting $k, r \longrightarrow \infty$ along the subsequence $\{n(m)\}$, we obtain

$$\limsup_{k,r \to \infty} \| x_k - x_r \|_{C(I,R^n)} \leq 2\varepsilon.$$

Since $\varepsilon(> 0)$ is arbitrary, it follows from this that from every sequence of G one can extract a subsequence which is a Cauchy sequence in $C(I, R^n)$. Since this is a Banach space with respect to the sup norm topology, there exists an element $x \in C(I, R^n)$ to which the sequence $\{x_k\}$ converges. This proves that the set G is relatively compact. Clearly if every t section of G is closed, the limit $x \in G$. Hence G is compact. This ends the proof. □

Remark. In fact Ascoli-Arzela theorem, as presented above, also holds for compact metric spaces (M, d) replacing compact interval I. For more general result See ([80] p153)

The space of continuous linear functionals on $C(I, R^n)$ has the representation

$$\ell(f) = \int_I < f(t), \mu(dt) >, f \in C(I, R^n)$$

where μ is an R^n valued measure defined on the Borel subsets of the set I. For example, note that for any $z \in R^n$ and the Dirac measure $\delta_s(dt)$, with mass concentrated at the point s, the element $\nu_{z,s}(dt) \equiv z\delta_s(dt)$ defines a continuous linear functional on $C(I, R^n)$. Indeed,

$$\ell_{z,s}(f) \equiv \int_I < f(t), \nu_{z,s}(dt) >= (f(s), z).$$

It is obvious that

$$|\ell_{z,s}(f)| \leq |z|_{R^n} \| f \| .$$

Hence the functional $\ell_{z,s}$ is a bounded linear (and hence continuous) functional on $C(I, R^n)$. Similarly for any sequence $J \equiv \{t_i\} \subset I$ and $Z \equiv \{z_i\} \subset R^n$, the measure $\nu_{Z,J} \equiv \sum z_i \delta_{t_i}(dt)$ defines a vector measure. The linear functional,

$$\ell_{Z,J}(f) \equiv \int_I < f(t), \nu_{Z,J}(dt) >= \sum_{t_i \in J} (f(t_i), z_i),$$

defines a bounded linear functional on $C(I, R^n)$ if and only if

$$\sum_{t_i \in J} |z_i| < \infty.$$

Two interesting facts emerge from this example. One, if J is a continuum and $|z_\tau| > 0$ for all $\tau \in J$, $\ell_{Z,J}$ defines an unbounded linear functional and hence is not an element of the (topological) dual of $C(I, R')$. Similarly if J is a countable set but $\sum |z_i|_{R^n} = \infty$, the functional $\ell_{Z,J}$ fails to define a continuous linear functional. From these elementary observations we conclude that the (topological) dual of the Banach space $C(I, R^n)$ is the space of countably additive R^n valued measures having bounded total variation on I given by

$$\| \mu \| \equiv \sup_{\pi} \sum_{\sigma \in \pi} |\mu(\sigma)|_{R^n},$$

where π is any finite partition of the set I by disjoint members from the class of Borel sets $\mathcal{B}_I$ of I. The supremum is taken over all such partitions. We denote this dual space by $\mathcal{M}(I, R^n)$ or $\mathcal{M}_c(I, R^n)$ to emphasize countable additivity. Furnished with the norm topology (total variation norm), as defined above, $\mathcal{M}(I, R^n)$ is a Banach space.

Note that any $g \in L_1(I, R^n)$ also induces a countably additive bounded vector measure through the mapping

$$\mu(\sigma) \equiv \int_{\sigma} g(t)dt.$$

Indeed, define

$$\ell_g(f) \equiv \int_I < f(t), \mu(dt) >= \int_I < f(t), g(t) > dt.$$

Clearly, it follows from this expression that for any $f \in C(I, R^n)$,

$$|\ell_g(f)| \leq \|f\| \int_I |g(t)|_{R^n} dt < \infty.$$

This shows that the map $e : g \longrightarrow \ell_g$ from $L_1(I, R^n)$ to $C(I, R^n)^* \equiv \mathcal{M}(I, R^n)$ is a continuous embedding. That is $e(L_1(I, R^n)) \subset \mathcal{M}(I, R^n)$. Thus the space of vector measures is much larger than the class of Lebesgue integrable vector valued functions. In the study of optimal controls involving vector measures as controls, some time we need to consider the question of compactness of subsets $\Gamma \subset \mathcal{M}(I, R^n)$. See the appendix for this.

1.6 Banach Fixed Point Theorems

Many problems of mathematical sciences lead to the question of existence and uniqueness of solutions of equations of the form $x = F(x)$ in appropriate metric spaces depending on the particular application. The question of existence and uniqueness of solutions of such problems are called fixed point problems.

Let $X \equiv (X, d)$ be a complete metric space and suppose $F : X \to X$.

Definition 1.6.1. A point $x^* \in X$ is said to be a fixed point of the map F if

$$x^* = F(x^*).$$

Let

$$Fix(F) \equiv \{x \in X : x = F(x)\},$$

denote the set of fixed points of the map F. If $Fix(F) = \emptyset$, the equation, $x = F(x)$, has no solution. If the set $Fix(F)$ consists of a single point in X only, then the equation, $x = F(x)$, has a unique solution. Otherwise the equation has multiple solutions.

Fixed point theorems are extremely important tools for proof of existence of solutions of linear and nonlinear problems. We present here only those we have used in the book.

Before we present fixed point theorems, for motivation we consider the following example. As we will see in later chapters, an input-output model often used in engineering, is given by a linear system governed by an integral operator like

$$y(t) \equiv \int_0^t K(t,s)u(s)ds, t \geq 0$$

where the input u may be provided by an expression like

$$u(t) = g(t, y(t)) + r(t), t \geq 0.$$

That is, the input has two components, one being the output feedback through a nonlinear device and the other is the direct input command. In this case we obtain an integral equation of the form

$$y(t) = w(t) + \int_0^t K(t,s)g(s, y(s))ds$$

where

$$w(t) = \int_0^t K(t,s)r(s)ds.$$

Depending on the functions r, g and the kernel K one can formulate this as an abstract fixed point problem

$$y = w + G(y) = F_w(y)$$

in appropriately chosen function spaces. This is precisely in the form we have presented earlier.

Theorem 1.6.2. (Banach Fixed Point Theorem) Let $X \equiv (X, d)$ be a complete metric space, and suppose $F : X \longrightarrow X$ is a contraction in the sense that there exists a number $\alpha \in [0, 1)$ such that

$$d(F(x), F(y)) \leq \alpha\, d(x, y) \;\; \forall x, y \in X.$$

Then F has a unique fixed point in X.

Proof. Let $x_0 \in X$ be any element. Define the sequence

$$x_{k+1} \equiv F(x_k), k \in N_0 \equiv \{0, 1, 2 \cdots\}. \tag{1.6.39}$$

Let $1 \leq p \in N_0$ and note that by repeated application of the triangle inequality we obtain

$$d(x_{n+p}, x_n) \leq \sum_{k=n}^{n+p-1} d(x_{k+1}, x_k) \leq \left(\sum_{k=n}^{n+p-1} \alpha^k \right) d(x_1, x_0). \tag{1.6.40}$$

Since $1 > \alpha \geq 0$, it is clear that

$$\sum_{k=n}^{n+p-1} \alpha^k \leq \alpha^n \left\{ \sum_{k=0}^{\infty} \alpha^k \right\} \leq (\alpha^n / (1 - \alpha)).$$

From (1.6.39), (1.6.40) and the above inequality it is clear that for any fixed $p(1 \leq p) \in N_0$ we have

$$d(x_{n+p}, x_n) \leq (\alpha^n/(1 - \alpha)) d(x_1, x_0)$$

and consequently

$$\lim_{n \to \infty} d(x_{n+p}, x_n) = 0.$$

Thus $\{x_n\}$ is a Cauchy sequence and hence bounded. Since (X, d) is a complete metric space, there exists a unique element $x^* \in X$ such that

$$x_n \xrightarrow{d} x^*$$

in the metric as indicated. Now we must show that x^* is the unique fixed point of the operator F. Again by use of the triangle inequality, the reader can easily verify that

$$\begin{aligned} d(x^*, F(x^*)) &\leq d(x^*, x_n) + d(F(x_{n-1}), F(x^*)) \\ &\leq d(x^*, x_n) + \alpha\, d(x_{n-1}, x^*), \;\; \forall \;\; n \geq 1. \end{aligned} \tag{1.6.41}$$

Letting $n \longrightarrow \infty$, it follows from this inequality that

$$d(x^*, F(x^*)) = 0$$

which proves that x^* is a fixed point of F. The uniqueness of solution is proved by contradiction. Suppose x^*, y^* are any two fixed points of F. Then

$$d(x^*, y^*) = d(F(x^*), F(y^*)) \leq \alpha d(x^*, y^*)$$

and since $0 \leq \alpha < 1$, it is clear that this inequality is false unless $x^* = y^*$. This completes the proof of the theorem. □

Corollary 1.6.3. Suppose X_0 is a closed subset of the metric space (X, d) and $F : X_0 \longrightarrow X_0$ is a contraction. Then F has a unique fixed point in X_0.

Proof. The proof is exactly the same. Here starting from any point $x_0 \in X_0$ one has the sequence $\{x_n\}$, as constructed above, confined in X_0. Since X_0 is closed the limit $x^* \in X_0$. The rest is obvious. □

Corollary 1.6.4. Suppose X_0 is a closed subset of the metric space (X, d) and the n-th iterate or power of F denoted by F^n has the property that $F^n : X_0 \longrightarrow X_0$ is a contraction. Then F^n has a unique fixed point in X_0 and it is also the unique fixed point of the operator F.

Proof. Since F^n is a contraction it follows from Theorem 1.6.2 that it has a unique fixed point, say, $x \in X_0$. We show that x is also the unique fixed point of F. Indeed,

$$\begin{aligned} d(F(x), x) &= d(F(x), F^n(x)) = d(F(F^n(x)), F^n(x)) \\ &= d(F^n(F(x)), F^n(x)) \leq \alpha\, d(F(x), x). \end{aligned} \tag{1.6.42}$$

Since $0 \leq \alpha < 1$, this is impossible unless x is a fixed point of F. Similarly one can verify that if x, y are two fixed points of F,

$$\begin{aligned} d(x, y) &= d(F(x), F(y)) = d(F(F^n(x)), F(F^n(y))) \\ &\leq \alpha\, d(F(x), F(y)) = \alpha\, d(x, y). \end{aligned}$$

This is impossible unless $x = y$ proving uniqueness. □

In the study of systems governed by differential inclusions modeling uncertain systems, that is systems with incomplete description or equivalently systems with parametric uncertainty, we need multi valued analysis in particular Banach fixed point theorem for multivalued maps. Let $X = (X, d)$ be a complete metric space and $c(X)$ denote the class of nonempty closed subsets of X. For each $A, B \in c(X)$ one can define a distance (metric) between them by the following expression

$$d_H(A, B) \equiv \max\{\sup\{d(A, y), y \in B\}, \sup\{d(x, B), x \in A\}\}.$$

It is not difficult to verify that

(MH1): $d_H(A, B) \geq 0, A, B \in c(X), d_H(A, B) = 0$ if and only if $A = B$.

(MH2): $d_H(A, B) = d_H(B, A)$, for $A, B \in c(X)$

(MH3): $d_H(A, B) \leq d_H(A, C) + d_H(C, B)$ for all $A, B, C \in c(X)$.

With this metric topology, $(c(X), d_H)$ is a complete metric space. The metric d_H is known as the Hausdorff metric. Now we are prepared to consider Banach fixed point theorem for multivalued maps.

Definition 1.6.5. Let X_0 be a closed subset of a complete metric space X and $G : X_0 \longrightarrow c(X_0)$ be a multivalued map. A point $z \in X_0$ is said to be a fixed point of the multifunction G if $z \in G(z)$.

Theorem 1.6.6. (Banach Fixed Point Theorem for Multi Functions) Let X_0 be a closed subset of a complete metric space X. Suppose G is a multi valued map from X_0 to $c(X_0)$ satisfying

$$d_H(G(x), G(y)) \leq \alpha\ d(x, y)\ \forall\ x, y \in X_0$$

with $\alpha \in [0, 1)$. Then G has at least one fixed point in X_0.

Proof. Take any $\rho \in (\alpha, 1)$, and $x_0 \in X_0$. Clearly $G(x_0)$ is a closed subset of X_0. Choose an element $x_1 \in G(x_0)$ such that $d(x_0, x_1) > 0$. Clearly if no such element exists then x_0 is already a fixed point of G and that will end the proof. Otherwise, it follows from the Lipschitz property that

$$d(x_1, G(x_1)) \leq d_H(G(x_0), G(x_1)) \leq \rho d(x_0, x_1).$$

Thus there exists an element $x_2 \in G(x_1) \subset X_0$, such that

$$d(x_1, x_2) \leq \rho d(x_0, x_1).$$

Proceeding in this manner we generate a sequence $\{x_n\}$, $n \in N_0 \equiv \{0, 1, 2, \cdots\}$ from the set X_0, satisfying the inclusions $x_{n+1} \in G(x_n), n \in N_0$. Clearly by the Lipschitz property of G, we have

$$d(x_2, x_1) \leq d_H(G(x_1), G(x_0)) \leq \rho d(x_1, x_0)$$
$$d(x_3, x_2) \leq d_H(G(x_2), G(x_1)) \leq \rho d(x_2, x_1) \leq \rho^2 d(x_1, x_0).$$

Thus by induction we have

$$d(x_{n+1}, x_n) \leq \rho^n d(x_1, x_0), \forall\, n \in N_0.$$

Hence for any integer $p \geq 1$, it follows from this inequality that

$$\begin{aligned} d(x_{n+p}, x_n) &\leq \left(\sum_{k=n}^{n+p-1} \rho^k \right) d(x_1, x_0) \\ &\leq \rho^n (1-\rho)^{-1} d(x_1, x_0). \end{aligned}$$

Since $\rho < 1$ this inequality implies that $\lim_{n\to\infty} d(x_{n+p}, x_n) = 0$ for any $p \geq 1$. Thus $\{x_n\}$ is a Cauchy sequence and since X is a complete metric space, there exists a $z \in X$ such that

$$\lim_{n\to\infty} d(x_n, z) = 0.$$

Since X_0 is a closed subset of X and all the elements of the sequence $\{x_n\}$ are contained in X_0, the limit $z \in X_0$. It remains to show that z is a fixed point of G. Note that

$$\begin{aligned} d(z, G(z)) &\leq d(z, x_{n+1}) + d(x_{n+1}, G(z)) \\ &\leq d(z, x_{n+1}) + d_H(G(x_n), G(z)) \\ &\leq d(z, x_{n+1}) + \rho\, d(x_n, z). \end{aligned}$$

This holds for all $n \in N$. Thus, upon letting $n \to \infty$, we find that $d(z, G(z) = 0$. Since $z \in X_0$, we have $G(z) \in c(X_0)$ and so the identity $d(z, G(z)) = 0$ implies that $z \in G(z)$. This proves that z is a fixed point of the multi valued map G thereby completing the proof. □

As a corollary of the above theorem we have the following result.

Corollary 1.6.7. Under the assumptions of Theorem 1.6.6, the fixed point set of the multivalued operator G denoted by

$$Fix(G) \equiv \{\xi \in X_0 : \xi \in G(\xi)\}$$

is a sequentially closed subset of X_0.

Proof. Let $\xi_n \in Fix(G)$ and suppose $d(\xi_n, \xi_0) \longrightarrow 0$ as $n \to \infty$. We show that $\xi_0 \in Fix(G)$. Clearly

$$\begin{aligned} d(\xi_0, G(\xi_0)) &\leq d(\xi_0, \xi_n) + d(\xi_n, G(\xi_0)) \\ &\leq d(\xi_0, \xi_n) + d_H(G(\xi_n), G(\xi_0)) \\ &\leq d(\xi_0, \xi_n) + \rho d(\xi_n, \xi_0). \end{aligned}$$

Letting $n \to \infty$, we obtain $d(\xi_0, G(\xi_0)) = 0$. Since $\xi_0 \in X_0$ and $G(x) \in c(X_0)$ for every $x \in X_0$, we have $\xi_0 \in G(\xi_0)$ and hence $\xi_0 \in Fix(G)$. Thus we have proved that $Fix(G)$ is a sequentially closed subset of X_0. □

1.7 Bibliographical Notes

Most of the materials presented here can be found in any book on real analysis and measure theory such as [44], [80], [69], [41], [19]. Results related to abstract analysis including characterization of the duals of L_p spaces, properties of reflexive Banach spaces and Banach fixed point theorems for single valued maps can be found in Dunford and Schwartz, Larsen and Yosida [33], [57], [98]. The question of vector measures will arise in several chapters. Here only a minor reference to this topic has been made. For detailed study of vector measures the classical book is due to Diestel and Uhl,Jr [30]. For our book however such a deep study is not essential. For Banach fixed point theorem for multi valued maps the most interesting references are Hu, Zeidler and Deimling [48], [100], and [27].

Chapter 2

Linear Systems

2.1 Introduction

Linear time invariant systems have the general form

$$dx(t)/dt \equiv \dot{x} = Ax(t) + Bu(t) + \gamma(t) \tag{2.1.1}$$

where $x(t)$ denotes the state of the system at time t and it is a vector of dimension say n, A is a $n \times n$ matrix representing the system matrix. This matrix characterizes the intrinsic behavior of the uncontrolled system. The matrix B is the control matrix and $u(t)$ is the control vector or input vector. The control matrix can be seen as the channel through which communication takes place between the external world and the internal (machine) world. This is the link through which external influence can be brought to bear upon the internal state of the system. Thus B also determines the extent by which external forces can change the course of evolution of the internal state x. The vector γ denotes the exogenous disturbance. A special case of this system is given by a $n-th$ order differential equation

$$\xi^{(n)} + a_{n-1}\xi^{(n-1)} + \cdots + a_1\xi^{(1)} + a_0\xi = u, \tag{2.1.2}$$

with constant or time varying coefficients independent of ξ or its derivatives. Defining $\{x_1 = \xi, x_2 \equiv \xi^{(1)}, x_3 \equiv \xi^{(2)}, \cdots, x_n \equiv \xi^{(n-1)}\}$, one can write this system in the canonical form

$$\dot{x} = Ax + bu, \tag{2.1.3}$$

where

$$A = \begin{pmatrix} 0 & 1 & 0\ldots & 0 \\ 0 & 0 & 1\ldots & 0 \\ \vdots & \vdots & \ddots & \vdots \\ -a_0 & -a_1 & \ldots & -a_{n-1} \end{pmatrix} \text{ and } b = \begin{pmatrix} 0 \\ 0 \\ \vdots \\ 1 \end{pmatrix}$$

and u is the scalar input. In many communication and control problems, one is simply interested in the solution ξ with the scalar control u. In that situation, one describes this system as a single input single output system (SISO). As opposed to this, in general, we have the multi input multi output systems (MIMO) given by (2.1.1).
Similarly Linear time varying systems have the general form

$$dx(t)/dt = A(t)x(t) + B(t)u(t) + \gamma(t) \tag{2.1.4}$$

where now both the system matrix A and the control matrix B may vary with time.

2.2 Some Examples

(E1) (Series RLC circuit): Consider a series RLC circuit driven by a voltage source. Using Kirchoff's current and voltage laws one can write the system equation as follows.

$$Ri + L\frac{d}{dt}i + v = V(t) \tag{2.2.5}$$

$$C\frac{d}{dt}v = i. \tag{2.2.6}$$

Define the state as $x = col(i, v)$, the system matrix as

$$A = \begin{pmatrix} -(R/L) & -(1/L) \\ (1/C) & 0 \end{pmatrix},$$

and the control matrix as

$$B = \begin{pmatrix} (1/L) \\ 0 \end{pmatrix}$$

with the control being the input voltage $u = V$. Thus the RLC circuit can be written in the standard canonical form

$$\dot{x}(t) = Ax(t) + Bu(t). \tag{2.2.7}$$

(E2) (Suspension Systems): The suspension system of a motor vehicle can be modeled by a system of mass, spring and damper as follows. Consider the static equilibrium position as the zero state and denote by ξ the displacement of the mass m from this rest state. The displaced mass experiences the following forces: inertial force given by $m\ddot{\xi}$, elastic force given by $k\xi$, force due to friction given by $b\dot{\xi}$. Let $u(t)$ denote the external force parallel to the spring. Then using one of the basic laws of Newtonian mechanics, which states that the algebraic sum of all the forces acting on the body m must equal zero, we have

$$m\ddot{\xi} + b\dot{\xi} + k\xi = u.$$

Defining the state x as

$$x = col(\xi, \dot{\xi}) \equiv col(x_1, x_2)$$

we can write the system model in the canonical form

$$(d/dt)\begin{pmatrix} x_1 \\ x_2 \end{pmatrix} = \begin{pmatrix} 0 & 1 \\ -(k/m) & -(b/m) \end{pmatrix}\begin{pmatrix} x_1 \\ x_2 \end{pmatrix} + \begin{pmatrix} 0 \\ (1/m) \end{pmatrix} u. \tag{2.2.8}$$

(E3) (Ecological System): Consider a habitat where only rabbits and foxes live. Rabbits are vegetarians and there are abundance of vegetation in the area. Foxes live by hunting rabbits. In the absence of rabbits the fox population will face starvation and eventual extinction. Letting $R(t)$ and $F(t)$ denote the population of rabbits and foxes at any time t, the system can be approximately modeled by the following system of linear differential equations

$$\begin{aligned} \dot{R} &= \alpha R - \beta F \\ \dot{F} &= -\gamma F + \delta R. \end{aligned} \tag{2.2.9}$$

The parameters $\{\alpha, \beta, \gamma, \delta\}$ are generally positive and have the following physical meaning. The parameter α denotes the intrinsic growth rate of rabbits, β the depletion rate per unit of fox population. In other words the actual depletion rate is linearly proportional to fox population, that is βF. The parameter γ denotes the mortality rate of foxes for lack of food and δ denotes the growth rate per unit of rabbit population. In other words the actual growth rate is linearly proportional to rabbit population, that is δR.

Defining the state x as $x = col(R, F) \equiv col(x_1, x_2)$, this equation can be written in the canonical form (2.1.1).

(E4) (Torsional Motion): Suppose one end of a cylindrical beam of length L is attached horizontally to a wall with the other end free to rotate around its central axis. The torsional (rotational) motion around the central axis is governed by a partial differential equation given by

$$J\frac{\partial^2}{\partial t^2}\phi + \frac{\partial^2}{\partial x^2}\left(EI\frac{\partial^2}{\partial x^2}\phi\right) = T, x \in (0, L), t \geq 0, \tag{2.2.10}$$

where J is the moment of inertia about the central axis, EI is the modulus of rigidity and ϕ is the angular displacement corresponding to the applied torque T. The parameters may be functions of x if the beam is not uniform. By use of modal expansion and retaining only the first n modes, this equation can be written as a system of n second order ordinary differential equations,

$$\mathcal{J}\frac{d^2}{dt^2}\Theta + K\Theta = \mathcal{T}, \tag{2.2.11}$$

where $\mathcal{J}$ is the matrix of mass-moment of inertia, K is the stiffness matrix and $\mathcal{T}$ is a n-vector of torques and Θ is the first n modes of angular displacement. Defining the state $x \equiv col(\Theta, \dot{\Theta})$, the system can be written in the standard canonical form

$$\dot{x} = Ax + Bu$$

where

$$A \equiv \begin{pmatrix} 0 & I \\ -\mathcal{J}^{-1}K & 0 \end{pmatrix} \text{ and } B \equiv \begin{pmatrix} 0 \\ \mathcal{J}^{-1} \end{pmatrix} \tag{2.2.12}$$

and $u = \mathcal{T}$.

In case we are only interested in the angular position of the free end of the beam one obtains a scalar differential equation $J\ddot{\theta} + k\theta = \tau$ where J is the mass moment of inertia about the central axis, k is the stiffness coefficient and τ is the torque.

(E5) (Machine Reliability-2 state): Consider a machine, for example an electrical motor, a generator or a microprocessor etc., which can reside in any one of two states: S1 and S0 at any given time.

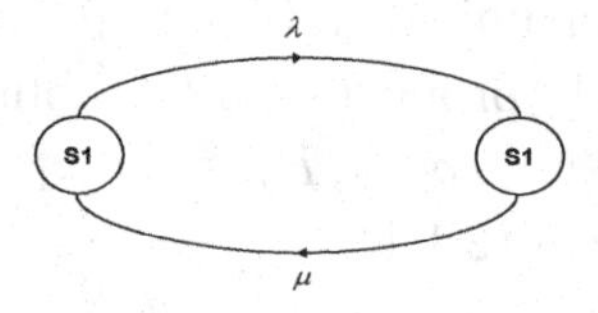

Figure 2.2.1: 2-State Diagram

S1 is the operating state and S0 denotes the failed state. Let λ denote the long term failure rate or transition rate from state S1 to state S0 and μ the repair rate from the failed state S0 to the operating state S1. If one looks at the history of operation of the machine, one will observe a random process taking values S0 and S1 alternately and making transitions at random times from one state to the other. By counting the number of times, $N_F(0,T)$, the machine has failed during the time interval $[0,T]$ and dividing by T one obtains

$$\lambda = \lim_{T\to\infty} (N_F(0,T)/T) \tag{2.2.13}$$

To compute the repair rate one must sum all the time intervals machine spent in failed state during the period $[0,T]$ and divide by T and take the limit. Let τ_{F_i} and τ_{R_i} denote the random times of the $i-th$ failure and $i-th$ repair. Then one can argue that the average long term repair rate is given by

$$\mu = \lim_{T\to\infty} (1/T) \sum_{i=1}^{N_F(0.T)} (\tau_{R_i} - \tau_{F_i}). \tag{2.2.14}$$

We are interested in the probabilistic dynamics of the machine. Let $P_1(t)$ denote the probability that the machine is in operating state at time t and $P_0(t)$ the probability that it is in the failed state. Then the probability that the machine is in state S1 at time $t+\Delta t$ is given by

$$P_1(t+\Delta t) = P_1(t)(1-\lambda\Delta t) + P_0(t)\mu\Delta t + o(\Delta t) \tag{2.2.15}$$

where the first term gives the probability that the machine was in state S1 at time t and that during the time interval $[t, t+\Delta t]$ it did not fail and the second term gives the probability that the machine was in failed sate S0 at time t and that it was repaired during the time interval $[t, t+\Delta t]$. The third term denotes the approximation error which is of small order in Δt in the sense that $\lim_{\Delta t\to 0}\{o(\Delta t)/\Delta t\} = 0$. From this one easily arrives at the following differential equation

$$\dot{P}_1(t) = -\lambda P_1(t) + \mu P_0(t), t \geq 0. \tag{2.2.16}$$

Similarly the probability that the machine is in failed state S0 at time $t+\Delta t$ is given by

$$P_0(t+\Delta t) = P_0(t)(1-\mu\Delta t) + P_1(t)\lambda\Delta t + o(\Delta t). \tag{2.2.17}$$

where the first term on the right gives the probability that the machine was in failed state at time t and that it did not get repaired during the time interval $[t, t+\Delta t]$. The second term denotes the probability that the

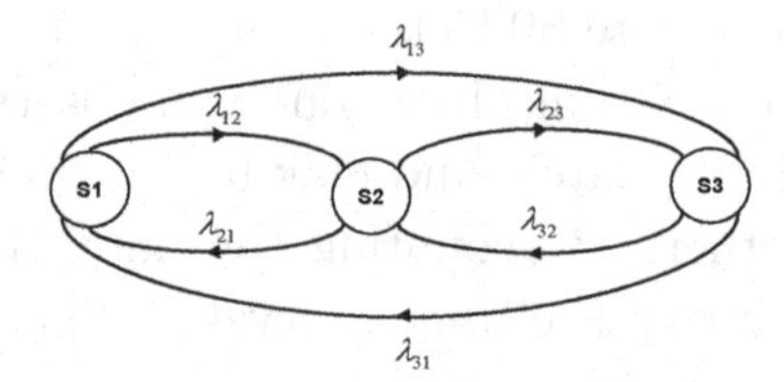

Figure 2.2.2: 3-State Diagram

machine was in operating state S1 at time t and that it failed during the interval $[t, t+\Delta t]$. The third term has the same meaning as stated earlier. This leads to the differential equation

$$\dot{P}_0(t) = \lambda P_1(t) - \mu P_0(t). \tag{2.2.18}$$

Thus finally we arrive at the probabilistic dynamics of the machine given by the following differential equations:

$$(d/dt)\begin{pmatrix} P_1 \\ P_0 \end{pmatrix} = \begin{pmatrix} -\lambda & \mu \\ \lambda & -\mu \end{pmatrix}\begin{pmatrix} P_1 \\ P_0 \end{pmatrix}. \tag{2.2.19}$$

(E6) (System Reliability-3 State): Suppose a production system can be in either one of the three states $\{S1, S2, S3\}$ where state $S1$ denotes the full capacity state, $S2$ denotes a partial capacity state and $S3$ denotes the failed state.

We may denote by $S(t)$ the state of the system at time t. Let $\{\lambda_{12}, \lambda_{13}, \lambda_{23}\}$ denote the transition (failure) rates from state $S1 \to S2$, $S1 \to S3$ and $S2 \to S3$ respectively. Similarly $\{\mu_{21}, \mu_{31}, \mu_{32}\}$ denote the transition (repair) rates from states $S2 \to S1$, $S3 \to S1$ and $S3 \to S2$. Let $\{P_1(t), P_2(t), P_3(t)\}$ denote the probabilities of the system being in states $\{S1, S2, S3\}$ respectively. Then using similar arguments as in the previous example we have

$$P_1(t+\Delta t) = P_1(t)(1-\lambda_{12}\Delta t)(1-\lambda_{13}\Delta t)+P_2(t)\mu_{21}\Delta t+P_3(t)\mu_{31}\Delta t+o(\Delta t). \tag{2.2.20}$$

Hence rearranging terms and letting $\Delta t \to 0$ we obtain the differential equation for P_1

$$\dot{P}_1(t) = -(\lambda_{12} + \lambda_{13})P_1(t) + \mu_{21}P_2(t) + \mu_{31}P_3(t). \tag{2.2.21}$$

Similarly for P_2 we have

$$P_2(t+\Delta t) = P_2(t)(1-\lambda_{23}\Delta t-\mu_{21}\Delta t)+P_1(t)\lambda_{12}\Delta t+P_3(t)\mu_{32}\Delta t+o(\Delta t). \tag{2.2.22}$$

Thus the corresponding differential equation is given by

$$\dot{P}_2(t) = -(\lambda_{23}+\mu_{21})P_2(t)+\lambda_{12}P_1(t)+\mu_{32}P_3(t). \tag{2.2.23}$$

Similarly for P_3 we have

$$P_3(t+\Delta t) = P_3(t)(1-\mu_{32}\Delta t-\mu_{31}\Delta t)+P_2(t)\lambda_{23}\Delta t+P_1(t)\lambda_{13}\Delta t+o(\Delta t). \tag{2.2.24}$$

Thus the corresponding differential equation is given by

$$\dot{P}_3(t) = -(\mu_{31}+\mu_{32})P_3(t)+\lambda_{23}P_2(t)+\lambda_{13}P_1(t). \tag{2.2.25}$$

Combining all the above we have the system equation in the standard canonical form

$$(d/dt)\begin{pmatrix} P_1 \\ P_2 \\ P_3 \end{pmatrix} = \begin{pmatrix} -(\lambda_{12}+\lambda_{13}) & \mu_{21} & \mu_{31} \\ \lambda_{12} & -(\lambda_{23}+\mu_{21}) & \mu_{32} \\ \lambda_{13} & \lambda_{23} & -(\mu_{31}+\mu_{32}) \end{pmatrix}\begin{pmatrix} P_1 \\ P_2 \\ P_3 \end{pmatrix}. \tag{2.2.26}$$

For a manufacturer who owns the system, it is important to have an estimate of the average daily production. Let C_1 denote the full capacity (production rate per unit time) when the system is in state $S1$ and $C_2(< C_1)$ is the capacity when the system is in state $S2$ and $C_3(= 0)$ is the capacity in state $S3$. Define the indicator function

$$1(S(t) = S) \equiv \begin{cases} 1 & \text{if } S(t) = S \\ 0 & \text{otherwise.} \end{cases}$$

Using the indicator function one may express the expected (mathematical average) production over the time interval $[0, T]$ as

$$\begin{aligned} Q_T &\equiv E\left\{\int_0^T \{C_1\ 1(S(t) = S1) + C_2\ 1(S(t) = S2) + C_3\ 1(S(t) = S3)\}dt\right\} \\ &= \int_0^T \{C_1\ P_1(t) + C_2\ P_2(t) + C_3\ P_3(t)\}dt \\ &= \int_0^T \{C_1\ P_1(t) + C_2\ P_2(t)\}dt. \end{aligned} \tag{2.2.27}$$

In general a multi state system is given by

$$(d/dt)P = AP, P(0) = P_0 \tag{2.2.28}$$

where P is a n-vector of probabilities with $P_i(t), i = 1, 2, \cdots, n$ denoting the probability that the system is in state Si at time t. The matrix A is the infinitesimal generator of the random process $S(t), t \geq 0$ which takes values from the set of states $\mathcal{S} \equiv \{S1, S2, S3, \cdots, Sn\}$ which constitutes the state space of the production system. For application to maintainable systems arising in manufacturing industry see [153].

It is interesting to note that the sum of the entries of any column of the infinitesimal generator A is zero. This is a unique characteristic of infinitesimal generators of Markov chains. In fact there is also a strong similarity with the Kirchoff's current law that states that the algebraic sum of currents in all the branches connected to any node must equal zero. Thus using this basic principle one can easily construct the generator matrix A just by inspection.

Now we consider the main objective of this chapter. We wish to give a complete mathematical representation of solutions of time invariant and time varying systems abbreviated by (TIS) and (TVS) respectively. First we consider the time invariant case.

2.3 Representation of Solutions for TIS

First we consider the time invariant homogeneous (or uncontrolled) system model

$$\dot{x} = Ax, x(0) = x_0 \in R^n. \tag{2.3.29}$$

Recall that if $n = 1$, both x and $A = a$ are scalars and the system reduces to an elementary first order ordinary differential equation written as $\dot{x} = ax$. The solution is then given by $x(t) = e^{ta}x_0 = x_0 e^{ta}$. The objective is to demonstrate that the solution of the multidimensional problem has similar appearance and it is given by

$$x(t) = e^{tA}x_0, t \geq 0, \tag{2.3.30}$$

where the exponential function e^{tA} has the meaning

$$e^{tA} \equiv \sum_{n=0}^{\infty}(1/n!)t^n A^n.$$

We demonstrate this using four different techniques.

(T1) Laplace Transform Technique: First we use the Laplace transform technique. Denote by $\hat{x}$ the Laplace transform of the vector valued function x given by

$$\hat{x}(s) \equiv \int_0^\infty e^{-st} x(t) dt.$$

Applying Laplace transform on either side of equation, (2.3.29), one obtains

$$(sI - A)\hat{x}(s) = x_0, s \in C.$$

Taking $Re(s)$ sufficiently large (larger than the largest eigen value of the matrix A or equivalently $Re(s) > \| A \|$) we can compute the inverse of the matrix $(sI - A)$ and obtain

$$\begin{aligned} \hat{x}(s) = (sI - A)^{-1} x_0 &= (1/s)(I - (1/s)A)^{-1} x_0 \\ &= (1/s) \sum_{n=0}^{\infty} (1/s^n) A^n x_0. \end{aligned} \tag{2.3.31}$$

This follows from the fact that, for $Re(s)$ sufficiently large, $(1/|s|) \| A \| < 1$ and hence the matrix $(I - (1/s)A)$ is invertible. Using the binomial expansion, $(1-a)^{-1} = 1 + a + a^2 + a^3 +, \cdots$, which is valid for $|a| < 1$, the result follows. Now Recalling that $\int_0^\infty \theta^n e^{-\theta} d\theta = n!$, we have

$$\int_0^\infty t^n e^{-st} dt = (n!/s^{n+1}).$$

Hence taking the inverse Laplace transform of (2.3.31), we arrive at the following expression

$$x(t) = \sum_{n=0}^{\infty} (1/n!) t^n A^n x_0 \equiv e^{tA} x_0. \tag{2.3.32}$$

An Example: Consider the machine reliability model of example (E5). This is a binary system and we have seen that the dynamics of evolution of the probability vector is given by equation (2.2.19). Using the Laplace transform technique, we reduce this equation into the following algebraic equation

$$s\hat{P} - A\hat{P} = P_0.$$

Inverting the matrix $(sI - A)$ we obtain

$$\begin{aligned}\hat{P} &= (sI - A)^{-1}P_0 \\ &= \begin{pmatrix} s+\lambda & -\mu \\ -\lambda & s+\mu \end{pmatrix}^{-1} \begin{pmatrix} P_{10} \\ P_{00} \end{pmatrix} \\ &= \begin{pmatrix} \frac{(s+\mu)}{s(s+\lambda+\mu)} & \frac{\mu}{s(s+\lambda+\mu)} \\ \frac{\lambda}{s(s+\lambda+\mu)} & \frac{s+\lambda}{s(s+\lambda+\mu)} \end{pmatrix} \begin{pmatrix} P_{10} \\ P_{00} \end{pmatrix}.\end{aligned}$$

From this we have

$$\begin{aligned}\hat{P}_1(s) &= P_{10}\frac{(s+\mu)}{s(s+\lambda+\mu)} + P_{00}\frac{\mu}{s(s+\lambda+\mu)} \\ \hat{P}_0(s) &= P_{10}\frac{\lambda}{s(s+\lambda+\mu)} + P_{00}\frac{s+\lambda}{s(s+\lambda+\mu)}.\end{aligned}$$

Taking the inverse transform we obtain the corresponding time functions

$$P_1(t) = \frac{\mu}{\lambda+\mu} + (\lambda P_{10} - \mu P_{00})\frac{e^{-(\lambda+\mu)t}}{\lambda+\mu} \tag{2.3.33}$$

$$P_0(t) = \frac{\lambda}{\lambda+\mu} + (\mu P_{00} - \lambda P_{10})\frac{e^{-(\lambda+\mu)t}}{\lambda+\mu}. \tag{2.3.34}$$

For a freshly bought machine, it is obvious that $P_{00} = 0$ and $P_{10} = 1$. In this case equations (2.3.33) and (2.3.34) reduce to

$$P_1(t) = \frac{\mu}{\lambda+\mu} + \frac{\lambda}{\lambda+\mu}e^{-(\lambda+\mu)t} \tag{2.3.35}$$

$$P_0(t) = \frac{\lambda}{\lambda+\mu} - \frac{\lambda}{\lambda+\mu}e^{-(\lambda+\mu)t}. \tag{2.3.36}$$

Notice that $P_1(t) + P_0(t) = 1$ for all $t \geq 0$, as expected. With time $P_1(t)$ decreases and $P_0(t)$ increases monotonically and converge to their long term values given by

$$\lim_{t\to\infty} P_1(t) = \frac{\mu}{\lambda+\mu} \tag{2.3.37}$$

$$\lim_{t\to\infty} P_0(t) = \frac{\lambda}{\lambda+\mu}. \tag{2.3.38}$$

(T2) Explicit Difference Approximations: We wish to compute the state $x(t)$ at an arbitrary but fixed time t. Partition the interval $[0, t]$ into n equal subintervals with each subinterval having width Δt so that $t = n\Delta t$.

Define

$$\begin{aligned} x(\triangle t) &\approx x_0 + \triangle t A x_0 = (I + \triangle t A) x_0 \\ x(2\triangle t) &\approx x(\triangle t) + \triangle t A x(\triangle t) = (I + \triangle t A) x(\triangle t) \\ &\approx (I + \triangle t A)^2 x_0. \end{aligned} \tag{2.3.39}$$

Thus for $x(t) \approx x(n\triangle t)$ we have

$$x(t) \approx x(n\triangle t) \approx (I + \triangle t A)^n x_0 \equiv x_n(t), \tag{2.3.40}$$

where we have denoted by $x_n(t)$ the n-th approximation of $x(t)$ at time t. Keeping t fixed, if we let $n \to \infty$, the difference approximation converges to the true solution and it is given by

$$x(t) \equiv \lim_{n\to\infty} x_n(t) = \lim_{n\to\infty} (I + (t/n)A)^n x_0 = e^{tA} x_0. \tag{2.3.41}$$

(T3) Implicit Difference Approximations: Here equation (2.3.39) is replaced by

$$\begin{aligned} x(\triangle t) &\approx x_0 + \triangle t A x(\triangle t) = (I - \triangle t A)^{-1} x_0 \\ x(2\triangle t) &\approx x(\triangle t) + \triangle t A\, x(2\triangle t) = (I - \triangle t A)^{-1} x(\triangle t) \\ &\approx (I - \triangle t A)^{-2} x_0. \end{aligned} \tag{2.3.42}$$

In general $x_n(t) \equiv x(n\triangle t)$ is given by

$$\begin{aligned} x_n(t) = x(n\triangle t) &= (I - \triangle t A)^{-n} x_0 \\ &= (I - (t/n)A)^{-n} x_0. \end{aligned} \tag{2.3.43}$$

Again one can use binomial expansion to verify that

$$\lim_{n\to\infty} (I - (t/n)A)^{-n} = e^{tA}.$$

Thus we have the exact solution given by the limit

$$x(t) = \lim_{n\to\infty} x_n(t) = \lim_{n\to\infty} (I - (t/n)A)^{-n} x_0 = e^{tA} x_0. \tag{2.3.44}$$

(T4) Successive Approximation Technique: Here we integrate the equation (2.3.29) obtaining

$$x(t) = x_0 + \int_0^t A x(\tau) d\tau, \quad t \geq 0. \tag{2.3.45}$$

This is an elementary linear integral equation where the unknown appears under the integral sign as well as outside of it. Our objective now is to find a solution of this integral equation. If the integrand is known the object on the left can be determined by evaluating the integral on the right, but to compute the integral one must know the object on the left. This cycle, analogous to "chicken-egg" situation, can be broken by introducing successive approximations. Define the following sequence of approximations:

$$\begin{aligned} x_1(t) &= x_0, \quad t \geq 0 \\ x_m(t) &= x_0 + \int_0^t A x_{m-1}(\tau) d\tau, \quad t \geq 0, m \in N. \end{aligned} \tag{2.3.46}$$

Substituting the first equation into the second and the second into the third and so on, one obtains

$$x_m(t) = \sum_{k=0}^{m-1} (t^k/k!) A^k x_0, \quad t \geq 0. \tag{2.3.47}$$

The reader can easily verify that $\{x_m\}$ is a Cauchy sequence in $C(I, R^n)$ and hence has a unique limit and it is given by the expression (2.3.32). That the series converges to the well defined limit (2.3.32) is also justified later under more general situations.

An Example: In fact this successive approximation technique is very powerful and can be applied to both time invariant as well as to time varying systems. For example, we consider the following problem

$$\dot{x}(t) = \beta(t) A x(t), x(0) = x_0, \tag{2.3.48}$$

where β is a scalar valued function. Again using the successive approximation method we have

$$\begin{aligned} x_1(t) &= x_0, \quad t \geq 0 \\ x_n(t) &= x_0 + \int_0^t \beta(\tau) A x_{n-1}(\tau) d\tau, \quad t \geq 0, n \geq 1. \end{aligned} \tag{2.3.49}$$

Define

$$h(t) \equiv \int_0^t \beta(\tau) d\tau.$$

Substituting the first equation into the second and the second into the third and so on, one can verify that

$$\begin{aligned} x_1(t) &= x_0, \quad t \geq 0 \\ x_n(t) &= \sum_{k=0}^{n-1} (h(t)^k/k!) A^k x_0, \quad t \geq 0, n \geq 1. \end{aligned} \tag{2.3.50}$$

The reader can easily verify that $\{x_n\}$ is a Cauchy sequence in $C(I, R^n)$ for any finite interval I. Hence, in the limit as $n \to \infty$, it follows from equation (2.3.50) that $x_n(t) \xrightarrow{u} x(t)$ on every finite interval I. Hence the solution to the problem (2.3.48) is given by

$$x(t) = e^{h(t)A}x_0 = e^{(\int_0^t \beta(\tau)d\tau)A}x_0. \tag{2.3.51}$$

Now we consider the general controlled system

$$\begin{aligned} \dot{x}(t) &= Ax(t) + Bu(t) + \gamma(t), \quad t \geq 0, \\ x(0) &= x_0, \end{aligned} \tag{2.3.52}$$

where $u(t)$ is the control vector and $\gamma(t)$ is the disturbance vector.

Theorem 2.3.1. Let u and γ be Lebesgue integrable functions. Then, a continuous function $\varphi(t), t \geq 0$, taking values from R^n, is a solution of equation (2.3.52) if and only if φ is given by

$$\varphi(t) = e^{tA}x_0 + \int_0^t e^{(t-s)A}Bu(s)ds + \int_0^t e^{(t-s)A}\gamma(s)ds, t \geq 0. \tag{2.3.53}$$

Proof. For the if part, we show that φ as given by the expression (2.3.53) satisfies the equations (2.3.52). Letting $t \downarrow 0$ it follows from the expression (2.3.53) due to integrability of the functions u and γ that $\varphi(0) = x_0$. Differentiating term by term the function φ as given, we have

$$\begin{aligned} \dot{\varphi}(t) &= Ae^{tA}x_0 + \left\{\int_0^t Ae^{(t-s)A}Bu(s)ds + Bu(t)\right\} \\ &\quad + \left\{\int_0^t Ae^{(t-s)A}\gamma(s)ds + \gamma(t)\right\}, \\ &= A\left(e^{tA}x_0 + \int_0^t e^{(t-s)A}Bu(s)ds + \int_0^t e^{(t-s)A}\gamma(s)ds\right) + Bu(t) + \gamma(t) \\ &= A\varphi(t) + Bu(t) + \gamma(t), \quad t > 0. \end{aligned} \tag{2.3.54}$$

Thus φ satisfies the equations (2.3.52). Suppose now that a function z satisfies the system equation (2.3.52). We must show that z equals φ as given. Since z satisfies the system equations (2.3.52) we have $z(0) = x_0$ and

$$\dot{z}(s) - Az(s) = Bu(s) + \gamma(s), \quad s > 0. \tag{2.3.55}$$

Multiplying the equation (2.3.55) on the left by the matrix e^{-sA} and integrating over the interval $[0, t]$ we obtain

$$\int_0^t e^{-sA}\Big(\dot{z}(s) - Az(s)\Big)ds = \int_0^t e^{-sA}\Big(Bu(s) + \gamma(s)\Big)ds. \tag{2.3.56}$$

Now using the fact that A commutes with any function of A, the expression on the left of equation (2.3.56) can be written as

$$\int_0^t e^{-sA}\Big(\dot{z}(s) - Az(s)\Big)ds = \int_0^t (d/ds)\Big(e^{-sA}z(s)\Big)ds$$
$$= e^{-tA}z(t) - z(0). \tag{2.3.57}$$

Hence it follows from equations (2.3.56) and (2.3.57) that

$$e^{-tA}z(t) - z(0) = \int_0^t e^{-sA}\Big(Bu(s) + \gamma(s)\Big)ds. \tag{2.3.58}$$

By multiplying this equation on the left by e^{tA} and recalling that $z(0) = x_0$, we arrive at the following expression

$$\begin{aligned} z(t) &= e^{tA}z(0) + \int_0^t e^{(t-s)A}\{Bu(s) + \gamma(s)\}ds \\ &= e^{tA}x_0 + \int_0^t e^{(t-s)A}\{Bu(s) + \gamma(s)\}ds \\ &= \varphi(t). \end{aligned} \tag{2.3.59}$$

Since t is arbitrary, this verifies that z coincides with φ. This ends the proof. □

Some special situations: Suppose an impulsive disturbance occurs at time $t_0 \geq 0$. That is, the disturbance γ is given by $\gamma(t) = \eta\ \delta(t - t_0)$ for some vector $\eta \in R^n$ where $\delta(t)$ denotes the Dirac delta function (actually measure). Then the response is given by

$$x(t) \equiv \begin{cases} e^{tA}x_0 + \displaystyle\int_0^t e^{(t-s)A}Bu(s)ds & \text{for } 0 \leq t < t_0 \\ e^{tA}x_0 + \displaystyle\int_0^t e^{(t-s)A}Bu(s)ds + e^{(t-t_0)A}\eta & \text{for } t \geq t_0\ . \end{cases} \tag{2.3.60}$$

This gives rise to a jump in state at t_0 given by

$$\Delta x(t_0) \equiv x(t_0+) - x(t_0-) = \eta.$$

Note that this solution is right continuous having left limit.

Discrete Time Model: A discrete time model of the general control system can be written as

$$\begin{aligned} x(k+1) &= Ax(k) + Bu(k) + \gamma(k),\ k \in N \equiv \{0, 1, 2, \cdot, \cdot\} \\ x(0) &= x_0, \end{aligned} \tag{2.3.61}$$

where A and B are suitable matrices. In the sequel we will have further occasions to discuss this model.

Transfer Function: For linear time invariant systems (only), one can use Laplace transform to construct the input-state and input-output transfer functions. In the multivariable case we have

$$\hat{x}(s) = (sI - A)^{-1}x_0 + (sI - A)^{-1}B\hat{u}(s). \tag{2.3.62}$$

Here the input-state transfer (matrix) function is given by

$$\hat{G}(s) \equiv (sI - A)^{-1}B$$

and in case the output is given by $y(t) = Hx(t)$, the input-output transfer function is given by

$$\hat{G}(s) = H(sI - A)^{-1}B.$$

In general the output is then given by the convolution integral

$$y(t) = \int_{-\infty}^{t} G(t-\tau)u(\tau)d\tau = \int_{-\infty}^{t} h(t-\tau)Bu(\tau)d\tau, \tag{2.3.63}$$

where G is the Laplace inverse of $\hat{G}(s)$ defining the so called input-output impulse response matrix and h is the Laplace inverse of $H(sI-A)^{-1}$, defining the disturbance-output impulse response. Note that if $u(t) \equiv 0$ for all $t < 0$ and an impulsive disturbance occurs at time $t = 0+$ so that $Bu(t) = x_0\delta(t)$, it follows from equation (2.3.63) that

$$\begin{aligned} y(t) &= \int_{-\infty}^{0} h(t-\tau)x_0\delta(\tau)d\tau + \int_{0}^{t} h(t-\tau)Bu(\tau)d\tau \\ &= h(t)x_0 + \int_{0}^{t} G(t-\tau)u(\tau)d\tau \\ &= He^{tA}x_0 + \int_{0}^{t} G(t-\tau)u(\tau)d\tau. \end{aligned}$$

2.4 Representation of Solutions for TVS

As in the previous section, first we consider the homogeneous or autonomous system

$$\begin{aligned} \dot{x}(t) &= A(t)x(t), t \geq s, \\ x(s) &= \xi, \end{aligned} \tag{2.4.64}$$

where now the system matrix A is time varying or a function of time t, s is the starting time and $\xi \in R^n$ is the starting state. Our objective now is to construct the transition operator $\Phi(t,s)$ that solves the above problem, that is,

$$x(t) \equiv \varphi(t, s, \xi) \equiv \Phi(t, s)\xi, t \geq s. \tag{2.4.65}$$

The very first question that one should ask is "can there be a function x whose derivative equals the product Ax at every (almost every) point of time?"

We wish to demonstrate that under certain mild assumptions on the matrix $A(t), t \in R$, equation (2.4.64) has indeed a unique solution and that the solution is continuously dependent on the initial state. In order to prove this result we need a very important inequality, known as the Gronwall inequality named after its author. This is one of the most celebrated result which has found broad applications in the theory of differential equations.

Lemma 2.4.1. Let $\phi(t), \alpha(t), K(t), t \geq 0$, be nonnegative functions with $\alpha \in L_\infty^{\ell oc}[0,\infty)$ and $K \in L_1^{\ell oc}[0,\infty)$, satisfying the integral inequality

$$\phi(t) \leq \alpha(t) + \int_0^t K(r)\phi(r)dr, t \geq 0.$$

Then $\phi \in L_\infty^{\ell oc}[0,\infty)$ and it is bounded above by the following expression:

$$\phi(t) \leq \alpha(t) + \int_0^t \exp\left\{\int_s^t K(r)dr\right\} K(s)\alpha(s)ds. \tag{2.4.66}$$

Further, if α is simply a nonnegative constant then this inequality reduces to

$$\phi(t) \leq \ \alpha \ \exp\left\{\int_0^t K(r)dr\right\}. \tag{2.4.67}$$

Proof. Define

$$\psi(t) \equiv \int_0^t K(r)\phi(r)dr.$$

Clearly ψ is absolutely continuous and its differential is given by $\dot{\psi} = K(t)\phi(t)$. Thus

$$\phi(t) \leq \alpha(t) + \psi(t). \tag{2.4.68}$$

Since $K(t) \geq 0$ for all $t \geq 0$, multiplying this on either side by $K(t)$ we have

$$\dot{\psi}(t) \leq (\alpha(t) + \psi(t))K(t), t \geq 0.$$

Solving this for ψ we obtain

$$\psi(t) \leq \int_0^t \exp\left\{\int_s^t K(r)dr\right\} \alpha(s)K(s)ds.$$

Substituting this in (2.4.68) we arrive at the inequality (2.4.66). The inequality (2.4.67) follows from (2.4.66). Indeed, it follows from (2.4.66) that

$$\phi(t) \leq \alpha \left\{1 + \exp\left\{\int_0^t K(r)dr\right\} \int_0^t \exp\left\{-\int_0^s K(r)dr\right\} K(s)ds\right\}.$$

Defining

$$\beta(s) \equiv \exp\left\{-\int_0^s K(r)dr\right\},$$

it follows from the above expression that

$$\begin{aligned}
\phi(t) &\leq \alpha \left\{1 + \exp\left\{\int_0^t K(r)dr\right\} \int_0^t \beta(s)K(s)ds\right\} \\
&\leq \alpha \left\{1 + \exp\left\{\int_0^t K(r)dr\right\} \left(\int_0^t -d\beta(s)\right)\right\} \\
&\leq \alpha \left\{1 + \exp\left\{\int_0^t K(r)dr\right\} (\beta(0) - \beta(t))\right\} \\
&\leq \alpha \left\{1 + \left(\exp\left\{\int_0^t K(r)dr\right\}\right) \left(1 - \exp\left\{-\int_0^t K(r)dr\right\}\right)\right\} \\
&\leq \alpha \exp\left\{\int_0^t K(r)dr\right\}.
\end{aligned}$$

Thus we have proved the inequality (2.4.67). This completes the proof. □

Remark. For constant α, one can derive the above result directly from the inequality

$$\dot{\psi}(t) \leq K(t)(\alpha + \psi(t))$$

by use of integration by parts. For more general results see (Q6)-Q(7).

Now we are ready to prove the following result.

Theorem 2.4.2. If the elements of the matrix $A(t), t \in R$, are locally integrable in the sense that $\int_I |a_{ij}|dt < \infty$ for all $i, j = 1, 2, \cdots, n$, and for every finite interval I, then the equation (2.4.64) has a unique (absolutely) continuous solution and that this solution is continuously dependent on the initial sate ξ.

Proof. For the proof use the successive approximation technique as discussed in the time invariant case. Consider the finite time interval $I \equiv [0, T]$

and the linear vector space $C(I, R^n)$ of continuous functions on I taking values from R^n. This is a complete normed space (Banach space) with respect to the norm defined by

$$\| x \| \equiv \sup_{t \in I} \| x(t) \| .$$

Define the following sequence of approximations as follows:

$$\begin{aligned} x_1(t) &= \xi, \quad t \geq s, s, t \in I, \\ x_{m+1}(t) &= \xi + \int_s^t A(\tau) x_m(\tau) d\tau, \quad t \geq s, m \in N. \end{aligned} \tag{2.4.69}$$

Define

$$h(t) \equiv \int_s^t \| A(\tau) \| d\tau. \tag{2.4.70}$$

Since the elements of the matrix are locally integrable, it is clear that the function h is nondecreasing, continuous, once differentiable and bounded for finite t. Note that

$$\begin{aligned} \| x_2(t) - x_1(t) \| &\leq \left(\int_s^t \| A(\tau) \| d\tau \right) \| \xi \| \\ &\leq h(t) \| \xi \| . \end{aligned} \tag{2.4.71}$$

Using the equations (2.4.69) and the inequality (2.4.71), it is easy to verify that

$$\| x_3(t) - x_2(t) \| \leq (h^2(t)/2) \| \xi \| . \tag{2.4.72}$$

Proceeding in this manner we arrive at the following inequality

$$\| x_{m+1}(t) - x_m(t) \| \leq (h^m(t)/m!) \| \xi \| . \tag{2.4.73}$$

Since for any positive integer p,

$$x_{m+p}(t) - x_m(t) = \sum_{k=m}^{m+p-1} (x_{k+1}(t) - x_k(t)), \tag{2.4.74}$$

we have

$$\begin{aligned} \| x_{m+p}(t) - x_m(t) \| &\leq \sum_{k=m}^{m+p-1} \| x_{k+1}(t) - x_k(t) \| \\ &\leq \sum_{k=m}^{m+p-1} (h^k(t)/k!) \| \xi \| \\ &\leq (\exp h(t))(h^m(t)/m!) \| \xi \| . \end{aligned} \tag{2.4.75}$$

It follows from this inequality that

$$\| x_{m+p} - x_m \| \equiv \sup_{t \in I}\{\| x_{m+p}(t) - x_m(t) \|\} \leq (\exp h(T))(h^m(T)/m!) \| \xi \| . \tag{2.4.76}$$

Since $h(T)$ is finite it follows from the above estimate that, for any integer p,

$$\lim_{m \to \infty} \| x_{m+p} - x_m \| = 0. \tag{2.4.77}$$

Thus $\{x_m\}$ is a Cauchy sequence in the Banach space $C(I, R^n)$ and hence there exists a unique $x \in C(I, R^n)$ to which x_m converges in the norm defined above. This means that

$$\lim_{m \to \infty} x_m(t) = x(t) \tag{2.4.78}$$

uniformly on the interval I. Hence letting $m \to \infty$ in the expression (2.4.69) we obtain

$$x(t) = \xi + \int_s^t A(\theta)x(\theta)d\theta, t \in I. \tag{2.4.79}$$

The absolute continuity follows from the integrability of $A(t), t \in I$. For continuity with respect to the initial data, let $\xi, \eta \in R^n$ and denote by x the solution corresponding to the initial state ξ and y the solution corresponding to η. Then it follows from equation (2.4.79) that

$$x(t) - y(t) = \xi - \eta + \int_s^t A(\theta)\{x(\theta) - y(\theta)\}d\theta, t \in I. \tag{2.4.80}$$

Clearly then by triangle inequality we have

$$\| x(t) - y(t) \| \leq \| \xi - \eta \| + \int_s^t \| A(\theta) \| \| x(\theta) - y(\theta) \| d\theta \tag{2.4.81}$$

and by Gronwall inequality (2.4.67) we obtain

$$\| x(t) - y(t) \| \leq \| \xi - \eta \| \exp h(t) \leq \exp h(T)\{\| \xi - \eta \|\} \leq c \| \xi - \eta \|, \tag{2.4.82}$$

where the constant $c = \exp h(T) < \infty$. Hence $\| x - y \|_{C(I,R^n)} \leq c \| \xi - \eta \|$. This shows that the solution is Lipschitz with respect to the initial data and hence continuously dependent on it. □

It is also clear from the above result that the solution is bounded above with the upper bound determined by the data $\{T, A, \xi\}$

$$\| x \|_{C(I,R^n)} \leq \{\exp h(T)\} \| \xi \| .$$

Now we are prepared to construct the transition matrix $\Phi(t,s), s \leq t$. Consider the system of equations

$$\begin{aligned} \dot{x}^i(t) &= A(t)x^i(t), \\ x^i(0) &= e_i, i = 1,2,3,\cdots,n, \end{aligned} \tag{2.4.83}$$

where e_i is the unit column vector with 1 at the i-th position and zero elsewhere. By virtue of the previous result, each one of these system of equations has a unique solution which we have denoted by $x^i(t), t \geq 0$. An equivalent way of writing the system of equations (2.4.83) is the matrix differential equation

$$\begin{aligned} \dot{X}(t) &= A(t)X(t), t > 0, \\ X(0) &= I, \end{aligned} \tag{2.4.84}$$

where I is the identity matrix. The matrix valued function $X(t), t \geq 0$, is called the fundamental solution. Using this fundamental solution we can construct the solution of the initial value problem

$$\dot{x}(t) = A(t)x(t), x(0) = x_0, \tag{2.4.85}$$

for any $x_0 \in R^n$, as

$$x(t) = X(t)x_0, t \geq 0. \tag{2.4.86}$$

Now we can show that

$$\Phi(t,s) = X(t)X^{-1}(s), s \leq t.$$

Clearly if $X(t), t \geq 0$, is nonsingular (or invertible) then we see that $x(t) \equiv \Phi(t,s)\xi, t \geq s$ is the solution of the general problem (2.4.64) since by differentiating this function we find that

$$\dot{x}(t) = A(t)X(t)X^{-1}(s)\xi = A(t)\Phi(t,s)\xi = A(t)x(t), t > 0$$

and that

$$\lim_{t \downarrow s} x(t) = \xi.$$

Hence it suffices to show that the fundamental matrix $X(t)$ is nonsingular. Consider the matrix differential equation

$$\begin{aligned} \dot{Y}(t) &= -Y(t)A(t), t \geq 0, \\ Y(0) &= I. \end{aligned} \tag{2.4.87}$$

Since this is a linear differential equation, with A locally integrable, it follows from Theorem 2.4.2 that this equation has a unique (absolutely) continuous solution, denoted by $Y(t), t \geq 0$. Now compute the derivative

$$\begin{aligned}(d/dt)(Y(t)X(t)) &= \dot{Y}(t)X(t) + Y(t)\dot{X}(t) \\ &= -Y(t)A(t)X(t) + Y(t)A(t)X(t) \\ &= 0. \end{aligned} \tag{2.4.88}$$

This last equation implies that the product $Y(t)X(t)$ is a constant matrix, that is, $Y(t)X(t) = C$. But we know that $Y(0)X(0) = I$ and hence $Y(t)X(t) = I$ for all $t \geq 0$. Hence

$$X^{-1}(t) = Y(t), t \geq 0 \tag{2.4.89}$$

proving that the fundamental solution matrix $X(t)$ is nonsingular and that the inverse is also a unique solution of a similar matrix differential equation, in this case, equation (2.4.87). Hence $t \longrightarrow X^{-1}(t)$ is absolutely continuous. Thus we have demonstrated the following result.

Corollary 2.4.3. The transition operator corresponding to the system (2.4.64), denoted by $\Phi(t,s), -\infty < s \leq t < \infty$, is given by

$$\Phi(t,s) = X(t)X^{-1}(s), -\infty < s \leq t < \infty. \tag{2.4.90}$$

Further, the solution of the autonomous (un controlled) system (2.4.64) is given by

$$x(t) = \Phi(t,s)\xi, t \geq s. \quad \bullet \tag{2.4.91}$$

The transition operator can be constructed also directly as follows. Using the successive approximation (2.4.69) it is easy to see that

$$\begin{aligned} x_n(t) \equiv \Phi_n(t,s)\xi \equiv \Big(I &+ \int_s^t dt_1 A(t_1) + \int_s^t dt_1 A(t_1) \int_s^{t_1} dt_2 A(t_2) \\ &+ \int_s^t dt_1 A(t_1) \int_s^{t_1} dt_2 A(t_2) \int_s^{t_2} dt_3 A(t_3) + \cdots, \\ &+ \int_s^t dt_1 A(t_1) \int_s^{t_1} dt_2 A(t_2) \int_s^{t_2} dt_3 A(t_3), \cdots, \int_s^{t_{n-2}} dt_{n-1} A(t_{n-1})\Big)\xi. \end{aligned} \tag{2.4.92}$$

The reader is encouraged to verify that

$$\| \Phi_n(t,s) \| \leq \{1 + h(t) + h^2(t)/2! + h^3(t)/3! + \cdots, + h^{n-1}(t)/(n-1)!\},$$

where h is given by the expression (2.4.70). Hence

$$\| \Phi_n(t,s) \| \leq \exp h(t), t \geq s.$$

For any finite T, define $\triangle \equiv \{0 \leq s \leq t \leq T\}$ and let $C(\triangle, R^{n\times n})$ denote the space of continuous functions with values in the space of $n \times n$ matrices with operator norm. Since h is bounded, $\{\Phi_n\}$ is bounded in $C(\triangle, R^{n\times n})$ and further one can verify that it is a Cauchy sequence. Hence there exists a unique $\Phi \in C(\triangle, R^{n\times n})$ so that

$$\lim_{n\to\infty} \Phi_n(t,s) = \Phi(t,s)$$

uniformly on $\triangle$. The limit $\Phi(t,s)$ is the transition operator.

Properties of Transition Operator: Now we state the basic properties of the transition operator $\Phi(t,s), s \leq t < \infty$. Using the expression (2.4.91) the reader can easily verify the following identities:

$$\begin{aligned}
&(P1): \Phi(t,t) = I,\\
&(P2): \Phi(t,\theta)\Phi(\theta,s) = \Phi(t,s), s \leq \theta \leq t, \text{ called the evolution property,}\\
&(P3): \partial/\partial t\Phi(t,s) = A(t)\Phi(t,s)\\
&(P4): \partial/\partial s\Phi(t,s) = -\Phi(t,s)A(s).
\end{aligned} \tag{2.4.93}$$

Controlled System: We have seen that the solution of the homogeneous system (2.4.64) is given by the expression (2.4.91). Now we consider the full control system subject to possibly an external perturbation (unwanted disturbance).

$$\begin{aligned}
\dot{x}(t) &= A(t)x(t) + B(t)u(t) + \gamma(t), t > s,\\
x(s) &= \xi.
\end{aligned} \tag{2.4.94}$$

Theorem 2.4.4. Suppose the elements of the matrix A are locally integrable and those of B are essentially bounded and let u and γ be integrable functions. Then, a continuous function $\varphi(t), t \geq 0$, taking values from R^n, is a solution of equation (2.4.94) if and only if φ is given by

$$\varphi(t) = \Phi(t,s)\xi + \int_s^t \Phi(t,\theta)B(\theta)u(\theta)d\theta + \int_0^t \Phi(t,\theta)\gamma(\theta)d\theta, t \geq s. \tag{2.4.95}$$

Proof. Clearly letting $t \downarrow s$ in equation (2.4.95), it follows from property P1 of the transition operator that $\lim_{t\downarrow s}\varphi(t) = \xi$. Differentiating φ and using the properties P1 and P3 of the transition operator Φ, we obtain

$$\begin{aligned}\dot{\varphi} &= A(t)\Phi(t,s)\xi + \int_s^t A(t)\Phi(t,\theta)[B(\theta)u(\theta) + \gamma(\theta)]d\theta \\ &\quad + \Phi(t,t)[B(t)u(t) + \gamma(t)] \\ &= A(t)\varphi(t) + B(t)u(t) + \gamma(t). \qquad (2.4.96)\end{aligned}$$

Thus the function φ as defined by equation (2.4.95) satisfies the initial value problem (2.4.94). This proves sufficiency. For the 'necessary condition, let x be a continuous function that satisfies the system equation (2.4.94). We show that x equals φ as defined by (2.4.95). If x satisfies equation (2.4.94), then

$$\dot{x}(\theta) - A(\theta)x(\theta) = B(\theta)u(\theta) + \gamma(\theta), \ a.e \ \ \infty > t \geq \theta > s$$

and hence for almost all θ in the interval indicated, we have

$$\Phi(t,\theta)[\dot{x}(\theta) - A(\theta)x(\theta)] = \Phi(t,\theta)[B(\theta)u(\theta) + \gamma(\theta)], s \leq \theta \leq t. \qquad (2.4.97)$$

Integrating this over the interval $[s,t]$ we obtain

$$\int_s^t \Phi(t,\theta)[\dot{x}(\theta) - A(\theta)x(\theta)]d\theta = \int_s^t \Phi(t,\theta)[B(\theta)u(\theta) + \gamma(\theta)]d\theta, t \geq s. \qquad (2.4.98)$$

Using the property (P4), the reader can easily verify that

$$\begin{aligned}\int_s^t \Phi(t,\theta)[\dot{x}(\theta) - A(\theta)x(\theta)]d\theta &= \int_s^t \partial/\partial\theta\{\Phi(t,\theta)x(\theta)\}d\theta \\ &= \Phi(t,t)x(t) - \Phi(t,s)x(s) \\ &= x(t) - \Phi(t,s)\xi. \qquad (2.4.99)\end{aligned}$$

Clearly it follows from the last two equations that

$$x(t) = \Phi(t,s)\xi + \int_s^t \Phi(t,\theta)[B(\theta)u(\theta) + \gamma(\theta)]d\theta. \qquad (2.4.100)$$

This shows that if x is a solution of the system equation (2.4.94) then it must be given by the expression (2.4.100) or (2.4.95). This completes the proof. □

2.5 Continuous Dependence of Solution on Data

We demonstrate here that the solution is continuously dependent on all the data such as the initial state, the input and even the system matrices. For simplicity of presentation we first consider the system

$$\dot{x}(t) = A(t)x(t) + f(t), x(s) = \xi. \qquad (2.5.101)$$

where $\xi \in R^n$ is the initial data and $f \in L_1([s,T], R^n)$ is the external input. We show that the solutions are continuously dependent on the data.

Corollary 2.5.1. Suppose the elements of A are locally integrable. Then the solution of the linear system (2.5.101) is Lipschitz and hence continuous from $R^n \times L_1([s,T], R^n)$ to $C([s,T], R^n)$.

Proof. Consider the same system (2.5.101) with the initial data ξ, η and inputs f, g. Let x, y denote the corresponding solutions. Define $e = (x - y)$. Then e satisfies the differential equation

$$\dot{e}(t) = A(t)e(t) + (f(t) - g(t)), e(s) = \xi - \eta.$$

Integrating this equation over the interval $[s,t]$ we have

$$e(t) = (\xi - \eta) + \int_s^t A(\tau)e(\tau)d\tau + \int_s^t (f(\tau) - g(\tau))d\tau.$$

Hence

$$\begin{aligned} \| e(t) \| &\leq \| (\xi - \eta) \| + \int_s^t \| A(\tau) \| \| e(\tau) \| d\tau + \int_s^t \| (f(\tau) - g(\tau)) \| d\tau \\ &\leq \| \xi - \eta \| + \int_s^T \| f(\tau) - g(\tau) \| d\tau + \int_s^t \| A(\tau) \| \| e(\tau) \| d\tau. \end{aligned} \tag{2.5.102}$$

Using Gronwall inequality (2.4.67) we obtain

$$\| e(t) \| \leq C_A \left\{ \| \xi - \eta \| + \int_s^T \| f(\tau) - g(\tau) \| d\tau \right\}, t \in [s,T] \tag{2.5.103}$$

where the constant C_A is given by

$$C_A \equiv \exp\left\{ \int_s^T \| A(\tau) \| d\tau \right\}.$$

Since the elements of A are locally integrable and T is finite, the constant C_A is finite. This shows that the map $\{\xi, f\} \longrightarrow x(t, \xi, f)$, from the space of data $R^n \times L_1([s,T], R^n)$ to the space of solutions $C([s,T], R^n)$, is Lipschitz continuous and hence continuous. This completes the proof.

The result of the above Corollary can be used to conclude that for every pair of data $\{\xi, f\} \in R^n \times L_1([s,T], R^n)$, the system (2.5.101) has a unique solution. This follows from the fact that if $\xi = \eta$ and $f = g$, then $e(t) \equiv 0$

and hence $x = y$. In other words there can not be two solutions for the same pair of data.

Very often we have to compare the responses of two systems one of which may be an approximation of the other or a perturbed version of the other. In other words we must look at the solution not only as a function of t but also as a functional of all the system parameters. Consider the control system

$$\dot{z} = A(t)z + B(t)u, z(s) = \xi \tag{2.5.104}$$

and consider $z \equiv z(t, \xi, A, B)$ as a functional of all the system parameters. Our objective is to study the continuity of this map with respect to all the parameters. We show that under mild conditions this map is continuous.

Theorem 2.5.2. Suppose the elements of A and B are locally integrable and $u \in L_\infty(I_s, R^d)$. Then the solution of the linear system (2.5.104) is Lipschitz and hence continuous from $R^n \times L_1([s,T], R^{n\times n}) \times L_1([s,T], R^{n\times d})$ to $C([s,T], R^n)$.

Proof. Let $(\xi, A, B) \in R^n \times L_1([s,T], R^{n\times n}) \times L_1([s,T], R^{n\times d})$ and $(\eta, \tilde{A}, \tilde{B}) \in R^n \times L_1([s,T], R^{n\times n}) \times L_1([s,T], R^{n\times d})$ and denote by $x(t) = z(t, \xi, A, B)$ and $y(t) = z(t, \eta, \tilde{A}, \tilde{B})$ the corresponding solutions. Define $e(t) = x(t) - y(t)$ and note that e satisfies the differential equation

$$\begin{aligned} \dot{e}(t) &= A(t)e(t) + (A(t) - \tilde{A}(t))y(t) + (B(t) - \tilde{B}(t))u(t), t \in I_s \\ e(s) &= \xi - \eta. \end{aligned}$$

Integrating this and using Gronwall inequality (2.4.66) one obtains

$$\begin{aligned} \| e(t) \| \leq \exp\Big(\int_s^T \| A(\theta) \| \, d\theta\Big)\Big\{\| \xi - \eta \| + \int_{I_s} \| A(\theta) - \tilde{A}(\theta) \| \| y(\theta) \| \, d\theta \\ + \int_{I_s} \| B(\theta) - \tilde{B}(\theta) \| \| u(\theta) \| \, d\theta\Big\}. \end{aligned} \tag{2.5.105}$$

Define

$$\begin{aligned} \| y \|_C &\equiv \sup\{\| y(t) \|, t \in I_s\}, \\ \| x \|_C &\equiv \sup\{\| x(t) \|, t \in I_s\}, \\ \| u \|_\infty &\equiv ess - \sup\{\| u(t) \|, t \in I_s\}. \end{aligned} \tag{2.5.106}$$

Since the solutions $x, y \in C(I_s, R^n)$ and $u \in L_\infty(I_s, R^d)$, and $A \in L_1(I_s, R^{n\times n})$, we have $\| x \|_C < \infty$, $\| y \|_C < \infty$, $\| u \|_\infty < \infty$ and $\exp\{\int_{I_s} \| A(t) \|$

$dt\} \equiv C_A < \infty$. Using these facts in equation (2.5.105) we arrive at the following inequality

$$\| e \|_C \leq \alpha \left\{ \| \xi - \eta \| + \int_{I_s} \| A(t) - \tilde{A}(t) \| \, dt + \int_{I_s} \| B(t) - \tilde{B}(t) \| \, dt \right\}, \tag{2.5.107}$$

where the constant α is given by

$$\alpha \equiv \max\{1, \| x \|_C, \| y \|_C, \| u \|_\infty\} C_A. \tag{2.5.108}$$

This proves that the data to solution map, $\{\xi, A, B\} \longrightarrow z(\cdot, \xi, A, B)$, is Lipschitz on $R^n \times L_1([s,T], R^{n\times n}) \times L_1([s,T], R^{n\times d})$ and hence continuous. This ends the proof. □

If the two systems are subjected to the same input and start from the same initial state, the inequality (2.5.107) simplifies to

$$\begin{aligned} \| e \|_C &\equiv \| z(A,B) - z(\tilde{A}, \tilde{B}) \| \\ &\leq \alpha \left\{ \int_{I_s} \| A(t) - \tilde{A}(t) \| \, dt + \int_{I_s} \| B(t) - \tilde{B}(t) \| \, dt \right\}. \end{aligned} \tag{2.5.109}$$

This is a measure of divergence of response of two distinct linear systems determined by the parameters (A, B) and $(\tilde{A}, \tilde{B})$. Difference in response at the terminal time can not exceed the limit as defined by the right hand side of the above expression.

Some Remarks: In view of the above result one can introduce an equivalence relation $\cong$ on the space of linear systems $\mathcal{L}_S$ determined by the pairs $(A, B) \in L_1(I_s, M(n \times n)) \times L_1(I_s, M(n \times d))$. This can be done simply by using the natural equivalence relation that already exists on L_1 spaces. Two elements, a, b of L_1 are equivalent, $a \cong b$, if and only if, the set on which they differ has Lebesgue measure zero. This equivalence relation can be transferred to the space of linear systems, $\mathcal{L}_S$. Two systems $S_1(A, B)$ and $S_2(C, D)$ are equivalent, $S_1 \cong S_2$, if and only if $A \cong C$ and $B \cong D$, as elements of some L_1 spaces.

2.6 Input-Output Stability and System Equivalence

Input-output Stability: Consider the linear time invariant system (2.3.63) with the input-output relation given by

$$y(t) = \int_0^\infty G(\tau) u(t - \tau) d\tau \tag{2.6.110}$$

where the input u is an R^d valued measurable function and G is the impulse response with values in the space of matrices $M(m \times d)$. Here we may state explicitly that the system is non anticipative and hence $G(t) = 0$ for all $t < 0$. We prove that, under some reasonable assumptions, the system is L_p stable in the sense that the output belongs to L_p whenever the input is in L_p. It is important to note that the input-output map given by an expression of the form (2.6.110) represents a much more general linear system than the one from which we have constructed it. In fact G does not have to be continuous, just measurability and integrability are sufficient.

Theorem 2.6.1. Consider the input-output system given by equation (2.6.110) and suppose $G \in L_1(R_0, M(m \times d))$. Then the system is L_p stable in the sense that for every input $u \in L_p(R, R^d)$, the output $y \in L_p(R, R^m)$, for all $1 \leq p \leq \infty$.

Proof. First take the case $p = 1$. Taking the vector norm on either side we obtain

$$\| y(t) \|_{R^m} \leq \int_0^\infty \| G(\tau) \|_M \quad \| u(t-\tau) \|_{R^d} \, d\tau. \tag{2.6.111}$$

Integrating on either side over the real line we have

$$\int_R \| y(t) \|_{R^m} \, dt \leq \int_0^\infty \| G(\tau) \|_M \quad \left\{ \int_R \| u(t-\tau) \|_{R^d} \, dt \right\} d\tau. \tag{2.6.112}$$

Since the integral

$$\int_R \| u(t-\tau) \|_{R^d} \, dt = \int_R \| u(r) \|_{R^d} \, dr,$$

it follows from (2.6.112) that

$$\| y \|_{L_1(R,R^m)} \leq \| G \|_{L_1(R_0,M)} \quad \| u \|_{L_1(R,R^d)} .$$

Thus L_1 input produces L_1 output. Now consider $p \in (1, \infty)$. Starting with (2.6.111), and using Holder inequality with q conjugate of p, we observe that

$$\begin{aligned}
\| y(t) \|_{R^m} &\leq \int_0^\infty \| G(\tau) \|_M \quad \| u(t-\tau) \|_{R^d} \, d\tau \\
&\leq \int_0^\infty \| G(\tau) \|_M^{1/p+1/q} \quad \| u(t-\tau) \|_{R^d} \, d\tau \\
&\leq \left(\int_0^\infty \| G(\tau) \|_M \, d\tau \right)^{1/q} \left(\int_{R_0} \| G(\tau) \|_M \| u(t-\tau) \|_{R^d}^p \, d\tau \right)^{1/p}.
\end{aligned}$$

Now integrating the $p-th$ power of the above inequality and using Fubini's theorem (Theorem A.6.6)we obtain

$$\begin{aligned}\int_R \| y(t) \|_{R^m}^p \, dt &\leq \| G \|_{L_1(R_0,M)}^{p/q} \int_{R_0} d\tau \, \| G(\tau) \|_M \left(\int_R \| u(t-\tau) \|_{R^d}^p \, dt \right) \\ &\leq \| G \|_{L_1(R_0,M)}^{(p/q)+1} \int_R \| u(\tau) \|^p \, d\tau \\ &= \| G \|_{L_1(R_0,M)}^p \| u \|_{L_p(R,R^d)}^p .\end{aligned}$$

Thus we have demonstrated that

$$\| y \|_{L_p(R,R^m)} \leq \| G \|_{L_1(R_0,M)} \| u \|_{L_p(R,R^d)} . \tag{2.6.113}$$

This shows that for any $p \in (1, \infty)$, L_p input produces L_p output. Now it remains to prove the case for $p = \infty$. But this follows immediately from the expression (2.6.111) by simply noting that if $u \in L_\infty(R, R^d)$, there exists a number $b < \infty$ such that the set on which $\| u(t) \|$ exceeds b is a set of Lebesgue measure zero. Hence

$$\| y \|_{L_\infty(R,R^m)} \leq \| G \|_{L_1(R_0,M)} \| u \|_{L_\infty(R,R^m)} .$$

This ends the proof. □

A similar result can be proved for time varying systems. Here the system, with input u and output y, is given by

$$\dot{x} = A(t)x + B(t)u, x(s) = \xi, t \geq s \tag{2.6.114}$$

$$y(t) = H(t)x(t), t \geq s. \tag{2.6.115}$$

It follows from Theorem 2.4.4, that the output is given by

$$y(t) = H(t)\Phi(t,s)\xi + \int_s^t H(t)\Phi(t,r)B(r)u(r)dr. \tag{2.6.116}$$

Letting $s \to -\infty$ and assuming that $\lim_{s\downarrow -\infty} \Phi(t,s)\xi = 0$, we have the input-output relation

$$y(t) = \int_{-\infty}^t K(t,\tau)u(\tau)d\tau, \text{ with } K(t,\tau) \equiv H(t)\Phi(t,\tau)B(\tau). \tag{2.6.117}$$

Theorem 2.6.2. Consider the input-output system given by equation (2.6.117) and suppose there exists an $h \in L_1[0, \infty)$ such that

$$\| K(t,\tau) \|_M \leq h(t-\tau), \ \forall \, t \geq \tau. \tag{2.6.118}$$

Then the time varying system is L_p stable in the sense that for every input $u \in L_p(R, R^d)$, the output $y \in L_p(R, R^m)$, for all $1 \le p \le \infty$.

Proof. Under the assumption (2.6.118), the proof is identical to that of Theorem 2.6.1. □

The reader may be interested to verify this in details.

System Equivalence: One interesting question about the input-output relations (2.6.110), (2.6.117) is; what makes two systems equivalent in the sense that the same input produces the same output? We introduce a formal definition for this. Let $\mathcal{U}$ and $\mathcal{Y}$ denote any two Banach spaces representing the input-output spaces. Let S_1 and S_2 be two linear systems with the corresponding operators being T_1 and T_2 which belong to the space of bounded linear operators denoted by $\mathcal{L}(\mathcal{U}, \mathcal{Y})$. Let $\mathcal{V}$ be a proper closed linear subspace of $\mathcal{U}$.

Definition 2.6.3. Two linear systems S_1 and S_2 with associated operators T_1 and T_2 are said to be input-output equivalent with respect to the pair of input-output spaces $(\mathcal{V}, \mathcal{Y})$, if and only if,

$$T_1 x = T_2 x \ \forall \ x \in \mathcal{V}.$$

This definition implicitly states that the two systems may not be equivalent with respect to the pair $(\mathcal{U}, \mathcal{Y})$.

Now considering the system (2.6.110), it is evident that if two such linear systems have the same kernel G (for time varying K), then the outputs are the same for the same inputs. The converse, however, is nontrivial. Does input-output equivalence imply that the kernels are identical? We answer this question in the following results.

Theorem 2.6.4. Any two linear systems of the form (2.6.110) or (2.6.117) are equivalent with respect to the pair of input-output spaces $(L_p(R, R^d)$, $L_p(R, R^m))$ if, and only if, the kernels $G_1, G_2 \in L_1([0,\infty), M(m \times d))$ (for time varying system K) of these integral operators are identical almost every where.

Proof. We prove this only for the time invariant system. The proof is similar for the time varying system. Consider the two systems S_1 and S_2 of the form (2.6.110) with the Kernels $G_1, G_2 \in L_1([0,\infty), M(m \times d))$. It is evident that if $G_1 \in L_1([0,\infty), M(m \times d))$ equals $G_2 \in L_1([0,\infty), M(m \times d))$ almost every where, then for every input $u \in L_p(R, R^d)$, the outputs y_1 and y_2 are identical elements of $L_p(R, R^m$ (more precisely belongs to the same equivalence class, that is, $y_1 \cong y_2$). For the converse, suppose the two systems S_1 and S_2 are equivalent with respect to the L_p spaces. We must show that in that case $G_1 \cong G_2$. Define $G = G_1 - G_2$. Since the two systems are equivalent, we have

$$\int_0^\infty G(\tau)u(t-\tau)d\tau = 0, \quad \text{for almost all } t \in R, \text{ and } \forall u \in L_p(R, R^d).$$

Let $J \subset R$ be any bounded measurable set. It is evident that

$$\int_J dt \Big(\int_0^\infty G(\tau)u(t-\tau)d\tau \Big) = 0, \ \forall u \in L_p(R, R^d).$$

By virtue of Fubini's theorem, one can interchange the order of integration giving

$$\int_0^\infty G(\tau)\Big(\int_J dt\ u(t-\tau)\Big)d\tau = \int_0^\infty G(\tau)v_J(\tau)d\tau = 0, \ \forall u \in L_p(R, R^d), \tag{2.6.119}$$

where

$$v_J(\tau) \equiv \int_J u(t-\tau)dt.$$

It follows from Holder inequality, that

$$\| v_J(\tau) \|_{R^d} \leq (\lambda(J))^{1/q} \| u \|_{L_p(R,R^d)} \ \forall \tau \in R$$

where q is the conjugate of p, and λ is the Lebesgue measure. Thus $v_J \in L_\infty([0,\infty), R^d)$. More precisely, $v_J \in B([0,\infty), R^d)$ which denotes the space of bounded (not only essentially bounded) measurable functions dense in $L_\infty([0,\infty), R^d)$. Hence it follows from (2.6.119) that, for any $\xi \in R^m$, we have

$$\int_0^\infty (v(\tau), G^{'}(\tau)\xi)d\tau = 0 \ \forall v \in L_\infty([0,\infty), R^d), \tag{2.6.120}$$

where $G^{'}(t)$ denotes the transpose of $G(t)$. Since the dual of L_1 space is L_∞ (see Chapter 1), the above identity also holds for

$$v(t) = sign(G^{'}(t)\xi)$$

where $sign(\eta) = (sign(\eta_1), sign(\eta_2), \cdots, sign(\eta_d))^{'}$. Substituting this in equation (2.6.120) we obtain

$$\int_0^\infty \| G^{'}(\tau)\xi \|_{R^d} \, d\tau = 0, \ \text{for } \xi \in R^m. \tag{2.6.121}$$

Since $\xi \in R^m$, is arbitrary, this implies that $G(t) = 0, a.e$ on $[0, \infty)$. This ends the proof. □

Remark. Two systems of the form (2.6.110) with kernels $G_1, G_2 \in L_1$ $([0,\infty), M(m \times d))$ which equal almost every where with respect to Lebesgue measure are not necessarily equivalent with respect to an input-space which contains Dirac measures and any output-space.

Here we present a case where this equivalence is recovered. For this we need the Stieltjes integral. As we have seen in the appendix A, any function of bounded variation can be represented by a signed measure. So we may consider our input space as the space of R^d-valued countably additive bounded vector measures. Denote by $\mathcal{M}(R, R^d)$ the space of such measures with bounded total variation.

Definition 2.6.5. Let $\mu \in \mathcal{M}(R, R^d)$ and $J \in B_R$, the Borel field of subsets of the real line R, and let $\Pi(J)$ denote the set of all partitions of the set J by a finite number of disjoint members $\{\sigma\}$ from B_R such that $\sigma \subseteq J$. The total variation of μ on J, denoted by $|\mu|(J)$, is defined as

$$|\mu|(J) = \sup_{\Pi(J)} \sum_{\sigma \in \Pi(J)} \parallel \mu(\sigma) \parallel_{R^d} .$$

The norm is given by $\parallel \mu \parallel_v \equiv |\mu|(R)$.

With respect to this norm $\mathcal{M}(R, R^d)$ is a Banach space.

A particular example is the sum of weighted Dirac measures

$$\mu(J) \equiv \sum_{t_i \in J} v_i \delta_{t_i}(J)$$

where $v_i \in R^d$ and $\delta_{t_i}(J) = 1$ if $t_i \in J$, otherwise zero. The total variation of this measure on J is given by

$$|\mu|(J) \equiv \sum_{t_i \in J} \parallel v_i \parallel_{R^d} .$$

Now we consider the system (2.6.110) driven by inputs from $\mathcal{U} \equiv \mathcal{M}$ (R, R^d). Note that this class of inputs is much broader than the class of measurable functions with values in R^d. In this case it is convenient to write this system as

$$y(t) \equiv \int_{-\infty}^{t} G(t-\tau)u(d\tau), u \in \mathcal{U}. \tag{2.6.122}$$

We must decide what the appropriate output space is. It is clear that if G is merely in $L_1([0,\infty), M)$, the output may not be defined at all since the control may have impulses exactly at the points where G has unbounded values and we end up with a sum of plus and minus infinities making no sense. So for G to be integrable with respect to all such control measures (including impulsive controls) we must put restriction on the class of kernels $\{G\}$. One choice is $C_b(R_0, M(m \times d))$, the Banach space of bounded continuous functions with values in $M(m \times d)$ furnished with the supremum norm. Taking the norm on either side of the expression (2.6.122), we have for any $u \in \mathcal{U}$

$$\begin{aligned} \| y(t) \| &\leq \int_{-\infty}^{t} \| G(t-\tau) \| \, |u|(d\tau) \\ &\leq \sup\{\| G(\theta) \|, \theta \in R_0\} \| u \|_v < \infty. \end{aligned} \tag{2.6.123}$$

Note that even though G is continuous, y is not, since u may consist of both impulsive and continuous signals. In this case the appropriate output space is $\mathcal{Y} = B(R, R^m)$, the space of bounded measurable functions on R taking values from R^m. Hence, for equivalence of two such systems, $G_1(t)$ must equal $G_2(t)$ for all $t \in R_0$. Thus in regard to systems with such inputs we have the following equivalence result.

Theorem 2.6.6. Any two linear systems of the form (2.6.122) or (2.6.117) are equivalent with respect to the pair of input-output spaces $(\mathcal{M}(R, R^d), B(R, R^m))$ if, and only if, the kernels of these integral operators $G \in C_b([0, \infty), M(m \times d))$ (for time varying system K) are identical every where on R_0.

Note that for the input-output model given by (2.6.122), the input is a vector valued measure while the kernel $G \in C_b(R_0, M)$. This role can be exchanged giving an input-output model of the form

$$y(t) = \int_0^{\infty} G(dr)u(t-r), \tag{2.6.124}$$

where $G(\cdot) \in \mathcal{M}(R_0, M(m \times d))$, the space of operator valued measures. The inputs belong to the Banach space of bounded continuous functions on R to R^d, denoted by $C_b(R, R^d)$. Again we may use the total variation norm for G which may be defined exactly as in Definition 2.6.5 giving $\| G \|_v = |G|(R_0)$. In this case the appropriate input-output spaces are $C_b(R, R^d)$ and $B(R, R^d)$ respectively.

Theorem 2.6.7. Any two linear systems of the form (2.6.124) are equivalent with respect to the pair of input-output spaces $(C_b(R, R^d), B(R, R^m))$ if, and only if, the corresponding kernels $G_1, G_2 \in \mathcal{M}(R_0, M(m \times d))$ and $\| G_1 - G_2 \|_v = 0$.

We give a somewhat imprecise outline of the proof of Theorem 2.6.7. The proof of Theorem 2.6.6 is entirely similar.

Proof. The proof of sufficiency is transparent. We prove the necessary condition. Our proof depends on the celebrated Riesz representation theorem which, loosely stated, says that every continuous linear functional on $C_b(\Omega)$ is representable by means of a suitable measure $\mu \in \mathcal{M}(\Omega)$ in the form

$$\ell(\phi) = \int_{\Omega} \phi(\omega)\mu(d\omega).$$

Suppose two systems S_1 and S_2 with the corresponding kernels G_1 and G_2 are equivalent with respect to the given pair of input-output spaces. Define $G = G_1 - G_2$ and note that, due to equivalence of the two systems, the function y given by

$$y(t) \equiv \int_{R_0} G(dr)u(t-r)$$

is zero every where for all $u \in C_b(R, R^d) \subset B(R, R^d)$. Clearly for any bounded set $J \in B_R$ we have, upon integration,

$$0 = \int_J y(t)dt = \int_{R_0} G(ds)w_J(s)$$

where $w_J(s) = \int_J u(t-s)dt$. Since $u \in C_b(R, R^d)$, it is locally integrable and so w_J is a bounded continuous function with values in R^d. Hence for any and every $\xi \in R^m$, we have

$$\int_{R_0} < G'(ds)\xi, w(s) > \ = 0 \ \ \forall\, w \in C_b(R_0, R^d).$$

Thus

$$\sup_{\|w\|_{C_b} \le 1} \left\{ \int_{R_0} < G'(ds)\xi, w(s) > \right\} = \ \| G'\xi \|_v = 0.$$

Since $\xi \in R^m$ is arbitrary, we have $\| G \|_v = 0$. Thus, for equivalence, it is necessary that the two kernels are equivalent with respect to the norm topology induced by the total variation norm for measures. This completes the proof. □

An interesting particular case of a linear system of the class (2.6.124) is given by a system with a kernel

$$G(dt) = G_0(t)dt + \sum_{\{i: t_i < t\}} G_i \delta_{t_i}(dt),$$

with the norm given by

$$\| G \|_v \equiv \int_{R_0} \| G_0(t) \|_M \, dt + \sum_{1 \leq i \leq \infty} \| G_i \|_M .$$

Remark. A more general linear time varying input-output system is given by

$$y(t) = \int_{-\infty}^{t} G(t, d\theta) u(\theta), t \in R, \tag{2.6.125}$$

where G is a function defined on $R \times B_R$ with values in the space of matrices $M(m \times d)$. In other words $G(t, \sigma)$ is a measurable function with respect to $t \in R$ and a measure (set function) on the Borel algebra of sets in R. For non anticipativeness, it is necessary that $G(t, \sigma) = 0$ for any $\sigma \notin B_{(-\infty, t]}$.

For this class of systems, the reader may like to prove an equivalence result similar to that of Theorem 2.6.7.

Example 1. An example of a system represented by (2.6.122) is found in inventory control problems. Consider a merchant who sells n distinct goods from his stores. He wants to maintain an appropriate supply of these goods so that he does not have to return his loyal customers. This means that he must control the inventory of these goods. Let $I = (I_1, I_2, \cdots, I_n)'$ denote the vector of inventory levels of these goods. A simple linear model for this system is given by

$$(d/dt)I_i = -\gamma_i I_i - \sum_{j \neq i} a_{ij} I_j + b_i u_i, i = 1, \cdots, n, \tag{2.6.126}$$

where γ_i is the rate of demand of the i-th good independent of other goods and a_{ij} is the demand of the same good induced by the demand of the j-th good, u_i is the supply rate with $b_i \in [0, 1]$ representing the spoilage factors. The coefficients $\gamma_i \geq 0, a_{ij} \geq 0$, while $a_{ij} = 0$ if the sale of the j-th good has no influence on the demand of the i-th good. The goods are supplied to the merchant at a set of instants of time on demand. Thus the control u is purely impulsive and can be represented by

$$u(dt) \equiv \sum_{k=1}^{m} u^k \delta_{t_k}(dt), \text{ with } u^k \equiv (u_1^k, u_2^k, \cdots, u_n^k)'$$

where $\delta_s(J) = 1$, if $s \in J$ and *zero* otherwise. Clearly this system can be written in the canonical form

$$dI = AI dt + Bu(dt). \tag{2.6.127}$$

This gives rise to an input-output model like (2.6.122).

The model (2.6.125) may arise from impulsive differential equations of the forms

$$\dot{x} = A(t)xdt + Bu(t), t \in R \setminus D \tag{2.6.128}$$
$$\triangle x(t_i) = A_i x(t_i), t_i \in D \equiv \{0 < t_1, t_2, \cdots, t_m, \cdots\} \tag{2.6.129}$$

where

$$\triangle x(t_i) \equiv x(t_i + 0) - x(t_i - 0) = x(t_i + 0) - x(t_i),$$

or equivalently,

$$\triangle x(t_i) \equiv x(t_i + 0) - x(t_i - 0) = x(t_i) - x(t_i -)$$

denotes the jumps at times $t_i \in D$ and $\{A_i\} \in M(n \times n)$ are constant matrices determining the size of the jumps. In the former case, solutions are continuous from the left having right hand limits, while for the later it is the opposite. Another system that gives rise to a similar model is given by

$$dx = A(t)xdt + B(dt)u(t), x(0) = x_0,$$

where $B \in \mathcal{M}(R, M(n \times d))$ is a matrix valued measure. Clearly the solution for this system is given by

$$x(t) = \Phi(t, 0)x_0 + \int_0^t \Phi(t, s)B(ds)u(s),$$

where Φ is the transition matrix corresponding to the matrix valued function $A(\cdot)$. If $y(t) \equiv H(t)x(t)$, the input-output model is

$$y(t) = g_0(t) + \int_0^t G(t, ds)u(s)$$

where $G(t, ds) = H(t)\Phi(t, s)B(ds)$ and $g_0(t) = H(t)\Phi(t, 0)x_0$.

2.7 Impulsive (Measure Driven) Systems

We have seen in the previous section some examples of impulsive systems arising in input-output models which are very popular in engineering sciences. The inventory control problem is another example. In this section we study more systematically systems governed by differential equations driven by measures. We shall call such systems as impulsive systems though they are much broader than systems merely subject to impulsive forces. We shall consider three broad classes: systems subject to impulsive inputs (controls), systems that experience impulsive structural perturbation and systems that combine both.

2.7.1 Systems Subject to Impulsive Inputs

The system model is given by

$$dx = A(t)xdt + B(t)u(dt), t \in I, x(0) = \xi, \tag{2.7.130}$$

where $I \equiv [0,T] \subset R$ is a bounded interval, $u \in \mathcal{M}(I, R^d)$ is a vector measure and $B \in C(I, M(n \times d))$ and $A \in L_1(I, M(n \times n))$.

Theorem 2.7.1. Suppose $A \in L_1(I, M(n \times n))$, $B \in C(I, M(n \times d))$ and Φ is the transition operator corresponding to the matrix valued function A. Then for every $\xi \in R^n$ and $u \in \mathcal{M}(I, R^d)$, equation (2.7.130) has a unique solution $x \in B(I, R^n)$ given by

$$x(t) \equiv \Phi(t,0)\xi + \int_0^t \Phi(t,s)B(s)u(ds), t \in I, \tag{2.7.131}$$

with $x \in BV(I, R^n)$ also. The solution map

$$(\xi, u) \longrightarrow x$$

is continuous from $R^n \times \mathcal{M}(I, R^d)$ to $B(I, R^n)$.

Proof. First we verify that x given by (2.7.131) is bounded and also a function of bounded variation on I. Since Φ is the transition operator corresponding to integrable A, there exists a constant $C > 0$ such that

$$\sup\{\| \Phi(t,s), 0 \le s \le t \le T\} \le C.$$

Then the boundedness follows from the inequality

$$\| x(t) \| \le C\Big\{\| \xi \| + \| B \|_0 \, |u_v(I)|\Big\}, \tag{2.7.132}$$

where

$$\| B \|_0 \equiv \sup\{\| B(t) \|, t \in I\}$$

and $|u|_v$ is the total variation norm of the vector measure u. Hence there exists a finite positive number b such that

$$\sup\{\| x(t) \|, t \in I\} \le b.$$

Thus $x \in B(I, R^n)$. We show that $x \in BV(I, R^n)$. For this it suffices to prove that there exists a constant $\gamma > 0$ such that for any $\varphi \in C(I, R^n)$

$$\left| \int_I (\varphi(t), dx(t)) \right| \le \gamma \| \varphi \|_0 \, .$$

This will mean that x has bounded variation. Again by direct computation one can verify that

$$\left|\int_I (\varphi, dx(t))\right| \leq \| \varphi \|_0 \left\{ b \int_I \| A(t) \| dt + \| B \|_0 |u|_v \right\}. \tag{2.7.133}$$

Thus taking

$$\gamma \equiv \left\{ b \int_I \| A(t) \| dt + \| B \|_0 |u|_v \right\}$$

we arrive at the desired inequality. This proves that $x \in BV(I, R^n)$. Clearly one can define a vector measure μ_x corresponding to x by the relation

$$\int_I (\psi(t), \mu_x(dt)) \equiv \int_I (\psi(t), dx(t))$$

for $\psi \in C(I, R^n)$. Two vector measures $\mu, \nu \in \mathcal{M}(I, R^n)$ are identical if and only if

$$\int_I (\varphi(t), \mu(dt)) = \int_I (\varphi(t), \nu(dt)) \ \forall \varphi \in C(I, R^n).$$

Now suppose $x \in B(I, R^n)$ satisfies the differential equation (2.7.130) in the sense of identity of vector measures as described above. Then we have

$$dx(s) - A(s)x(s)ds = B(s)u(ds), s \in I.$$

Multiplying this on either side by the transition operator and integrating we obtain

$$\int_0^t \Phi(t,s)(dx(s) - A(s)x(s)ds) = \int_0^t \Phi(t,s)B(s)u(ds), t \in I. \tag{2.7.134}$$

Integrating this by parts and using the properties of the transition operator Φ we arrive at the expression (2.7.131). Thus if x solves the differential equation (2.7.130) in the sense of equivalence of any two measures, it is given by the expression (2.7.131). Now we must prove that if x is given by the expression (2.7.131) then it is the solution of equation (2.7.130) in the sense of identity of measures. Rigorous proof of this requires regularization of the vector measure u by C_0^∞ mollifiers giving smooth functions approximating the measure as follows. Define

$$W(t) \equiv \int_0^t u(ds), t \geq 0.$$

Since u is a countably additive bounded vector measure, W is a function of bounded variation. Using this notation we can also express the system in the form

$$dx(t) = A(t)x(t)dt + B(t)dW(t), t \in I.$$

For approximation, we define

$$W_n(t) \equiv \int_{R_+} \rho_n(t-s)W(s)ds, t \in I,$$

where $\{\rho_n\}$ are nonnegative C^∞ functions with compact supports satisfying

$$\int_{R_+} \rho_n(t)dt = 1, \ \forall \ n \in N.$$

The sequence of functions $\{W_n\}$, so constructed, are also C^∞ functions having W as the weak (pointwise) limit. Let $\{x_n\}$ denote the corresponding approximating solutions of (2.7.130) given by

$$x_n(t) = \Phi(t,0)\xi + \int_0^t \Phi(t,s)B(s)dW_n(s), t \in I.$$

Clearly this is differentiable and it satisfies the differential equation

$$\dot{x}_n(t) = A(t)x_n(t) + B(t)\dot{W}_n(t),$$

for almost all $t \in I$ and also in the sense of identities in the space of measures,

$$dx_n(t) = A(t)x_n(t)dt + B(t)dW_n(t).$$

Thus for any $\varphi \in C(I, R^n)$ we have

$$\int_I (\varphi(t), dx_n(t)) = \int_I (A^*(t)\varphi(t), x_n(t))dt + \int_I (B^*(t)\varphi(t), dW_n(t)). \quad (2.7.135)$$

Using the representation of solutions based on transition operators, one can easily verify that as W_n converges weakly to W, x_n converges strongly to x in $B(I, R^n)$. Clearly for any $\varphi \in C_0^\infty(I, R^n)$, ($C^\infty$ functions with compact support) we have

$$\int_I (\varphi(t), dx_n(t)) = -\int_I (d\varphi(t), x_n(t)) \longrightarrow -\int_I (d\varphi(t), x(t)) = \int_I (\varphi(t), dx(t)).$$

Then we can justify taking the limit ($n \to \infty$) in equation (2.7.135) to arrive at the expression

$$\int_I (\varphi(t), dx(t)) = \int_I (A^*(t)\varphi(t), x(t))dt + \int_I (B^*(t)\varphi(t), dW(t)). \quad (2.7.136)$$

Since $\varphi \in C_0^\infty(I, R^n)$ is arbitrary, it follows from this that x, given by the expression (2.7.131), satisfies the identity (2.7.130) in the sense of distribution. The uniqueness of solution follows from the inequality (2.7.132).

Indeed, it follows from this that if x and y are any two solutions of the system (2.7.130) corresponding to the data $\{\xi, u\} \in R^n \times \mathcal{M}(I, R^d)$ and $\{\zeta, v\} \in R^n \times \mathcal{M}(I, R^d)$ respectively, then

$$\sup\{\| x(t) - y(t) \|, t \in I\} \leq C\Big\{\| \xi - \zeta \| + \| B \|_0 \, |u - v|_v(I)\Big\}. \quad (2.7.137)$$

Thus there can be at most one solution corresponding to any given data and further it follows from the same inequality that the solution map

$$(\xi, u) \longrightarrow x$$

is not only continuous from $R^n \times \mathcal{M}(I, R^d)$ to $B(I, R^n)$ but also Lipschitz. This completes the proof. □

2.7.2 Systems Subject to Impulsive Structural Perturbation

Many physical systems may experience and undergo impulsive changes in their structural configuration or parameter values. This class of systems can be described by using the following fundamental system of impulsive differential equations:

$$dx(t) = A_0(t)x(t)dt + A_1(dt)x(t-), x(0+) = \xi, t \in I. \quad (2.7.138)$$

We wish to develop sufficient conditions that guarantee the existence of a transition operator corresponding to the above model or solutions of the above equations corresponding to any given initial state $\xi \in R^n$.

Theorem 2.7.2. Suppose $A_0 \in L_1(I, M(n \times n))$ and A_1 is a matrix or operator valued measure, more precisely, $A_1 \in \mathcal{M}_c(I, M(n \times n))$. Then for every $\xi \in R^n$ the equation (2.7.138) has a unique solution $x \in B(I, R^n)$ with $x \in BV(I, R^n)$ and there exists a measurable transition operator $\Phi(t, s), 0 \leq s \leq t \leq T$ so that

$$x(t) = \Phi(t, 0)\xi. \quad (2.7.139)$$

Proof. Since $A_0 \in L_1(I, M(n \times n))$ it has a unique transition operator which we denote by $\Phi_0(t, s)$. Thus we can write the differential equation (2.7.138) as an integral equation

$$x(t) = \Phi_0(t, 0)\xi + \int_0^t \Phi_0(t, s)A_1(ds)x(s-), t \in I. \quad (2.7.140)$$

We consider the general situation. Let $x(s+) = \xi$ be given ; then we may write the system as an integral equation starting from time $s \in (0, T)$ as follows

$$x(t) = \Phi_0(t,s)\xi + \int_s^t \Phi_0(t,\tau)A_1(d\tau)x(\tau-), t \in (s,T]. \quad (2.7.141)$$

Thus for the proof of existence of a measurable transition operator for the system (2.7.138), it suffices to prove the existence of a unique (measurable) solution $x = x(\cdot, s, \xi) \in B((s,T], R^n)$ of the integral equation (2.7.141) and that the map

$$\xi \longrightarrow x(\cdot, s, \xi)$$

is a bounded linear map. For any $\sigma \in \mathcal{B}_I$, define

$$\mu_1(\sigma) = |A_1|_v(\sigma) \equiv \sup_\pi \sum_{J \in \pi} \| A_1(\sigma \cap J) \|_{\mathcal{L}(R^n,R^n)} \quad (2.7.142)$$

where π denotes any partition of the interval I into a finite number of disjoint members from $\mathcal{B}_I$. Since A_1 is a countably additive bounded vector measure having bounded total variation, μ_1 is a countably additive bounded positive measure. Define the operator G by

$$(Gf)(t) \equiv \Phi_0(t,s)\xi + \int_s^t \Phi_0(t,r)A_1(dr)f(r), t \in [s,T]. \quad (2.7.143)$$

Since Φ_0 is continuous in both the variables and $A_1 \in \mathcal{M}_c(I, M(n \times n))$, it is clear that, for every $f \in B([s,T], R^n)$, the function,

$$t \longrightarrow \int_s^t \Phi_0(t,r)A_1(dr)f(r), t \in [s,T],$$

is measurable and uniformly (not essentially) bounded, that is, it is an element of $B([s,T], R^n)$ and hence, for every $f \in B([s,T], R^n)$, $(Gf) \in B([s,T], R^n)$. Thus G maps $B([s,T], R^n)$ into itself. We prove that G has a unique fixed point in $B([s,T], R^n)$. For $x, y \in B([s,T], R^n)$ with $x(s+) = y(s+) = \xi$, we have

$$\begin{aligned}(Gx)(t) - (Gy)(t) &= \int_s^t \Phi_0(t,r)A_1(dr)(x(r-) - y(r-)) \\ &\equiv \int_{(s,t]} \Phi_0(t,r)A_1(dr)(x(r-) - y(r-)), t \in [s,T].\end{aligned}$$

Taking the norm we obtain

$$\| (Gx)(t) - (Gy)(t) \| \leq M_0 \int_0^t \| x(r-) - y(r-) \| \mu_1(dr), t \in [s,T]. \quad (2.7.144)$$

Define

$$d_t(x,y) \equiv \sup\{\| x(r) - y(r) \| \; s \le r \le t\}$$

and set $d(x,y) = d_T(x,y)$. Since $B([s,T],R^n)$ is a Banach space, B furnished with this metric topology is a complete metric space. Using this metric, the inequality (2.7.144) and the fact that $t \longrightarrow d_t(x,y)$ is a monotone nondecreasing function, the reader can easily verify that

$$d_t(Gx,Gy) \;\le M_0 \int_s^t d_{r-}(x,y)\mu_1(dr) \le M_0 \int_s^t d_r(x,y)\mu_1(dr), t \in [s,T]. \tag{2.7.145}$$

Define

$$\alpha(t) \equiv \mu_1((0,t]).$$

Since μ_1 is a bounded positive measure, the function α is a positive monotone (nondecreasing) right continuous function of its argument having bounded total variation. Thus the inequality (2.7.145) is equivalent to the following inequality,

$$d_t(Gx,Gy) \le M_0 \int_s^t d_r(x,y)d\alpha(r), t \in [s,T]. \tag{2.7.146}$$

Using this inequality and repeated substitution of this into itself (after two iterations) we obtain

$$\begin{aligned} d_t(G^2x,G^2y) &\le M_0 \int_s^t d_r(Gx,Gy)d\alpha(r), t \in [s,T] \\ &\le M_0^2 \; d_t(x,y) \int_s^t \alpha(r)d\alpha(r) \\ &\le (M_0^2/2)\alpha^2(t)d_t(x,y). \end{aligned}$$

The last identity follows from the fact that α is a (positive) nondecreasing right continuous function. After m iterations we arrive at the following inequality,

$$d_t(G^mx,G^my) \le M_0^m(\frac{\alpha^m(t)}{\Gamma(m+1)})d_t(x,y). \tag{2.7.147}$$

Using the metric d, it follows from the above inequality that

$$d(G^mx,G^my) \le M_0^m(\frac{\alpha^m(T)}{\Gamma(m+1)})d(x,y). \tag{2.7.148}$$

Since $\alpha(T) < \infty$ it follows from this estimate that for m sufficiently large, say m_0, the operator G^{m_0} is a contraction on the metric space $(B([s,T], R^n), d)$. In other words,

$$\left(M_0^{m_0}(\alpha(T))^{m_0}/\Gamma(m_0+1)\right) \equiv \rho < 1$$

and

$$d(G^{m_0}x, G^{m_0}y) \leq \rho d(x,y) \ \ \forall \ x,y \in B([s,T], R^n).$$

Hence by Banach fixed point theorem G^{m_0} has a unique fixed point, say, $x^* \in B([s,T], R^n)$, that is $x^* = G^{m_0}x^*$. Hence

$$d(x^*, Gx^*) \leq d(G^{m_0}x^*, G(G^{m_0}x^*)) = d(G^{m_0}x^*, G^{m_0}(Gx^*)) \leq \rho d(x^*, Gx^*).$$

Since $\rho < 1$, this inequality is true if and only if $d(x^*, Gx^*) = 0$. Thus x^* is also the unique fixed point of the operator G. Hence we have

$$\begin{aligned} x^*(t,s,\xi) &= (Gx^*)(t,s,\xi) \\ &\equiv \Phi_0(t,s)\xi + \int_s^t \Phi_0(t,r)A_1(dr)x^*(r-,s,\xi), t \in [s,T]. \quad (2.7.149) \end{aligned}$$

We show now that the solution is bounded (in norm). Define $\psi(t) \equiv d_t(x^*, 0)$. Using this it follows from the above identity that

$$\begin{aligned} \psi(t) &\leq M_0 \parallel \xi \parallel + M_0 \int_s^t \psi(r-)\mu_1(dr) \\ &\leq M_0 \parallel \xi \parallel + M_0 \int_s^t \psi(r)\mu_1(dr) \end{aligned}$$

Since μ_1 is a countably additive bounded positive measure having bounded total variation on I, it follows from the generalized Gronwall type inequality that

$$\psi(t) \leq M_0 \parallel \xi \parallel Exp\left\{\int_s^t \mu_1(dr)\right\} \ \ \forall \ t \in [s,T].$$

Hence

$$\sup\{\parallel x^*(t,s,\xi) \parallel, t \in [s,T]\} \leq M_0 e^{\alpha(T)} \parallel \xi \parallel. \quad (2.7.150)$$

We leave it as an easy exercise for the reader to verify that for any $\beta, \gamma \in R$ and any $\xi, \eta \in R^n$

$$x^*(t,s,\beta\xi + \gamma\eta) = \beta x^*(t,s,\xi) + \gamma x^*(t,s,\eta), t \in [s,T]. \quad (2.7.151)$$

Thus the map

$$\xi \longrightarrow x(t,s,\xi), t \geq s$$

is linear and it follows from equation (2.7.150) that it is also bounded. Measurability of the map

$$t \longrightarrow x^*(t,s,\xi)$$

follows from the fact that $x^* \in B([s,T],R^n)$. Hence there exists a bounded measurable operator valued function or equivalently a $M(n\times n)$ valued function $\Phi(t,s), 0 \leq s \leq t \leq T$ such that

$$x(t,s,\xi) = \Phi(t,s)\xi.$$

The uniqueness of solution implies that

$$\Phi(t,s) = \Phi(t,r)\Phi(r,s), 0 \leq s \leq r \leq t < \infty. \tag{2.7.152}$$

This proves the existence and uniqueness of a bounded measurable transition operator. The existence and uniqueness of a solution x^* for the initial value problem (2.7.138) now follows as a special case for $s = 0+$. To complete the proof we must show that $x^* \in BV(I,R^n)$. The proof of this is entirely similar to that given in Theorem 2.7.1. This completes the proof of the theorem. □

Remark. One may introduce the convention that whenever there is a jump in the solution, this event would be automatically included in the definition of the transition operator. Thus the transition property (2.7.152) may be rewritten as

$$\Phi(t+,s) \equiv \Phi(t+,r)\Phi(r+,s), 0 \leq s \leq r \leq t < \infty.$$

Remark. Note that the solutions are not only elements of the Banach space $B(I,R^n)$ (uniformly bounded measurable functions), they are also regular enough to have bounded total variation. In other words the solutions are elements of the space $BV(I,R^n)$.

Examples. Some interesting special cases of the system (2.7.138) are obtained by using possible representations of the operator valued measure $A_1(\cdot)$. Letting $\delta_\tau(\cdot)$ denote the Dirac measure with mass concentrated at τ, two such representations are given by

$$A_1(\sigma) \equiv \sum_{k=1}^{\ell} C_k \delta_{t_k}(\sigma) \quad C_k \in M(n \times n), \tag{2.7.153}$$

$$A_1(\sigma) \equiv \left\{ \int_\sigma K(s)\beta(ds) + \int_\sigma L(s)ds \right\}, \tag{2.7.154}$$

for any $\sigma \in \mathcal{B}_I$ and any countably additive measure β having bounded total variation with $K \in L_1(\beta, M(n \times n))$ and $L \in L_1(dt, M(n \times n))$. Clearly the first example is equivalent to the classical model described by

$$\begin{aligned}\dot{x} &= A_0(t)x(t), x(0+) = \xi, t \in I \setminus D \\ \triangle x(t_k) &= C_k x(t_k), t_k \in D \equiv \{t_k, k = 1, 2 \cdots, \ell\} \subset I.\end{aligned}$$

as presented in Section 2.6.

2.7.3 Controlled Impulsive Systems

Now we consider the most general model given by

$$dx = A_0(t)xdt + A_1(dt)x(t-) + B(t)u(dt) + f(t)dt, t \in I, x(0+) = \xi. \tag{2.7.155}$$

We prove the following result.

Theorem 2.7.3. Let A_0, A_1 satisfy the assumptions of Theorem 2.7.2 and $B \in C(I, M(n \times d))$ bounded. Then, for every $\xi \in R^n$, control measure $u \in \mathcal{M}(I, R^d)$ and $f \in L_1(I, R^n)$, equation (2.7.155) has a unique solution $x \in B(I, R^n)$ and further, the map

$$(\xi, u, f) \longrightarrow x(\cdot, \xi, u, f)$$

from $R^n \times \mathcal{M}(I, R^d) \times L_1(I, R^n)$ to $B(I, R^n)$ is bounded and Lipschitz continuous.

Proof. Under the given assumptions, it follows from the previous theorem (Theorem 2.7.2) that the pair $\{A_0, A_1\}$ generates a unique bounded measurable transition operator $\Phi(t, s), 0 \leq s \leq t \leq T$. Define

$$\sup\{\| \Phi(t, s) \|, 0 \leq s \leq t \leq T\} \leq M.$$

Clearly this is finite. Using this transition operator we can immediately write down the solution explicitly as follows

$$x(t) = \Phi(t, 0)\xi + \int_0^t \Phi(t, r)B(r)u(dr) + \int_0^t \Phi(t, r)f(r)dr, t \in I. \tag{2.7.156}$$

One can verify, as in Theorem 2.7.1, that this is the only solution of equation (2.7.155) and that it belongs to $B(I, R^n)$. For the Lipschitz continuity, let

$x_i \in B(I, R^n)$ denote the solution corresponding to the data $(\xi_i, u_i, f_i) \in R^n \times \mathcal{M}(I, R^d) \times L_1(I, R^n)$, $i = 1, 2$. Then one can easily derive that

$$\| x_1(t) - x_2(t) \| \leq M\left\{ \| \xi - \eta \| + \int_0^t \| B(s) \| \, |u - v|(ds) + \int_0^t \| f_1(s) - f_2(s) \| \, ds \right\}, t \in I. \tag{2.7.157}$$

Hence we have

$$\| x_1 - x_2 \|_{B(I,R^n)} \leq M\left\{ \| \xi - \eta \| + \| B \|_0 \, |u - v|(I) + \| f_1 - f_2 \|_{L_1(I,R^n)} \right\} \tag{2.7.158}$$

where $|u - v|(I) \equiv \| u - v \|_{\mathcal{M}(I,R^d)}$ denotes the standard total variation norm. Since $B \in C(I, M(n \times d))$, its supnorm is finite. Thus from the above inequality we conclude that the solution of equation (2.7.155) is Lipschitz in the data, and that the solution is unique. This completes the proof. □

Remark. Notice that in both the Theorems 2.7.1 and 2.7.3, we made the assumption that $B \in C(I, M(n \times d))$ and, as a consequence, we found that the linear operator L defined by

$$(Lu)(t) \equiv \int_0^t \Phi(t, s) B(s) u(ds)$$

is bounded from $\mathcal{M}(I, R^d)$ to the Banach space $B(I, R^n)$. For any fixed $u \in \mathcal{M}(I, R^d)$, we can also consider a similar operator Γ that maps $C(I, M(n \times d))$ to $B(I, R^n)$ given by

$$(\Gamma B)(t) \equiv \int_0^t \Phi(t, s) B(s) u(ds).$$

Clearly this is a bounded linear operator and

$$\| (\Gamma B) \|_{B(I,R^n)} \leq (M |u|_v) \, \| B \|_{C(I,M(n \times d))} \, .$$

This ineqiality remains valid if $C(I, M(n \times d))$ is replaced by $B(I, M(n \times d))$. Thus the map Γ can be continuously extended from $C(I, M(n \times d))$ to $B(I, M(n \times d))$. Thus all the results presented in this section also hold for general $B \in B(I, M(n \times d))$.

2.8 Exercises

(Q1): Construct the infinitesimal generator of a 4-state machine (manufacturing system) reliability model using the fact that the column sum is zero.

(Q2): Show that for an irreparable 2-state machine the transition probabilities are given by $P_1(t) = e^{-\lambda t}, P_0(t) = 1 - e^{-\lambda t}$.
(Q3): Let $\alpha \geq 0$, $K \in L_1^{\ell oc}$ and nonnegative, $\psi(t) \geq 0$ for all $t \geq 0$ satisfying

$$\dot{\psi}(t) \leq K(t)(\alpha + \psi(t)), t \geq 0.$$

Show that

$$\psi(t) \leq \alpha \exp\left\{\int_0^t K(s)ds\right\}.$$

(Q4) Consider the fundamental solutions X and Y of the matrix differential equations (2.4.84) and (2.4.87). It was shown in the text that $d/dt(Y(t)X(t)) = 0$. Prove that $d/dt(X(t)Y(t)) = 0$ also.
(Q5): Develop a theory of equivalence of input-output systems described by

$$y(t) = \int_0^t G(t, ds)u(s), t \in I,$$

where $G : I \times \mathcal{B}_I \longrightarrow M(d \times n)$ is measurable in t and a countably additive measure on $\mathcal{B}_I$ having bounded total variation on I.
(Q6): Verify the identity (2.7.151).
(Q7): Verify the second Remark following Theorem 2.7.2.
(Q8): Let h, φ, K be nonnegative functions on $[0, \infty)$ satisfying

$$\varphi(t) \leq h(t) + \int_0^t K(s)\varphi(s)\beta(ds), t \geq 0,$$

where β is a countably additive bounded positive measure having bounded variation on bounded subsets of $[0, \infty)$ possessing no atom at $\{0\}$, h continuous, and $K \in L_1^{\ell oc}([0, \infty), \beta)$. Prove that

$$\varphi(t) \leq h(t) + \int_0^t \exp\left(\int_s^t K(r)\beta(dr)\right)h(s)K(s)\beta(ds), t \in [0, \infty).$$

Verify that this inequality is also valid for bounded measurable h.
(Q9): From the previous result verify: if $0 \leq h(t) \leq a$ for all $t \geq 0$, then

$$\varphi(t) \leq a \exp\left\{\int_0^t K(r)\beta(dr)\right\}, t \geq 0.$$

2.9 Bibliographical Notes

One of the reasons for linear systems to be so popular in engineering is due its explicit input-output representation. There are many books on this topic with different levels of materials presented. Only books and papers easily available to the author have been included resulting in exclusion of many excellent contributions. An incomplete list is presented here: [2],[3], [8], [9],[17],[23], [31],[47],[62],[83]. Linear systems arising from partial differential equations, for example, Euler beam equation, linear wave equations such as Maxwell equation, equations of flexible rods, elastic strings etc. are not considered in this book. By use of modal expansion, as seen in example (E4) of this chapter, these systems can be approximated by linear ordinary differential equations which can then be used to construct input-output models. For a brief survey on such systems and their control see [126].

Chapter 3

Nonlinear Systems

3.1 Introduction

Most of the physical systems encountered in practice are nonlinear. This is the overriding reason why we must study nonlinear systems. The canonical forms for nonlinear time invariant and time varying systems are given, respectively, by the following equations,

$$\dot{x} = f(x, u) \tag{3.1.1}$$

$$\dot{x} = f(t, x, u), \tag{3.1.2}$$

where x is the state vector of dimension n and $f = col(f_1, f_2, f_3, \cdots, f_n)$ is a n-vector which is a function of the state and control variables in the time invariant case while in the time varying case f is also explicitly dependent on t. Impulsive nonlinear systems are described by similar equations in association with another equation that describes the evolution of jumps. This is further generalized by including nonlinear systems which are governed by differential equations and inclusions driven by vector measures. This is introduced in Section 3.6. Here we study the questions of existence of solutions and their regularity properties which are used in control theory in later chapters.

3.2 Some Examples

For motivation, we begin with some simple yet useful examples.
(E1) (Ecological System): The prey-predator model given in example (E3) of Chapter 2 was assumed to be linear. A more realistic model is given by the logistic equation known as the Lotka-Volterra equation. Here one assumes that the effective growth rate of rabbits is linearly proportional to

the fox population (in the negative sense). That is, $\alpha_e \equiv (\alpha - \beta F)$. Similarly the effective growth rate of the fox population is again linearly proportional (in the positive sense) to the rabbit population given by $\gamma_e \equiv (-\gamma + \delta R)$. Based on these hypotheses, the population dynamics is given by

$$\begin{aligned}\dot{R} &= \alpha_e R = (\alpha - \beta F)R = \alpha R - \beta F R \\ \dot{F} &= \gamma_e F = (-\gamma + \delta R)F = -\gamma F + \delta F R.\end{aligned} \tag{3.2.3}$$

The parameters $\{\alpha, \beta, \gamma, \delta\}$ are generally positive and have the following physical meaning. The parameter α denotes the intrinsic growth rate of rabbits, β the depletion rate per unit of fox population. In other words the actual depletion rate is linearly proportional to fox population, that is βF. The parameter γ denotes the mortality rate of foxes for lack of food and δ denotes the growth rate per unit of rabbit population. In other words the actual growth rate is linearly proportional to rabbit population, that is δR.

Defining the state x as

$$x = \begin{pmatrix} R \\ F \end{pmatrix} = \begin{pmatrix} x_1 \\ x_2 \end{pmatrix},$$

this equation can be written in the canonical form (3.1.1) giving

$$\begin{aligned}\dot{x}_1 &= f_1(x_1, x_2) \equiv \alpha x_1 - \beta x_1 x_2 \\ \dot{x}_2 &= f_2(x_1, x_2) \equiv -\gamma x_2 + \delta x_1 x_2.\end{aligned} \tag{3.2.4}$$

(E2) (Nonlinear Circuit): A nonlinear parallel RLC circuit is connected to a current source I. The resistor is nonlinear in the sense that its $i - v$ characteristic is given by $i = g(v)$ where g is a nonlinear function. The energy storage elements of the circuit are L and C. Hence the appropriate state variables are the current through the inductor i and the voltage across the capacitor v. Using Kirchoff laws one can write the system of equations as

$$\begin{aligned}(d/dt)i &= (1/L)v \\ (d/dt)v &= -(1/C)i - (1/C)g(v) + (1/C)I.\end{aligned} \tag{3.2.5}$$

Here the state is given by x as

$$x = \begin{pmatrix} i \\ v \end{pmatrix} = \begin{pmatrix} x_1 \\ x_2 \end{pmatrix}.$$

(E3) (Nonlinear time varying Circuit): Consider the same circuit. In case the capacitor is time varying, the appropriate state variables to choose are the inductor current and the charge on the capacitor, (i, q). Using the basic Kirchoff's current and voltage laws one can again derive the system model as

$$\begin{aligned} (d/dt)i &= (1/LC(t))q \\ (d/dt)q &= -i - g(q/C(t)) + I. \end{aligned} \tag{3.2.6}$$

Again, here the state is given by x as

$$x = \begin{pmatrix} i \\ q \end{pmatrix} = \begin{pmatrix} x_1 \\ x_2 \end{pmatrix}.$$

3.3 Existence and Uniqueness of Solutions

First we consider the system model

$$\dot{x}(t) = f(t, x(t)), x(0) = x_0. \tag{3.3.7}$$

Suppose that f of equation (3.3.7) is defined for all $(t, x) \in I \times R^n$ in the sense that it is single valued and that $\| f(t, x) \| < \infty$ for each $(t, x) \in I \times R^n$. We use finite difference scheme to construct an approximate solution as follows. Take the interval $[0, T]$ and partition this into n equal subintervals of length Δt so that $T = n\Delta t$. Then define the following approximations

$$\begin{aligned} x(\Delta t) &\approx x_0 + \Delta t f(0, x_0) \\ x(2\Delta t) &\approx x(\Delta t) + \Delta t f(\Delta t, x(\Delta t)) \\ x(k\Delta t) &\approx x((k-1)\Delta t) + \Delta t f((k-1)\Delta t, x((k-1)\Delta t)), k = 1, 2, \cdots, n. \end{aligned} \tag{3.3.8}$$

Note that this is simply a linear extrapolation giving a piecewise linear approximation of actual solution. For each choice of $n \in N \equiv \{0, 1, 2, 3 \cdots\}$ we have one such approximation which we may denote by $x_n \equiv \{x_n(t), t \in [0, T]\}$. Under some additional assumptions on the vector valued function f one can show that the sequence x_n converges uniformly on the interval $I = [0, T]$ to a function x which is a solution. This is an explicit difference scheme. One can also construct the solution using an implicit difference scheme

$$x(k\Delta t) \approx x((k-1)\Delta t) + \Delta t f(k\Delta t, x(k\Delta t)), k = 1, 2, \cdots, n, \tag{3.3.9}$$

as discussed for linear systems. Solving this nonlinear algebraic equation we obtain

$$x(k\triangle t) \approx (I - \triangle t F_k)^{-1} x((k-1)\triangle t), k = 1, 2, \cdots, n. \tag{3.3.10}$$

where F_k is the nonlinear operator from R^n to R^n defined by

$$F_k(x) = f(k\triangle t, x).$$

Note that if for each k, F_k is continuous and maps bounded sets of R^n into bounded sets of R^n and the range of the operator $(I - \triangle t F_k)$ coincides with R^n, then for sufficiently small $\triangle t$ the operator $(I - \triangle t F_k)$ is invertible and the approximate solution is well defined. In particular, if the operator F_k is Lipschitz, then, for sufficiently small $\triangle t$, $\triangle t F_k$ is a contraction. Then by virtue of Banach fixed point theorem, a unique solution exists. If F_k is only locally Lipschitz, one can prove existence and uniqueness of solutions locally in time possibly with finite blow up time. We shall study this further in the following sections.

Below we present some sufficient conditions for existence and uniqueness of solutions. In general, in engineering literature existence questions are never asked. The argument is that a physical system has always a solution. However when dealing with inexact mathematical models of physical systems, the question is crucial. A solution of the mathematical model, if one exists, requires to be close to the true response of the physical system. If in the first place existence is not assured, the question of closeness is a far cry.

In fact there is no general theory giving necessary and sufficient conditions for existence of solutions. Most of the existence results give sufficient conditions. The implication is, if for a given system these conditions are not satisfied it does not mean that the system has no solution.

Theorem 3.3.1. If $f(t, x)$ is continuous on $[0, \infty) \times R^n$ and bounded on bounded sets, the system (3.3.7) has a solution $x \in C(I_{t_b}, R^n)$ (not necessarily unique) where $I_{t_b} = [0, t_b)$ is the maximal interval of existence of solutions and t_b is the blow up time.

Proof. (See [2], Theorem 3.1.17). □

Some Examples. For illustration we consider some simple examples.

(E1): Consider the scalar equation

$$\dot{z} = -z^2, z(0) = z_0.$$

By simply integrating this equation it is easy to verify that the solution is given by

$$z(t) = \frac{z_0}{(1 + tz_0)}.$$

Clearly if $z_0 > 0$, the solution is defined for all $t \geq 0$ and vanishes as $t \to \infty$. If, on the other hand, $z_0 < 0$, the solution blows up at time $t_b = (1/|z_0|)$. If the sign of the right hand side is changed from $-z^2$ to $+z^2$, the solution is defined for all time $t \geq 0$ provided $z_0 < 0$ and blows up in time $t_b = (1/z_0)$ if $z_0 > 0$. Similar conclusions hold for $\dot{z} = \alpha z^{2n}, z(0) = z_0$, for $\alpha = -1$ or $+1$. Note that here f is locally Lipschitz but does not satisfy the (at most) linear growth condition (A1) as seen below.

(E2): Consider the scalar equation

$$\dot{z} = -z^3, z(0) = z_0.$$

Integrating this equation one finds that the solution is given by

$$z(t) = (+\,-)\frac{z_0}{\sqrt{1 + 2tz_0^2}}, \forall t \geq 0.$$

There are two solutions and they are defined for all $t \geq 0$. On the other hand with the sign changed to $+z^3$, the solution is defined locally and is given by

$$z(t) = (+\,-)\frac{z_0}{\sqrt{1 - 2tz_0^2}}, t \in [0, t_b)$$

where t_b is the blow up time given by $t_b = (1/2z_0^2)$.

Similar conclusions hold for $\dot{z} = \alpha z^{2n+1}, z(0) = z_0, \alpha = -1$ or $+1$.

The continuity of f in t is not essential. It suffices if f is (measurable and) locally integrable in t. For this generality, one must impose more stronger regularity condition on f with respect to the state variable $x \in R^n$.

The simplest such existence result is based on the following assumptions. We consider the time varying system (3.3.7). The function f is said to satisfy the linear growth condition and Lipschitz condition with respect to the state variable in R^n if there exist functions $K, L \in L_1^+(I)$ such that

$$(A1): \parallel f(t,x) \parallel \leq K(t)\{1 + \parallel x \parallel\} \tag{3.3.11}$$

$$(A2): \parallel f(t,x) - f(t,y) \parallel \leq L(t) \parallel x - y \parallel. \tag{3.3.12}$$

Theorem 3.3.2. Under the assumptions (A1) and (A2) as given above, for each $x_0 \in R^n$, the system (3.3.7) has a unique solution $x \in C(I, R^n)$.

Proof. Integrating equation (3.3.7) we obtain the integral equation

$$x(t) = x_0 + \int_0^t f(s, x(s))ds, t \in I \equiv [0, T]. \tag{3.3.13}$$

We use the successive approximation technique. Define

$$\begin{aligned} x_1(t) &= x_0, t \in I \\ x_2(t) &= x_0 + \int_0^t f(s, x_1(s))ds, t \in I, \end{aligned}$$

and by induction,

$$x_{n+1}(t) = x_0 + \int_0^t f(s, x_n(s))ds, t \in I, n \in N. \tag{3.3.14}$$

We show that $\{x_n\}$ is a Cauchy sequence in $C(I, R^n)$ and that it converges to a unique limit which is the solution of the integral equation and eventually a solution of the system equation (3.3.7). First note that

$$\begin{aligned} \| x_2(t) - x_1(t) \| &\leq \int_0^t \| f(s, x_0) \| ds \\ &\leq (1 + \| x_0 \|) \int_0^t K(s)ds. \end{aligned} \tag{3.3.15}$$

Hence

$$\| x_2 - x_1 \|_C \equiv \sup_{t \in I} \| x_2(t) - x_1(t) \| \leq (1 + \| x_0 \|) \int_0^T K(s)ds \equiv C. \tag{3.3.16}$$

Using the Lipschitz condition (A2) and the sequence (3.3.14) we obtain

$$\begin{aligned} \| x_3(t) - x_2(t) \| &\leq \int_0^t \| f(s, x_2(s)) - f(s, x_1(s)) \| ds \\ &\leq \int_0^t L(s) \| x_2(s) - x_1(s) \| ds \\ &\leq C \int_0^t L(s)ds \equiv C\ell(t), \end{aligned} \tag{3.3.17}$$

where $\ell(t) \equiv \int_0^t L(s)ds$. Clearly this is a nonnegative and nondecreasing function of t and $\ell(T) < \infty$ for any finite T. Following this procedure successively one finds that

$$\begin{aligned} \| x_4(t) - x_3(t) \| &\leq \int_0^t L(s) \| x_3(s) - x_2(s) \| ds \\ &\leq \int_0^t L(s)C\ell(s)ds = C \int_0^t \ell(s)d\ell(s) \\ &\leq (C/2)\ell^2(t), t \in I, \end{aligned} \tag{3.3.18}$$

and

$$\| x_5(t) - x_4(t) \| \leq (C/3!)\ell^3(t), t \in I, \tag{3.3.19}$$

$$\| x_6(t) - x_5(t) \| \leq (C/4!)\ell^4(t), t \in I, \tag{3.3.20}$$

and so on. Thus by induction

$$\| x_{n+1}(t) - x_n(t) \| \leq (C/(n-1)!)\ell^{n-1}(t), \quad n \geq 1, t \in I. \tag{3.3.21}$$

Since

$$x_{n+p}(t) - x_n(t) = \sum_{k=n}^{n+p-1} (x_{k+1}(t) - x_k(t)) \tag{3.3.22}$$

we have

$$\begin{aligned} \| x_{n+p}(t) - x_n(t) \| &\leq \sum_{k=n}^{n+p-1} \| x_{k+1}(t) - x_k(t) \| \\ &\leq (C/(n-1)!)\ell^{n-1}(t) \exp \ell(t), t \in I. \end{aligned}$$

Hence

$$\| x_{n+p} - x_n \|_C \equiv \sup_{t \in I} \| x_{n+p}(t) - x_n(t) \| \leq \{C \exp \ell(T)\}(\ell^{n-1}(T)/(n-1)!). \tag{3.3.23}$$

Note that the expression on the right hand side of this estimate is independent of p. Thus letting $n \to \infty$ we see that

$$\lim_{n \to \infty} \| x_{n+p} - x_n \| = 0$$

for any integer p. Hence $\{x_n\}$ is a Cauchy sequence in the Banach space $C(I, R^n)$ and consequently there exists a unique limit, say, $x \in C(I, R^n)$. That is

$$\lim_{n \to \infty} x_n(t) = x(t), \text{ uniformly on } I.$$

Since $\xi \longrightarrow f(t, \xi)$ is continuous on R^n, it follows from the above result that

$$\lim_{n \to \infty} f(t, x_n(t)) = f(t, x(t)) \quad a.a \quad t \in I. \tag{3.3.24}$$

Further, since every Cauchy sequence is bounded, there exists a finite positive number b so that

$$\| f(t, x_n(t)) \| \leq K(t)(1 + \| x_n(t) \|) \leq (1 + b)K(t) \ a.e. \tag{3.3.25}$$

By virtue of the two conditions (3.3.24) and (3.3.25), Lebesgue dominated convergence theorem holds and hence

$$\lim_{n\to\infty}\int_0^t f(s,x_n(s))ds = \int_0^t f(s,x(s))ds, \text{ for each } t \in I.$$

Thus letting $n \to \infty$, it follows from (3.3.14) that

$$x(t) = x_0 + \int_0^t f(s,x(s))ds, \ t \in I. \tag{3.3.26}$$

In other words the limit x is a solution of the integral equation (3.3.13). Since f is Lipschitz in x, this is the unique solution. Indeed, if there is another solution, say $y \in C(I,R^n)$ that satisfies the integral equation, we have

$$\| x(t) - y(t) \| \leq \int_0^t K(s) \| x(s) - y(s) \| ds, t \in I.$$

By Gronwall inequality this implies that $x(t) = y(t)$ for all $t \in I$. So far we have shown that the limit x is the unique solution of the integral equation. We must show that it is also the solution of our original system equation (3.3.7). Since $g(t) \equiv f(t,x(t))$, $t \in I$ is an integrable function (the reader may justify this), its integral $\int_0^t g(s)ds \equiv \int_0^t f(s,x(s))ds, t \in I$ is an absolutely continuous function of time t and hence it is differentiable almost every where and the derivative equals $g(t) = f(t,x(t))$. Thus differentiating either side of equation (3.3.26) we obtain the differential equation (3.3.7). This completes the proof. □

Remark. Examining the last part of the proof we note that, under the given assumptions on f, $x \in C(I,R^n)$ is a solution if it is absolutely continuous and the identity (3.3.7) holds for almost all $t \in I$. This is a relaxed notion of solution. In contrast, if f is also continuous in t, we have a C^1 solution known as the classical solution. The reader can easily check that for $n \in N_0 \equiv \{0,1,2\cdots\}$, if $f \in C^n$ in both the arguments, $x \in C^{n+1}$.

A more general result is given by the following theorem. This theorem requires only local Lipschitz condition and, as in Theorem 3.2.2, the global linear growth condition. Let

$$B_R \equiv \{\xi \in R^n : \| \xi \| \leq R\}$$

denote the closed ball of radius R centered at the origin.

Theorem 3.3.3. Suppose there exists a function $K \in L_1^+$ and, for each positive number R, a function $L_R \in L_1^+$ so that the following inequalities hold

$$(A1) : \| f(t,x) \| \leq K(t)\{1+ \| x \|\} \quad \forall x \in R^n, \tag{3.3.27}$$

$$(A2) : \| f(t,x) - f(t,y) \| \leq L_R(t) \| x - y \| \quad \forall x, y \in B_R. \tag{3.3.28}$$

Then for each $x_0 \in R^n$, the system (3.3.7) has a unique solution $x \in C(I, R^n)$.

Proof. We use the previous theorem to prove this result. From the linear growth assumption it follows that a solution of the system equation (3.3.7), if one exists, remains confined in a ball

$$B_{\tilde{R}} \equiv \{\xi \in R^n : \| \xi \| \leq \tilde{R}\}$$

of radius $\tilde{R}$ around the origin where $\tilde{R}$ is possibly dependent on T and $\| x_0 \|$. Indeed, let $x \in C(I, R^n)$ be a solution of equation (3.3.7), or equivalently the integral equation

$$x(t) = x_0 + \int_0^t f(s, x(s))ds. \tag{3.3.29}$$

Then it follows from the assumption (A1) that

$$\| x(t) \| \leq \| x_0 \| + \int_0^t K(s)[1+ \| x(s) \|]ds, t \in I. \tag{3.3.30}$$

Using Gronwall inequality, it follows from this estimate that

$$(1+ \| x(t) \|) \leq (1+ \| x_0 \|) \exp \int_0^T K(s)ds \equiv \tilde{R}. \tag{3.3.31}$$

In other words if x is a solution of the problem, then $x(t) \in B_{\tilde{R}}$ for all $t \in I$. Now choose $R > \tilde{R}$ and define

$$\phi_R(x) = \begin{cases} x & \text{if } x \in B_R \\ (R/ \| x \|)x & \text{for } x \notin B_R. \end{cases} \tag{3.3.32}$$

The function ϕ_R is called the retraction. It maps every point out side the ball B_R on to the sphere $S_R = \partial B_R$. Let $\tilde{x} = \phi_R(x)$ and $\tilde{y} = \phi_R(y)$. Geometrically it is clear that $\| \tilde{x} - \tilde{y} \| \leq \| x - y \|$. This is because the distance between any two points out side the ball is always greater than the distance between their retracts. We leave it to the reader to verify that this inequality can be also derived purely algebraically using Cauchy inequality. Define $f_R(t,x) \equiv$

$f(t, \phi_R(x))$. Note that f_R has of course the linear growth property (A1) and further it is also globally Lipschitz, that is,

$$\| f_R(t,x) - f_R(t,y) \| \leq L_R(t) \| x - y \|, \ \forall x, y \in R^n. \tag{3.3.33}$$

Hence it follows from the previous theorem (Theorem 3.2.1) that the initial value problem

$$\dot{\psi}(t) = f_R(t, \psi(t)), t \in I, t > 0,$$
$$\psi(0) = x_0$$

or equivalently the integral equation

$$\psi(t) = x_0 + \int_0^t f_R(s, \psi(s))ds, t \in I, \tag{3.3.34}$$

has a unique solution which we denote by $x_R(t), t \in I$. Since f_R satisfies the same growth condition (A1), $x_R(t) \in B_{\tilde{R}} \subset B_R$ for all $t \in I$. Hence

$$x_R(t) = x_0 + \int_0^t f_R(s, x_R(s))ds,$$
$$= x_0 + \int_0^t f(s, x_R(s))ds, t \in I. \tag{3.3.35}$$

Thus x_R is independent of R for all $R > \tilde{R}$. Thus we may denote the solution by x without the subscript R. This ends the proof. □

Theorem 3.3.4. Let $I \equiv [0, T]$, $T \in (0, \infty)$ and suppose that $f(\cdot, 0) \in L_1(I, R^n)$ and that, for every finite positive number r, there exists a function $L_r \in L_1^+(I)$ so that

$$\| f(t,x) - f(t,y) \| \leq L_r(t) \| x - y \| \quad \forall x, y \in B_r. \tag{3.3.36}$$

Then for each $x_0 \in R^n$, there exists a maximal interval $I(x_0) = [0, \tau(x_0)) \subset I$ such that the system (3.3.7) has a unique solution $x \in C(I(x_0), R^n)$. Further, if $\tau(x_0) \in \text{Int } I$, then $\lim_{t \to \tau(x_0)} \| x(t) \| = +\infty$. If $\tau(x_0) = T$, the solution may be continued beyond I till it blows up.

Proof. Define $r_0 \equiv \| x_0 \|$ where $x_0 \in R^n$ is the given initial state. Take any finite $r > r_0$, and note that $x_0 \in \text{Int}B_r$, where B_r is the closed ball of radius r around the origin of R^n. Now observe that

$$\| f(t,x) \| \leq \| f(t,0) \| + L_r(t) \| x \|$$
$$\leq K_r(t)(1 + \| x \|), x \in B_r, \tag{3.3.37}$$

where $K_r(t) \equiv \sup\{\| f(t,0) \|, L_r(t)\}$. Since $f(\cdot,0) \in L_1(I, R^n)$ and $L_r \in L_1^+(I)$ we have $K_r \in L_1^+(I)$. Thus f satisfies the assumptions (A1) and (A2) of Theorem 3.3.2 in B_r and hence there exists a time $T_r \leq T$ such that the system (3.3.7) has a unique solution $x \in C([0, T_r), R^n)$ till the time T_r at which it hits the boundary $\partial B_r \equiv \{\xi \in R^n : \| \xi \| = r\}$. If it never hits the boundary, set $T_r = T$, to end the process and we may conclude that we have a unique solution. If $T_r < T$, we start all over again with the initial state $x(T_r)$. Clearly $\| x(T_r) \| = r$, which we denote by $r_1 = r$. Again we choose a positive number, say, $r_2 > r_1$, and note that $x(T_{r_1}) \in \text{Int}B_{r_2}$. By our assumption on f, we can again find functions $L_{r_2}, K_{r_2} \in L_1^+(I)$ so that f has the growth and Lipschitz properties with respect to these functions on B_{r_2}, thereby guaranteeing the existence and uniqueness of a solution till time, say, T_{r_2} at which time the solution hits the boundary of the ball B_{r_2}. If $T_{r_2} = T$, the process is complete and we have a unique solution; if $T_{r_2} < T$ we repeat the process. This way we choose an increasing sequence of positive numbers $\{r_k\}$ with $r_k \to \infty$ and a corresponding sequence of $\{T_{r_k}\} \leq T$ so that we have unique solutions on each of the sequence of intervals $[T_{r_k}, T_{r_{k+1}})$. If $\lim_{k\to\infty} T_{r_k} = \tau < T$, the solution blows up at time τ and we have a unique solution locally on the interval $[0, \tau)$, with τ being the blow up time. If on the other hand $\tau = T$ we have a unique global solution. This completes the proof. □

Remark. This result tells us that a system like (3.3.7) with f only locally Lipschitz has always a unique solution till a blow up time. With little reflection, the reader may observe that Theorem 3.3.4 covers all functions f, having components $\{f_i\}$ which are polynomials in the variables $\{x_i, i = 1, 2, 3, \cdots n\}$. Some simple examples are (3.2.4) and the examples (E1) and (E2) of this section.

3.4 Properties of Solution Operator

In view of the above results we can define a solution operator $\{T_{t,s}(\cdot), t \geq s \geq 0\}$, that maps the state ξ given at time s to the current state, $x(t)$, at time t. That is,

$$x(t) = x(t, s, \xi) \equiv T_{t,s}(\xi), t \geq s \geq 0. \tag{3.4.38}$$

Theorem 3.4.1. Like the transition operator for linear systems, the family of solution operators $\{T_{t,s}(\cdot), t \geq s \geq 0, \}$ satisfies the evolution (semigroup) property in the sense given bellow:

(T1): $T_{s,s} = I$, $\forall s \geq 0$.
(T2): $\lim_{t \downarrow s} T_{t,s}(\xi) = \xi, t > s \geq 0$, $\forall\, \xi \in R^n$
(T3): $T_{t,\tau}(T_{\tau,s}(\xi)) = T_{t,s}(\xi)$, $\forall\, t \geq \tau \geq s \geq 0$, $\forall\, \xi \in R^n$

Proof. The proof follows from continuity of solution with respect to $t \geq s \geq 0$ and the uniqueness. □

In case the system is time invariant, the evolution operator reduces to a family of nonlinear semigroups of operators. This is stated in the following corollary.

Corollary 3.4.2. Like the transition operator for linear systems, the family of solution operators $\{T_t(\cdot), t \geq 0\}$, satisfies the semigroup property in the sense given bellow:
(T1): $T_0 = I$,
(T2): $\lim_{t \downarrow 0} T_t(\xi) = \xi, t > 0$, $\forall\, \xi \in R^n$
(T3): $T_t(T_\tau(\xi)) = T_{t+\tau}(\xi)$, $\forall\, t, \tau > 0$, $\forall\, \xi \in R^n$

Some Physical Examples. Here we present some simple examples.

(E1): The differential equation describing the motion of a forced pendulum is given by an equation of the form

$$\ddot{\theta} + k \sin\theta = u, \tag{3.4.39}$$

where k is a constant and $u = u(t)$ is the external force. Defining $x_1 = \theta, x_2 = \dot{\theta}$ we write this in the canonical form giving

$$\begin{aligned} \dot{x}_1 &= x_2 \\ \dot{x}_2 &= -k \sin x_1 + u. \end{aligned} \tag{3.4.40}$$

Note that k is negative for inverted pendulum. For any integrable function u, and any given initial state, it follows from Theorem 3.2.2 that this equation has a unique solution.

(E2): Consider a parallel RLC circuit connected to a current source. The capacitor is nonlinear and its charge voltage characteristic is given by $v = g(q)$ where q is the charge and v is the corresponding voltage. Letting $I(t)$ denote the current source, q the charge on the capacitor, and i the current in the inductor, it follows from Kirchoff's current and voltage laws, that the system is governed by the following set of equations

$$\begin{aligned} (dq/dt) &= -(1/R)g(q) - i + I(t) \\ (di/dt) &= (1/L)g(q). \end{aligned} \tag{3.4.41}$$

The state of this system is naturally given by the charge on the capacitor and the current in the inductor. Usually the voltage-charge characteristics g is Lipschitz continuous with $g(0) = 0$. Again by Theorem 3.2.2, this system has unique solution for a given initial state.

(E3) Here we consider a simple example that satisfies the local Lipschitz property as seen in Theorem 3.3.3. Let $\mathcal{R}_i, i = 1, 2, 3, \cdots, m$ denote a finite number of open regions in R^n so that $R^n = \cup_{i=1}^m \bar{\mathcal{R}}_i$, where $\bar{\mathcal{R}}_i$ denotes the closure of the region $\mathcal{R}_i$. Let $\{g_i, A_i, i = 1, 2, \cdots, m\}$ be a family of vectors and matrices which are functions of time with elements locally integrable. Consider the system

$$\begin{aligned} \dot{x}(t) &= A_i(t)x(t) + g_i(t) \equiv f_i(t, x(t)), t \geq 0, \quad \text{for } x \in \mathcal{R}_i, i = 1, 2, \cdots, m \\ x(0) &= x_0. \end{aligned} \tag{3.4.42}$$

This is a system that satisfies the conditions of Theorem 3.3.3 and hence has a unique solution in each region. Note that this is a locally or piecewise linear system and some authors call it multiplant system. If $x_0 \in \mathcal{R}_i$, a unique solution is defined $x_i(t, x_0)$. This solution may reside in $\mathcal{R}_i$ for ever or may exit at some finite time τ_i and enter a region that borders with $\mathcal{R}_i$. Let $\mathcal{R}_j$ denote the region that has common border with $\mathcal{R}_i$, that is, $\partial\mathcal{R}_i \cap \partial\mathcal{R}_j \neq \emptyset$. For exit it is necessary that $(f_i(\tau_i, x_i(\tau_i)), \nu_i(x_i(\tau_i))) < 0$ where $\nu_i(z)$ denotes the inward normal at the point $z \in \partial\mathcal{R}_i$.

3.5 Continuous Dependence of Solutions

The subject of continuous dependence of solutions is vital in applications where one expects small changes in the parameters to produce small changes in the outcome. This expectation is not valid in all situations. For example, a system may be stable for some values of the parameters, but a small change may destabilize the system. Here we are interested in continuous dependence of solutions with respect to certain parameters. Consider the system governed by a differential equation

$$\dot{x} = f(t, x, \alpha), x(0) = \xi \in R^n, t \in I \equiv [0, T], \ T < \infty, \tag{3.5.43}$$

where α takes values from a metric space Λ. Suppose $f : I \times R^n \times \Lambda \longrightarrow R^n$ is measurable in t on I and continuous in the rest of the variables on $R^n \times \Lambda$.

Theorem 3.5.1. Consider the system (3.5.43). Suppose there exists a $K_\Lambda \in L_1^+(I)$ such that the following conditions hold

$$\| f(t, x, \alpha) \| \leq K_\Lambda(t)(1 + \| x \|) \ \forall \ \alpha \in \Lambda, x \in R^n \tag{3.5.44}$$

$$\| f(t,x,\alpha) - f(t,y,\alpha) \| \leq K_\Lambda(t) \| x - y \| \; \forall \; \alpha \in \Lambda, x, y \in R^n. \quad (3.5.45)$$

Then the solution map $\{\xi, \alpha\} \longrightarrow x \equiv x(\cdot, \xi, \alpha)$ is sequentially continuous.

Proof. Let $\{\xi_k, \alpha_k\} \in R^n \times \Lambda$ be any sequence that converges to an element $\{\xi, \alpha\} \in R^n \times \Lambda$ and let $x_k \in C(I, R^n)$ and $x \in C(I, R^n)$ denote the corresponding solutions of equation (3.5.43) respectively. We must show that $x_k \longrightarrow x$. Note that under the given assumptions it follows from Theorem 3.3.2 that these solutions do exist and are unique. Then the reader can easily verify that

$$\begin{aligned} \| x(t) - x_k(t) \| \; &\leq \; \| \xi - \xi_k \| + \int_0^t K_\Lambda(s) \| x(s) - x_k(s) \| \, ds \\ &+ \int_0^T \| f(s, x(s), \alpha) - f(s, x(s), \alpha_k) \| \, ds, t \in I. \end{aligned} \quad (3.5.46)$$

Hence by using the elementary Gronwall inequality we obtain

$$\| x - x_k \|_C \; \leq \; \beta \left(\| \xi - \xi_k \| + \int_0^T \| f(s, x(s), \alpha) - f(s, x(s), \alpha_k) \| \, ds \right) \quad (3.5.47)$$

where $\beta \equiv exp\{\int_0^T K_\Lambda(s) ds\}$ is a finite positive number. Since x is the solution corresponding to the initial state ξ and the parameter α we have

$$\| x(t) \| \leq \; \| \xi \| + \int_0^t K_\Lambda(s)(1+ \| x(s) \|) ds.$$

By use of Gronwall inequality once again, it follows from this that

$$\sup_{t \in I} \| x(t) \| \; \leq \; \beta \left(\| \xi \| + \int_0^T K_\Lambda(s) ds \right).$$

Hence there exists a constant $b > 0$ such that

$$\| x(t) \| \; \leq \; b \, \forall \, t \in I.$$

Now, by virtue of the growth condition for f (uniformly on Λ) we have

$$\begin{aligned} \| f(t, x(t), \alpha) \| \; &\leq K_\Lambda(t) \; (1+ \| x(t) \|) \; \forall \; t \in I, \\ \| f(t, x(t), \alpha_k) \| \; &\leq K_\Lambda(t) \; (1+ \| x(t) \|) \; \forall \; t \in I. \end{aligned}$$

Thus the integrand of (3.5.47) is dominated by an integrable function as follows

$$\| f(t, x(t), \alpha) - f(t, x(t), \alpha_k) \| \; \leq 2(1 + b) K_\Lambda(t), t \in I.$$

Further, since f is continuous in α on Λ, it is clear that

$$f(t, x(t), \alpha_k) \longrightarrow f(t, x(t), \alpha) \ a.e$$

as $k \to \infty$. In other words, the integrand converges to zero a.e on I. Thus by virtue of Lebesgue dominated convergence theorem, it follows from the expression (3.5.47) that

$$\lim_{k \to \infty} \| x - x_k \|_C = 0.$$

This shows that the map $\{\xi, \alpha\} \longrightarrow x(\cdot, \xi, \alpha)$ is sequentially continuous on $R^n \times \Lambda$. This completes the proof. □

Remark. Since $R^n \times \Lambda$ is a metric space, the sequential continuity is equivalent to continuity. In other words $(\xi, \alpha) \longrightarrow x(\cdot, \xi, \alpha)$ is continuous from $R^n \times \Lambda \longrightarrow C(I, R^n)$.

Later in the sequel we will be concerned with continuous dependence of solutions also with respect to controls or inputs. Consider the controlled system

$$\dot{x} = f(t, x, u), \ x(0) = \xi, t \in I, \tag{3.5.48}$$

where the controls take values from a closed subset $U \subset R^d$. For admissible controls we consider the metric space (M, d) (see example E4 section 1.5 of Chapter 1). Here M is the space of all measurable functions on I taking values from the set U. Recall that the metric d was defined as follows

$$d(u, v) \equiv \lambda\{t \in I : u(t) \neq v(t)\},$$

where λ is the Lebesgue measure. It is shown in the appendix A (Theorem A.5.3) that this is a metric and that M furnished with this metric, is a complete metric space. We wish to prove a result asserting continuous dependence of solutions with respect to this metric.

Theorem 3.5.2. Consider the system (3.5.48) and suppose f is measurable in t on I and continuous in the variables $x \in R^n$ and $u \in U$ and there exists a $K \in L_1^+(I)$ such that

$$\| f(t, x, u) \| \leq K(t)(1 + \| x \|) \ \forall \ u \in U, x \in R^n \tag{3.5.49}$$

$$\| f(t, x, u) - f(t, y, u) \| \leq K(t) \| x - y \| \ \ \forall \ u \in U, x, y \in R^n. \tag{3.5.50}$$

Then the solution x is a continuous map from the metric space $R^n \times (M, d)$ to the Banach space $C(I, R^n)$.

Proof. Take any pair of inputs $u, v \in (M, d)$ and initial data $\xi, \eta \in R^n$. Let $x(\cdot, \xi, u) \equiv x(\xi, u)) \in C(I, R^n)$ and $x(\cdot, \eta, v) \equiv x(\eta, v) \in C(I, R^n)$ denote the corresponding solutions of the system equation (3.5.48). Note that existence and uniqueness of solutions follow from Theorem 3.3.3. Clearly

$$x(t, \xi, u) = \xi + \int_0^t f(s, x(s, \xi, u), u(s))ds, t \in I,$$
$$x(t, \eta, v) = \eta + \int_0^t f(s, x(s, \eta, v), v(s))ds, t \in I.$$

Thus taking the difference and using the Lipschitz condition, we arrive at the following inequality

$$\begin{aligned} &\| x(t, \xi, u) - x(t, \eta, v) \| \\ &\quad \leq \ \| \xi - \eta \| + \int_0^t \| f(s, x(s, \eta, v), u(s)) - f(s, x(s, \eta, v), v(s)) \| ds \\ &\quad + \int_0^t K(s) \| x(s, \xi, u) - x(s, \eta, v) \| ds \qquad (3.5.51) \\ &\quad \leq \ \| \xi - \eta \| + \int_0^T \| f(s, x(s, \eta, v), u(s)) - f(s, x(s, \eta, v), v(s)) \| ds \\ &\quad + \int_0^t K(s) \| x(s, \xi, u) - x(s, \eta, v) \| ds. \qquad (3.5.52) \end{aligned}$$

Using the elementary Gronwall inequality, it follows from the above expression that

$$\begin{aligned} &\sup_{t \in I} \| x(t, \xi, u) - x(t, \eta, v) \| \\ &\quad \leq C\{\| \xi - \eta \| + \int_I \| f(s, x(s, \eta, v), u(s)) - f(s, x(s, \eta, v), v(s)) \| ds\} \end{aligned} \qquad (3.5.53)$$

where the constant C is given by $C \equiv exp\{\int_0^T K(t)dt\}$. By virtue of the growth condition (3.5.49), it follows again from Gronwall inequality that

$$\| x(\eta, v) \|_C \equiv \sup_{t \in I} \| x(t, \eta, v) \| \equiv b \leq \ C \left(\| \eta \| + \int_I K(s)ds \right) < \infty.$$

Note that

$$\begin{aligned} J &\equiv \{t \in I : f(t, x(t, \eta, v), u(t)) \neq f(t, x(t, \eta, v), v(t))\} \\ &= \{t \in I : u(t) \neq v(t)\}, \end{aligned}$$

and hence $d(u,v) = \lambda(J)$. Using these facts we deduce from (3.5.53) that

$$\| x(\xi,u) - x(\eta,v) \|_C \leq C\{\| \xi - \eta \| + 2(1+b)\int_J K(s)ds\}$$
$$\leq C\{\| \xi - \eta \| + 2(1+b)\mu_K(J)\}, \qquad (3.5.54)$$

where μ_K is the measure defined by

$$\mu_K(\sigma) \equiv \int_\sigma K(s)\lambda(ds) = \int_\sigma K(s)ds.$$

Since $K \in L_1^+(I)$, it is clear from this that $\mu_K(J) \longrightarrow 0$ whenever $d(u,v) \equiv \lambda(J) \longrightarrow 0$. Thus it follows from the expression (3.5.54) that the solution map $x : R^n \times M \longrightarrow C(I, R^n)$ is continuous. This completes the proof. □

Remark. In case K is a constant, the solution map $(\xi, u) \longrightarrow x(\xi, u)$ is Lipschitz,

$$\| x(\xi,u) - x(\eta,v) \|_C \ \leq \ \tilde{C}\{\| \xi - \eta \| + d(u,v)\},$$

with Lipschitz constant $\tilde{C} \equiv max\{C, 2(1+b)K\}$.

In the above result we have assumed that f is Lipschitz in $x \in R^n$ uniformly with respect to the control space U. This is relaxed in the following result. Let U be a closed but not necessarily bounded subset of R^d. For each $r \in N \equiv \{1, 2, \cdots\}$, define $U_r \equiv \{\zeta \in U : \| \zeta \| \leq r\}$. Now define $M_r = M(I, U_r)$ the metric space of measurable functions on I with values in the closed ball U_r with the induced metric given by

$$d_r(u,v) = d(u,v) \text{ for } u, v \in M_r.$$

Using this family of semi-metrics $\{d_r, r \in N\}$ we may now define a metric ρ on $M = M(I, U)$ as follows:

$$\rho(u,v) = \sum_{r=1}^{\infty} (1/2^r) \min\{d_r(u,v), 1\}.$$

With respect to the metric ρ, (M, ρ) is a complete metric space.

Theorem 3.5.3. Consider the system (3.5.48) and suppose f is measurable in t on I and continuous in the variables $x \in R^n$ and $u \in U$ and for each $r \in N$, there exists a $K_r \in L_1^+(I)$ such that

$$\| f(t,x,u) \| \leq K_r(t)(1 + \| x \|) \ \forall \ u \in U_r, x \in R^n \qquad (3.5.55)$$
$$\| f(t,x,u) - f(t,y,u) \| \leq K_r(t) \| x - y \| \ \ \forall \ u \in U_r, x, y \in R^n. \qquad (3.5.56)$$

Then the solution map x is locally continuous from the metric space $R^n \times (M, \rho)$ to the Banach space $C(I, R^n)$.

Proof. Consider the retraction of the ball $U_r, r \in N$, as follows

$$\varphi_r(u) \equiv \begin{cases} u, & \text{if } u \in U_r; \\ (r/ \parallel u \parallel)u, & \text{otherwise.} \end{cases}$$

Let $u, v \in M$ and define $u_r(t) = \varphi_r(u(t))$ and $v_r(t) = \varphi_r(v(t))$ and let $x(\xi, u_r)$ and $x(\eta, v_r)$ denote the solutions of the system (3.5.48) corresponding to the pairs $\{\xi, u_r\}$ and $\{\eta, v_r\}$ respectively. Then it follows from the previous theorem, in particular the inequality (3.5.54), that

$$\parallel x(\xi, u_r) - x(\eta, v_r) \parallel_C \; \leq \; C_r\{\parallel \xi - \eta \parallel + 2(1 + b_r)\mu_r(J_r)\}, \tag{3.5.57}$$

where

$$C_r \equiv \exp\left\{\int_I K_r(s)ds\right\}, \quad b_r \equiv \parallel x(\eta, v_r) \parallel_C$$

$$J_r \equiv \{t \in I : u_r(t) \neq v_r(t)\} \quad \mu_r(\Gamma) \equiv \int_\Gamma K_r(s)ds,$$

for any measurable set $\Gamma \subset I$.

Since $K_r \in L_1^+(I)$, the measure μ_r vanishes on every measurable set on which the Lebesgue measure λ does. In other words μ_r is absolutely continuous with respect to the Lebesgue measure. Hence it follows from the inequality (3.5.57) that the solution map $x : R^n \times M_r \longrightarrow C(I, R^n)$ is continuous. This is true for each finite $r \in N$ and hence for every finite $r \in R_0 = [0, \infty)$. Thus the solution map is locally continuous. □

3.6 Impulsive Systems

So far we have considered regular nonlinear systems. There are situations where a system may be forced, from time to time, to leave its current state abruptly and jump into another state determined by an external agency. For example, an orbiting satellite may be hit by meteorites form time to time causing an abrupt change of state. In the absence of such forces, the satellite executes its motion according to its normal attitude dynamics. Similar examples can be found in the study of demography where due to war there is large scale migration of population in a very short span of time. Similarly epidemic may cause abrupt changes in the demography.

Further a system may also be labeled impulsive, if the control forces are impulsive. Such examples arise in the study of traffic flow control in computer communication networks.

3.6.1 Classical Models

A most widely studied impulsive system is described by the following model,

$$\begin{aligned} \dot{x} &= f(t,x), t \in I \setminus D, \qquad (3.6.58) \\ \triangle x(t_i) &= G(t_i, x(t_i)), t_i \in D \equiv \{0 = t_0 < t_1 < t_2, \cdots, < t_m < T\}, t_i \in I, \end{aligned}$$

where $\triangle$ denotes the jump operator given by

$$\triangle x(t_i) \equiv x(t_i + 0) - x(t_i - 0) \equiv x(t_i + 0) - x(t_i).$$

In other words the system evolves normally according to the first equation during each of the time intervals $[t_{i-1}, t_i)$, and at time $t_i + 0$ it makes a jump to the state

$$x(t_i + 0) = x(t_i) + G(t_i, x(t_i)).$$

Clearly the jump size is determined by the function G. Note that G could be viewed also as a feedback control (law) that imparts control actions at a set of predetermined points of time according to the state of the system at those instances. One may also visualize even more general systems of the form

$$\dot{x} = f(t, x, u), t \in I \setminus D, \qquad (3.6.59)$$

$$\triangle x(t_i) = G(t_i, x(t_i), v_i), t_i \in D, \qquad (3.6.60)$$

where u is a regular control and v is an impulsive control. Before we discuss the questions of existence, uniqueness and regularity properties of solutions of such models, for motivation we may consider a practical example arising from construction engineering.

(E1) (Piling): For construction of buildings, civil engineers use piling to reinforce the soil beneath the construction site. Piling is also used to construct supporting pillars by pouring concrete mixture in a hollow steel pipe which is hammer driven into the ground. A massive hammer is used to drop under gravity on a supporting plate placed on the pipe. We are interested to estimate the depth of penetration after each strike. This is an impulsive force and the system may be considered as an impulsive system. Letting $x(t)$ denote the depth of penetration at time t measured downward from the surface, m the mass of the pipe, M the mass of the hammer, r the radius of the pipe, β the coefficient of friction per unit surface area, one can verify that x satisfies the second order differential equation given by

$$m\ddot{x}(t) + (\beta 4\pi r x(t))\dot{x}(t) = F(t), \qquad (3.6.61)$$

where F is the force imparted to the standing pipe by the hammer. Note that the resistance to penetration is given by $R = \beta(4\pi r x)$ where $(4\pi r x)$ is the surface area of the pipe in contact with the surrounding ground. Here we have assumed that after each strike the hammer bounces back. Define the set

$$D \equiv \{0 < t_1 < t_2 < t_m < T\} \subset I$$

to be the set of time instances at which the hammer strikes the supporting plate. We shall give an approximate solution of this problem after the first strike and over the time interval $[0, t_2)$. Note that the velocity of the pipe before the first strike is 0 and immediately after the strike is $\sqrt{2gH}$ where H is the height of fall. Thus the force imparted to the pipe is an impulse one given by $F(t) = M\sqrt{2gH}\delta(t-t_1)$. Assuming the initial position and velocity to be $x(0) = x_0, \dot{x}(0) = 0$ and that the friction is approximately constant in between strikes, we can approximate the dynamics as follows

$$m\ddot{x}(t) + (4\beta\pi r x_0)\dot{x}(t) = M\sqrt{2gH}\delta(t-t_1), t \in [0, t_2). \tag{3.6.62}$$

Solving this equation, we find that $x(t) = x_0$ for $0 \le t < t_1$ and

$$x(t) = x_0 + \frac{M\sqrt{2gH}}{4\beta\pi r x_0}\Big(1 - exp\{-(4\beta\pi r x_0/m)(t-t_1)\}\Big), \text{ for } t_1 \le t < t_2. \tag{3.6.63}$$

It is clear from this expression that penetration increases with the increase of the weight of the hammer or the height of fall, while it decreases with the increase of radius of the pipe, friction coefficient β, and the initial depth x_0. If we wish to consider the result of only one strike, we must set $F(t) \equiv 0$ for $t > t_1$. In that case the maximum possible penetration due to first strike is given by

$$x_1 = x_0 + \left(\frac{M\sqrt{2gH}}{4\beta\pi r x_0}\right).$$

Assuming the time difference between the strikes to be sufficiently large, we can take $x(t_2-) = x_1$ and $\dot{x}(t_2-) = 0$ as the initial state for the second strike. In that case, one can write the solution approximately as

$$x(t) = x_1 + \frac{M\sqrt{2gH}}{4\beta\pi r x_1}\Big(1 - exp\{-(4\beta\pi r x_1/m)(t-t_2)\}\Big), t_2 \le t < t_3.$$

This process can be continued add infinitum to construct an approximate solution and the depth of penetration can be estimated after any number of strikes.

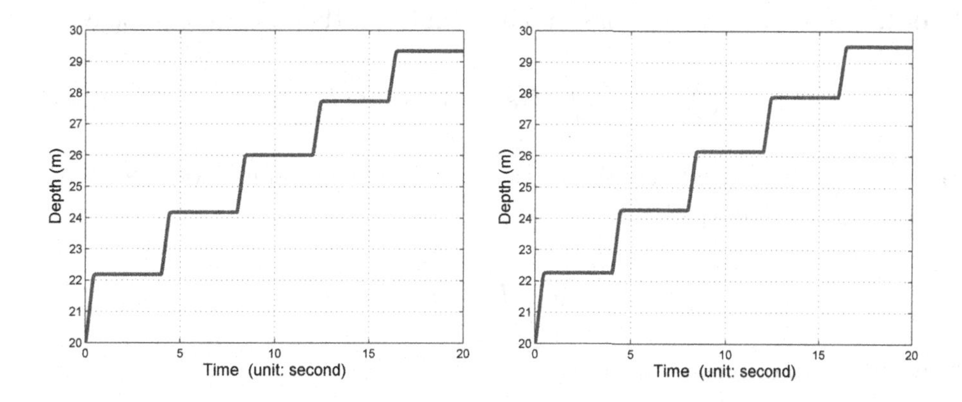

Figure 3.6.1: Pilling

We leave it as an exercise for the reader to compute the solution for multiple strikes with the force given by

$$F(t) \equiv \sum_{i=1}^{m} M\sqrt{2gH_i}\ \delta(t - t_i),$$

where H_i may vary from one strike to another. For illustration, some numerical results based on exact solution and approximate ones are presented below. The parameters used are $m = 2.5 \times 10^3 Kg, M = 7.85 \times 10^3 Kg, \beta = 0.4, r = 0.6m, H = 20m, g = 9.8m/sec^2$. The graph on the left corresponds to the exact solution and that on the right is based on approximate solution. Notice that the approximate solution overestimates the depth of penetration slightly. Clearly this is due to assumption of constancy of the friction during the process of penetration for each strike.

Here we present some results on the questions of existence of solutions of Impulsive systems given by equation (3.6.58). Let $PWC(I, R^n)$ denote the class of piece wise continuous functions on I with values in R^n, and $PWC_\ell(I, R^n)(PWC_r(I, R^n))$ those that are left (right) continuous with right (left) hand limits.

Theorem 3.6.1. Suppose f satisfy the assumptions of Theorem 3.3.3 and $G : I \times R^n \longrightarrow R^n$ is continuous and bounded on bounded subsets of $I \times R^n$. Then for every initial state $x_0 \in R^n$, the system (3.6.58) has a unique solution in $PWC_\ell(I, R^n)$.

Proof. Under the assumptions of Theorem 3.3.3, the integral equation

$$x(t) = x_0 + \int_0^t f(s, x(s))ds, 0 \leq t < t_1,$$

has a unique (absolutely continuous) solution. By continuity we extend this solution up to time t_1 giving

$$x(t_1) = x_0 + \int_0^{t_1} f(s, x(s))ds.$$

Now the jump occurs at t_1 giving

$$x(t_1 + 0) = x(t_1) + G(t_1, x(t_1)).$$

Note that the jump size is determined by the value of G just prior to jump. Since G is continuous and bounded on bounded sets of $I \times R^n$, this is uniquely defined. Starting with this as the initial state for the next interval $(t_1, t_2]$ we solve the integral equation

$$x(t) = x(t_1 + 0) + \int_{t_1}^t f(s, x(s))ds, t \in [t_1, t_2].$$

Again by virtue of Theorem 3.3.3, this equation has a unique solution $x \in AC([t_1, t_2], R^n)$. It is clear from these steps that we can construct a unique solution for the system (3.6.58) on the entire interval I piece by piece. Also note that the solution so constructed is continuous from the left and it has limits from the right. Thus $x \in PWC_\ell(I, R^n)$. This proves the statement of the theorem. □

This result can be further extended to the case in which f is only locally Lipschitz. This is given in the following theorem.

Theorem 3.6.2. Suppose f satisfy the assumptions of Theorem 3.3.4 and $G : I \times R^n \longrightarrow R^n$ is continuous and bounded on bounded subsets of $I \times R^n$. Then for every initial state $x_0 \in R^n$, the system (3.6.58) has a unique solution in $PWC_\ell(I, R^n)$.

Proof. Since the proof is very similar to that of the previous theorem, we give only an outline. Let $I \equiv [0, T]$. Since $x_0 \in R^n$, there exists a positive number r_0 such that $x_0 \in B_{r_0}$. Now choose $r_1 > r_0$ and note that by our assumption there exists $L_{r_1} \equiv L_1 \in L_1^+(I)$ such that

$$\| f(t, x) - f(t, y) \| \leq L_1(t) \| x - y \|, x, y \in B_{r_1}.$$

Then we solve the integral equation

$$x(t) = x_0 + \int_0^t f(s, x(s))ds, 0 \leq t < T_1 \wedge t_1,$$

where T_1 denotes the first time the solution hits the boundary ∂B_{r_1}. The existence and uniqueness of a solution follows from Theorem 3.3.4. If $T_1 > t_1$, to continue, define

$$x(t_1 + 0) = x(t_1) + G(t_1, x(t_1)).$$

If $x(t_1 + 0) \in IntB_{r_1}$ solve the integral equation

$$x(t) = x(t_1 + 0) + \int_{t_1}^t f(s, x(s))ds, t_1 \leq t \leq T_1 \wedge t_2.$$

If $x(t_1 + 0) \notin B_{r_1}$, take a ball of radius $r_2 > r_1$ so that $x(t_1 + 0) \in IntB_{r_2}$ and repeat the process. Note that $\{T_k\} \in I$ is a nondecreasing sequence. If it converges to a point $\tau < T$ as $k \to \infty$, the solution blows up at time τ and we have a unique solution $x \in PWC_\ell([0, \tau_0], R^n)$ for any $\tau_0 < \tau$. This completes the outline of our proof. □

Again the question of continuous dependence of solutions on controls for impulsive systems given by (3.6.59) and (3.6.60) is crucial in applications to control theory. Let $U \subset R^d$ and $V \subset R^k$ be compact sets and let $M(I \setminus D, U)$ denote the class of measurable functions defined on $I \setminus D$ with values in U and $F(D, V)$ denote the class of functions defined on D with values in V respectively. For the class of admissible controls, we choose the complete metric space $M(I \setminus D, U) \times F(D, V) \equiv \mathcal{W}$ which is furnished with the metric topology given by

$$\begin{aligned} \rho(w, \tilde{w}) \equiv \rho(\{u, v\}, \{\tilde{u}, \tilde{v}\}) &\equiv \lambda\{t \in I \setminus D : u(t) \neq \tilde{u}(t)\} \\ &\quad + sup_{1 \leq i \leq N} \parallel v_i - \tilde{v}_i \parallel \\ &\equiv \lambda\{t \in I : u(t) \neq \tilde{u}(t)\} + sup_{1 \leq i \leq N} \parallel v_i - \tilde{v}_i \parallel, \end{aligned}$$

for $w, \tilde{w} \in \mathcal{W}$. It is clear that $(\mathcal{W}, \rho)$ is a complete metric space. We prove the following theorem.

Theorem 3.6.3. Consider the control system governed by the impulsive differential equations (3.6.59) and (3.6.60) with admissible controls $\mathcal{W}$ and suppose the following assumptions hold. There exist a $K \in L_1^+(I)$ and a finite positive number L so that

(Af): $f : I \times R^n \times U \longrightarrow R^n$ is measurable in the first variable and continuous with respect to the rest satisfying

$$\begin{aligned}&\| f(t,x,u) \| \leq K(t)(1+ \| x \|), \forall x \in R^n, \forall u \in U\\&\| f(t,x,u) - f(t,y,u) \| \leq K(t)(\| x-y \|), \forall x,y \in R^n, \forall u \in U;\end{aligned}$$

(AG): $G : I \times R^n \times V \longrightarrow R^n$ is continuous in all the variables satisfying

$$\begin{aligned}&\| G(t,x,v) \| \leq L(1+ \| x \|), \forall t \in D, x \in R^n, \forall v \in V\\&\| G(t,x,v) - G(t,y,v) \| \leq L(\| x-y \|), \forall t \in D, x,y \in R^n, \forall v \in V.\end{aligned}$$

Then the solution map $x : \mathcal{W} \longrightarrow PWC(I, R^n)$ denoted by $x(w)$ is continuous with respect to the respective topologies.

Proof. Let $w \equiv \{u,v\}, \tilde{w} \equiv \{\tilde{u},\tilde{v}\}$ be any two control policies from the class $\mathcal{W}$. Under the given assumptions, it follows from Theorem 3.6.1 that there corresponds a unique solution in $PWC_\ell(I, R^n)$ to each of the control policies w and $\tilde{w}$ respectively. Let these solutions be denoted by $x = x(w)$ and $\tilde{x} \equiv x(\tilde{w})$ with values $x(t), \tilde{x}(t) \in R^n$. It is also clear from Theorem 3.6.1 that there exists a finite positive number b such that $x(t), \tilde{x}(t) \in B_b(R^n)$ for all $t \in I$. According to the definition of the set D, we introduce the intervals $\{I_i \equiv [t_{i-1}, t_i), 1 \leq i \leq m+1\}$ where $I_{m+1} \equiv (t_m, t_{m+1}] \equiv (t_m, T]$. For convenience of notation, define

$$\begin{aligned}&J_i \equiv \{t \in I_i : u(t) \neq \tilde{u}(t)\}, 1 \leq i \leq m+1\\&C \equiv exp\left\{\int_I K(t)dt\right\}, \quad \text{and } \mu_K(J_i) \equiv \int_{J_i} K(t)dt.\end{aligned}$$

For $t \in I_1$, the reader can easily check that

$$\begin{aligned}\| x(t) - \tilde{x}(t) \| &\leq exp\left\{\int_0^{t_1} K(s)ds\right\}\left\{2(1+b)\int_{J_1} K(s)ds\right\}\\&\leq 2(1+b)C\int_{J_1} K(s)ds \equiv 2(1+b)C\mu_K(J_1). \qquad (3.6.64)\end{aligned}$$

By extension by continuity we have

$$\| x(t_1-) - \tilde{x}(t_1-) \| = \| x(t_1) - \tilde{x}(t_1) \| \leq 2(1+b)C\mu_K(J_1). \qquad (3.6.65)$$

Now considering the jump at t_1 we have

$$x(t_1+) = x(t_1) + G(t_1, x(t_1), v_1) \qquad (3.6.66)$$

$$\tilde{x}(t_1+) = \tilde{x}(t_1) + G(t_1, \tilde{x}(t_1), \tilde{v}_1). \qquad (3.6.67)$$

Using the Lipschitz property of G and the inequality (3.6.65), it follows from simple computation that

$$\begin{aligned} \| x(t_1+) - \tilde{x}(t_1+) \| &\leq (1+L) \| x(t_1) - \tilde{x}(t_1) \| \\ &\quad + \| G(t_1, \tilde{x}(t_1), v_1) - G(t_1, \tilde{x}(t_1), \tilde{v}_1) \| \\ &\leq 2(1+b)C(1+L)\mu_K(J_1) + \| G_1(v_1) - G_1(\tilde{v}_1) \|, \end{aligned} \tag{3.6.68}$$

where, for convenience of notation, we have used

$$\| G_i(v_i) - G_i(\tilde{v}_i) \| \equiv \| G(t_i, \tilde{x}(t_i), v_i) - G(t_i, \tilde{x}(t_i), \tilde{v}_i) \| .$$

Similarly for the interval I_2, we have

$$\begin{aligned} \| x(t_2+) - \tilde{x}(t_2+) \| &\leq 2(1+b)C^2(1+L)^2\mu_K(J_1) + 2(1+b)C(1+L)\mu_K(J_2) \\ &\quad + C(1+L) \| G_1(v_1) - G_1(\tilde{v}_1) \| + \| G_2(v_2) - G_2(\tilde{v}_2) \| . \end{aligned}$$

Again for convenience of notation we let $\alpha \equiv (1+L)C$. Using this notation and repeating the above process for all the intervals, one finds that

$$\sup\{\| x(t) - \tilde{x}(t) \|, t \in I\}$$

$$\leq C\Big\{2(1+b)\sum_{r=1}^{m+1} \alpha^{m+1-r}\mu_K(J_r) + \sum_{r=1}^{m} \alpha^{m-r} \| G_r(v_r) - G_r(\tilde{v}_r) \| \Big\}. \tag{3.6.69}$$

Since $K \in L_1^+(I)$, it is clear that the measure μ_K is absolutely continuous with respect to the Lebesgue measure λ. Further, it follows from our hypothesis on G that $v \longrightarrow G_i(v)$ is continuous from V to R^n for each i with $1 \leq i \leq m$. Thus it follows from the above inequality that $w \longrightarrow x(w)$ is continuous from $(\mathcal{W}, \rho)$ to $PWC(I, R^n)$. □

If the assumptions (Af) and (AG) are slightly modified we can prove the following result.

Corollary 3.6.4. Suppose the general assumptions of Theorem 3.6.3 hold with the following additional properties: f is uniformly Lipschitz in the second argument with Lipschitz constant K and G is Lipschitz also in the third argument. Then the map $w \longrightarrow x(w)$ is Lipschitz from $(\mathcal{W}, \rho)$ to $PWC(I, R^n)$ and there exists a positive constant γ so that

$$\| x(w) - x(\tilde{w}) \|_{PWC} \leq \gamma \, \rho(w, \tilde{w}).$$

Remark. The reader is encouraged to give an estimate of the Lipschitz constant γ.

Remark. Continuous dependence of solutions on other parameters such as the time instances at which impulsive controls are to be applied is also very important in control theory. The reader may like to develop such results.

3.6.2 General Impulsive Models(Driven by Vector Measures)

The model for impulsive systems given above is classical and has been extensively studied by Lakshmikantham, Bainov and Simeonov [56]. Silva and Vinter [196] have studied differential equations and inclusions driven by vector measures. It was substantially generalized by the author in several of his papers in recent years [105, 106, 107, 112, 113, 114, 118, 122, 124, 127, 128, 129, 130, 132, 134, 141]. These models can be described as follows:

$$dx = f(t,x)dt + G(t,x)\nu(dt), x(0) = x_0, t \in I \equiv [0,T], \quad (3.6.70)$$

where $f : I \times R^n \longrightarrow R^n, G : I \times R^n \longrightarrow M(n \times m)$ and $\nu \in \mathcal{M}(I, R^m)$ is a vector measure. If the measure ν is purely atomic given by a sum of Dirac measures, we revert to the classical model (3.6.58). Thus the system model given by (3.6.70) is very general. For the system (3.6.70), here we consider the questions of existence and uniqueness including other regularity properties of solutions. First we point out that the system (3.6.70) should be interpreted as

$$dx(t) = f(t,x(t))dt + G(t,x(t-))\nu(dt), x(0) = x_0, t \in I \equiv [0,T], \quad (3.6.71)$$

where, in case of a jump at the instant of time $\{t\}$ due to ν being atomic at this point, the state immediately after the jump is given by

$$x(t+) \equiv x(t) = x(t-) + G(t, x(t-))\nu(\{t\}).$$

This interpretation is intuitively appealing and physically reasonable since it says that if the system has made an impulsive change of state it must have done so from the immediate past state. This is also in conformity with impulsive models of the simple form given by (3.6.58).

If ν is purely atomic and has bounded variation on bounded sets, the atoms are countable and may be enumerated as $a(\nu) \equiv \{t_i > 0, i \in N\}$ where N denotes the set of positive integers. In this case the system (3.6.71) reduces to the classical case treated in Section 2. Strictly speaking we have

$$\dot{x} = f(t,x), x(0) = x_0, t \in I \setminus D \quad (3.6.72)$$

$$x(t_i+) \equiv x(t_i) = x(t_i-) + G(t_i, x(t_i-)\nu(\{t_i\}), t_i \in D, \quad (3.6.73)$$

where $D \equiv a(\nu) \cap I$. Fundamentally, this system is identical to the one given by (3.6.58) with only a minor change of notation. Thus the system described by equation (3.6.71) is much more general and contains the model (3.6.58) as a special class. This is the class of systems for which we now discuss the questions of existence,uniqueness and regularity properties of solutions. The first question we must answer is what we mean by a solution of equation (3.6.71). This is given in the following definition. First recall that $B(I, R^n)$ denotes the Banach space of bounded measurable functions from I to R^n furnished with the sup norm topology.

Definition 3.6.5. An element $x \in B(I, R^n)$ is said to be a solution of equation (3.6.71) if $x(0) = x_0$ and x satisfies the following integral equation

$$x(t) = x_0 + \int_0^t f(s, x(s))ds + \int_0^t G(s, x(s-))\nu(ds), t \in I. \tag{3.6.74}$$

It is important to justify the definition of solution as given above. Suppose t_1 is an atom of the measure ν. Then at t_1 we expect a jump in the solution. This may be given by either of the following two expressions:

$$x(t_1+) \equiv x(t_1) = x(t_1-) + G(t_1, x(t_1-))\nu(\{t_1\})$$
$$x(t_1) = x(t_1-) + G(t_1, x(t_1))\nu(\{t_1\}).$$

Clearly the first expression is explicit; the state at time t_1 is given entirely in terms of the state just before the jump and the size of the atom. This is compatible with the class of basic impulsive systems given by equation (3.6.58). On the other hand the second expression is implicit; the state at time t_1 is a result of its interaction with itself, the state just before the jump and the size of the atom. This leads to a fixed point problem of the form

$$\xi = \eta + G(t_1, \xi)\nu(\{t_1\})$$

where $\eta \in R^n$ is given, we must solve for ξ. This equation may have no solution, one solution or even multiple solutions leading to ambiguities, where as the first choice offers unique solution. As an example consider the scalar equation

$$dy = (1 - y^2)\delta_{t_1}(dt), t \geq 0,$$

where the measure $\nu(dt) = \delta_{t_1}(dt)$, that is the Dirac measure with mass 1 at t_1 and zero elsewhere. If one chooses the implicit scheme, the reader can easily verify that for $y(0) = 1$, there are two real solutions given by

$$y(t) = \begin{cases} 1, \text{ for } t \geq 0 \\ 1, \text{ for } 0 \leq t < t_1, \text{and } \; -2 \text{ for } t \geq t_1. \end{cases}$$

In case $y(0) = -3$, there are no real solutions. In fact there are two complex solutions which clearly do not make physical sense for differential equations with real coefficients. In contrast, if explicit scheme is used, we have unique solutions for both the initial conditions. For $y(0) = 1$, the solution is $y(t) = 1, t \geq 0$, and for $y(0) = -3$ the solution is given by

$$y(t) = \begin{cases} -3, & \text{for } 0 \leq t < t_1 \\ -11 & \text{for } t \geq t_1. \end{cases}$$

These examples and the compatibility of the general models with the standard impulsive differential equations described by equation (3.6.58), justify the definition 3.6.5.

For scalar valued positive measures a notion of robust solutions based on so called "graph completion technique" was developed by Dal Maso and Rampazzo [168] and Silva and Vinter [196]. This technique does not appear to admit extension to systems driven by vector measures considered here. For detailed discussion of this topic interested reader is referred to [151].

Theorem 3.6.6. Consider the system (3.6.71) and suppose $\nu \in \mathcal{M}(I, R^m)$ is a countably additive bounded vector measure having bounded total variation on I and f and G are measurable in $t \in I$ and continuous in $x \in R^n$ and there exist two functions $K \in L_1^+(I) \equiv L_1^+(I, \lambda)$ (λ = Lebesgue measure) and $L \in L_1^+(I, |\nu|)$ such that

(F): $\| f(t,x) \| \leq K(t)\{1 + \| x \|\}$, $\quad \| f(t,x) - f(t,y) \| \leq K(t)\{\| x - y \|\}, x, y \in R^n$.

(G): $\| G(t,x) \| \leq L(t)\{1 + \| x \|\}$, $\quad \| G(t,x) - G(t,y) \| \leq L(t)\{\| x - y \|\}, x, y \in R^n$.

Then for every $x_0 \in R^n$, equation (3.6.71) has a unique solution $x \in B(I, R^n)$ and further it is piecewise continuous.

Proof. First we prove an a priori bound, that is, if x is a solution of equation (3.6.71) then there exists a finite positive number r such that

$$\| x \| \equiv \sup\{\| x(t) \|, t \in I\} \leq r.$$

Since by definition x must satisfy equation (3.6.74), it follows from this that

$$\| x(t) \| \leq \left(\| x_0 \| + \int_0^t K(s)ds + \int_0^t L(s)|\nu|(ds)\right) + \int_0^t K(s) \| x(s) \| ds + \int_0^t L(s) \| x(s-) \| |\nu|(ds), \qquad (3.6.75)$$

for all $t \in I$. Define $\varphi(t) \equiv \sup\{\| x(s) \|, 0 \leq s \leq t\}$ and the measure μ by

$$\mu(\sigma) \equiv \int_\sigma K(s)ds + \int_\sigma L(s)|\nu|(ds), \sigma \in \mathcal{B}.$$

Since $|\nu|$ is a countably additive bounded positive measure and $K \in L_1^+(I)$ and $L \in L_1^+(I,|\nu|)$, it is clear that μ is also a countably additive bounded positive measure having bounded total variation on I. Defining

$$C \equiv \{\parallel x_0 \parallel + \mu(I)\},$$

and noting that $\varphi(t), t \in I$, is a nonnegative nondecreasing function we have $\varphi(t-) \leq \varphi(t), t \in I$ and it follows from (3.6.75) that

$$\varphi(t) \ \leq \ C + \int_0^t \varphi(s)\mu(ds). \tag{3.6.76}$$

By virtue of a generalized Gronwall type inequality, it follows from this expression that

$$\varphi(t) \leq C exp\{\mu([0,t])\}, t \in I.$$

Hence we conclude that

$$\parallel x \parallel \ \equiv \ \sup\{\parallel x(t) \parallel\} \ \leq \ C \ exp\{\mu(I)\}.$$

Thus taking $r \equiv C exp\{\mu(I)\}$, we have the a priori bound. As a consequence of this result we may also conclude that any solution of equation (3.6.71), if one exists, must belong to the Banach space $B(I, R^n)$ and be bounded. Now we consider the question of existence of solutions. Define the operator $\mathcal{G}$ by

$$(\mathcal{G}x)(t) \equiv x_0 + \int_0^t f(s,x(s))ds + \int_0^t G(s,x(s-))\nu(ds), t \in I. \tag{3.6.77}$$

Clearly it follows from the a priori bound that

$$\mathcal{G} : B(I,R^n) \longrightarrow B(I,R^n),$$

and for $x, y \in B(I, R^n)$,

$$\begin{aligned} \parallel (\mathcal{G}x)(t) - (\mathcal{G}y)(t) \parallel \leq & \int_0^t K(s) \parallel x(s) - y(s) \parallel ds \\ & + \int_0^t L(s) \parallel x(s-) - y(s-) \parallel |\nu|(ds), t \in I. \end{aligned} \tag{3.6.78}$$

Now defining

$$d_t(x,y) \equiv \sup\{\parallel x(s) - y(s) \parallel, 0 \leq s \leq t\}$$

and noting that $d_{t-}(x,y) \leq d_t(x,y)$ for all $x, y \in B(I, R^n)$ it follows from the expression (3.6.78) that

$$d_t(\mathcal{G}x, \mathcal{G}y) \leq \int_0^t d_s(x,y)\mu(ds), \ t \in I, \ x,y \in B(I,R^n). \tag{3.6.79}$$

Define

$$v(t) \equiv \mu([0,t]) \equiv \int_0^t \mu(ds), t \in I.$$

Since μ is a countably additive positive measure having bounded total variation (on bounded sets), the function v, as defined above, is a nonnegative, nondecreasing function of bounded variation (on bounded sets) having at most a countable number of discontinuities. Without loss of generality, we may assume that 0 is not an atom of the measure ν. Then 0 is not an atom of μ either and consequently, $v(0) = 0$. Using this v we can rewrite the expression (3.6.79) as

$$d_t(\mathcal{G}x, \mathcal{G}y) \leq \int_0^t d_s(x,y)dv(s), \ t \in I, \ x,y \in B(I,R^n). \tag{3.6.80}$$

Repeating this for the second iterate, we have

$$\begin{aligned} d_t(\mathcal{G}^2x, \mathcal{G}^2y) &\leq \int_0^t d_s(\mathcal{G}x, \mathcal{G}y)dv(s) \\ &\leq \int_0^t d_s(x,y)v(s)dv(s) \\ &\leq d_t(x,y)(v^2(t)/2), \ t \in I, \ x,y \in B(I,R^n). \end{aligned} \tag{3.6.81}$$

Repeating this iterative process n times we arrive at the following inequality

$$d_t(\mathcal{G}^n x, \mathcal{G}^n y) \leq d_t(x,y)(v^n(t)/n!), t \in I.$$

Define $d(x,y) \equiv d_T(x,y)$ for $x,y \in B(I,R^n)$. Note that $(B(I,R^n),d)$ is a complete metric space. Then it follows from the above inequality that

$$d(\mathcal{G}^n x, \mathcal{G}^n y) \leq (v^n(T)/n!)d(x,y). \tag{3.6.82}$$

Since $0 < v(T) < \infty$ for finite T, it follows from the above inequality that for some $m \in N$ the m-th iterate of the operator $\mathcal{G}$, given by $\mathcal{G}^m$, is a contraction. Since $(B(I,R^n),d)$ is a complete metric space, it follows from this and Banach fixed point theorem that $\mathcal{G}^m$ and hence $\mathcal{G}$ itself has a unique fixed point $x^o \in B(I,R^n)$, that is,

$$x^o = \mathcal{G}x^o.$$

Hence x^o is the unique solution of the integral equation (3.6.74). This completes the proof. □

Remark. Since the measure ν is countably additive having bounded total variation on I, the solution x^o can have at most a finite number of discontinuities on the interval I. Thus the solution x^o is also piece wise continuous. This is summarized by the statement $x^o \in B(I, R^n) \cap PWC(I, R^n)$.

Following result establishes the continuous dependence of solution with respect to the initial state and the measure input.

Corollary 3.6.7. Suppose the assumptions of Theorem 3.6.6 hold with L being uniformly bounded by a constant $\ell \geq 0$. Then the solution map is continuous and Lipschitz with respect to the initial state and the measure input in the norm topologies. More precisely, $(x_0, u) \longrightarrow x(\cdot, x_0, u)$ from $R^n \times \mathcal{M}(I, R^m)$ to $B(I, R^n)$ is continuous and Lipschitz.

Proof. For every $x_0 \in R^n$ and $u \in \mathcal{M}(I, R^m)$, we have

$$\begin{aligned} \| x(t) \| \leq \| x_0 \| &+ \int_0^t K(s)(1+ \| x(s) \|)ds \\ &+ \int_0^t L(s)(1+ \| x(s-) \|)|u|(ds), t \in I, \end{aligned}$$

where, by $|u|(\sigma)$ we mean the measure induced by the variation of the vector measure u. As seen before, this leads to the following estimate:

$$\sup\{\| x(t) \|, t \in I\} \leq C \exp\{\mu(I)\},$$

where

$$C = \{\| x_0 \| + \mu(I)\}, \quad \mu(I) = \int_I K(s)ds + \int_I L(s)|u|(ds).$$

Thus $x \in B(I, R^n)$. Let $x(\xi, u), x(\zeta, v)$ denote the solutions corresponding to initial conditions $x(\xi, u)(0) = \xi, x(\zeta, v)(0) = \zeta$ and control measures $u, v \in \mathcal{M}(I, R^m)$ respectively. Then by simple computation, we obtain

$$\| x(\xi, u) - x(\zeta, v) \| \leq \exp\{\mu(I)\}\Big\{\| \xi - \zeta \| + \int_I L(s)|(u-v)|(ds)\Big\}. \tag{3.6.83}$$

Since L is bounded above by $\ell \geq 0$, it follows from the above estimate that

$$\begin{aligned} \| x(\xi, u) - x(\zeta, v) \| &\leq \exp\{\mu(I)\}\Big\{\| \xi - \zeta \| + \ell \| (u-v) \|_V\Big\} \\ &\leq c\{\| \xi - \zeta \| + \| u - v \|_V\}, \end{aligned} \tag{3.6.84}$$

with $c \equiv max\{1, \ell\} \exp \mu(I)$ where $\| u \|_V$ denotes the total variation norm. This verifies Lipschitz continuity of the solution map from the product space

$R^n \times \mathcal{M}(I, R^m)$ to the Banach space $B(I, R^n)$ with Lipschitz constant c. This completes the proof. □

Remark. For any nonnegative bounded measurable function $L \not\equiv 0$ defined on the interval I, introduce the set

$$M_L \equiv \left\{\mu \in \mathcal{M}(I, R^m) : \int_I L(t)|\mu|(dt) < \infty\right\}.$$

It is easy to see that it is a nontrivial linear subspace of the Banach space $\mathcal{M}(I, R^m)$. We introduce the norm topology on it by defining

$$\| \mu \|_L \equiv \int_I L(t)|\mu|(dt).$$

Completing M_L with respect to this norm topology we obtain a closed linear subspace of the Banach space $\mathcal{M}(I, R^m)$. Hence M_L itself is a Banach space. However note that $\| \mu \|_L = 0$ does not imply that the total variation norm $\| \mu \| = 0$. This simply means that the support of μ is contained in $I_0 \equiv \{t \in I : L(t) = 0\}$. In fact $\mu, \nu \in \mathcal{M}(I, R^m)$ are equivalent with respect to the L-norm topology if and only if $\mu, \nu \in M_L$ and $\| \mu - \nu \|_L = 0$. Using this topology, one can consider the continuous dependence of solutions in the form

$$\| x(\xi, u) - x(\zeta, v) \| \leq \exp\{\mu(I)\}\left\{\| \xi - \zeta \| + \| (u - v) \|_L\right\}.$$

This is precisely the expression (3.6.83) and clearly it is a more accurate estimate of the modulus (Lipschitz) of continuity though it is dependent on the particular system under consideration. Thus the uniform boundedness assumption of Corollary 3.6.7 for L is not necessary at all. The reader can appreciate the significance of this new topology if one assumes that the set I_0 is nonempty and has positive Lebesgue measure. On this set $|\nu|(\sigma), \sigma \subset I_0$, can assume the value $+\infty$ without affecting the conclusion.

3.7 Differential Inclusions

In this section we wish to study systems governed by differential inclusions. Any dynamic system with incomplete description or incomplete knowledge of the fundamental parameters can be described by differential inclusions of the form

$$\dot{x}(t) \in F(t, x(t)), x(0) = x_0, t \in I, \tag{3.7.85}$$

where F is a suitable multi valued or set valued function. Here we give some examples where differential inclusions are the appropriate models.

(E1): A classical example is control under parametric uncertainty. Consider a controlled nonlinear system given by

$$\dot{x} = f(t, x, u; \alpha), x(0) = x_0$$

where α is a vector of parameters that determines the vector field f. If all the parameters are exactly known then we have an exact differential equation governing the dynamics. In many physical problems these parameters are fundamental and are determined experimentally using basic sciences giving an approximate range of values this may take. Letting $\mathcal{P}$ denote this range we have the multifunction $F(t, x, u) \equiv f(t, x, u, \mathcal{P})$ leading to the controlled differential inclusion

$$\dot{x} \in F(t, x, u), x(0) = x_0.$$

(E2): A closely related problem arises in the study of controlled nonlinear variational inequalities of the form

$$< \dot{x}(t) - f(t, x(t), u(t)), y - x(t) > \ \leq \ \psi(y) - \psi(x(t)), y \in R^n, t \in I.$$

If ψ is a convex function from R^n to the reals, possessing subdifferential $\partial\psi$, this is equivalent to the differential inclusion

$$\dot{x} - f(t, x, u) \ \in \ \partial\psi(x).$$

Defining the multifunction $F(t, x, u) \equiv f(t, x, u) + \partial\psi(x)$, we have the controlled differential inclusion $\dot{x} \in F(t, x, u)$.

(E3): Another interesting example comes from the class of systems governed by differential-algebraic equations of the form

$$\dot{x} = f(x, y), g(x, y) = 0$$

where $f : R^n \times R^m \longrightarrow R^n$ and $g : R^n \times R^m \longrightarrow R^m$. Define the multifunction $G : R^n \longrightarrow 2^{R^m} \setminus \emptyset$ by setting

$$G(x) \equiv \{y \in R^m : g(x, y) = 0\}.$$

Then an equivalent formulation of the diff-algebraic system is given by the differential inclusion $\dot{x} \in F(x)$ where $F(x) \equiv f(x, G(x))$ provided $G(x) \neq \emptyset$.

In order to study differential inclusions, we need certain basic regularity properties of the multifunctions. Let 2^{R^n} denote the power set of R^n and $c(R^n) \subset 2^{R^n} \setminus \emptyset$ denote the set of all nonempty closed subsets of R^n. The most important properties are upper semi continuity (usc), lower semi continuity (lsc) and measurability.

Definition 3.7.1. (usc) Let $(X, d), (Y, \rho)$ be two metric spaces. A multi function $F : X \longrightarrow 2^Y \setminus \emptyset$ is said to be upper semi continuous at $x \in X$ if, for every open set $V \subset Y$ with $F(x) \subset V$, there exists an open neighborhood U of x such that $F(x) \subset V$ for all $x \in U$. The multi function F is said to be upper semi continuous on X if it is so for every point of X.

Note that if F is a single valued map the above definition is precisely the definition of continuity.

Definition 3.7.2. (lsc) A multi function $F : X \longrightarrow 2^Y \setminus \emptyset$ is said to be lower semi continuous at $x \in X$, if for every open set $\mathcal{O} \subset Y$ for which $F(x) \cap \mathcal{O} \neq \emptyset$, there exists an open neighborhood $N(x)$ of x such that $F(\xi) \cap \mathcal{O} \neq \emptyset$ for all $\xi \in N(x)$.

A multi function $F : X \longrightarrow 2^Y \setminus \emptyset$ is said to be continuous on X if it is both usc and lsc on X. A multi function is said to be Hausdorff continuous on X if it is continuous with respect to the Hausdorff metric d_H introduced in Chapter 1.

Definition 3.7.3. (measurability) Let $(\Omega, \mathcal{B})$ be a measurable space and $Y = (Y, \rho)$ a metric space. A multifunction $F : \Omega \longrightarrow 2^Y \setminus \emptyset$ is said to be measurable if, for every open set $\mathcal{O} \subset Y$, the set

$$F^-(\mathcal{O}) \equiv \{\omega \in \Omega : F(\omega) \cap \mathcal{O} \neq \emptyset\} \in \mathcal{B}.$$

An equivalent definition of measurability is: the multifunction F is measurable if and only if the function $\omega \longrightarrow \rho(y, F(\omega))$ is $\mathcal{B}$ measurable for every $y \in Y$.

Following result on existence of solutions of differential inclusion is elementary but very useful in the study of optimal controls as seen in Chapter 8.

Theorem 3.7.4. Consider the system $\dot{x} \in F(t, x), x(0) = x_0$ and suppose the multifunction F satisfy the following assumptions:

(F1): $F : I \times R^n \longrightarrow c(R^n)$ is measurable in t on I for each fixed $x \in R^n$, and, for almost all $t \in I$, it is upper semicontinuous (usc) on R^n,

(F2): for every finite positive number r, there exists an $\ell_r \in L_1^+(I)$ such that

$$\inf\{\| z \|, z \in F(t,x)\} \leq \ell_r(t), \ \forall\, t \in I, \ \forall\, x \in B_r,$$

(F3): there exists an $\ell \in L_1^+(I)$ such that

$$d_H(F(t,x), F(t,y)) \leq \ell(t) \| x - y \|, \ \forall\, x, y \in R^n, t \in I.$$

Then, for each initial state $x(0) = x_0 \in R^n$, the system (3.7.85) has at least one solution $x \in C(I, R^n)$. Further $x \in AC(I, R^n)$.

Proof. Define the map $\mathcal{R}$ by

$$\mathcal{R}(f)(t) \equiv \mathcal{R}_t(f) \equiv x_0 + \int_0^t f(s)ds, t \in I.$$

Clearly $\mathcal{R}$ is an affine continuous (bounded) map from $L_1(I, R^n)$ to AC $(I, R^n) \subset C(I, R^n)$. Define the multifunction $\hat{F}$ from $L_1(I, R^n)$ to $L_1(I, R^n)$ as follows. For each $g \in L_1(I, R^n)$, set

$$\hat{F}(g) \equiv \{f \in L_1(I, R^n) : f(t) \in F(t, \mathcal{R}_t(g)), \ a.e. \ t \in I\}.$$

Assuming momentarily that this is a nonempty set, it suffices to show that $\hat{F}$ has a fixed point in $L_1(I, R^n)$. Indeed, suppose that $f \in L_1(I, R^n)$ is a fixed point of $\hat{F}$, that is, $f \in \hat{F}(f)$. Then by definition this means that $f(t) \in F(t, \mathcal{R}_t(f)), a.e\ t \in I$. Defining x by

$$x(t) = \mathcal{R}_t(f) \equiv x_0 + \int_0^t f(s)ds, t \in I,$$

we have $\dot{x}(t) \in F(t, x(t)), t \in I$ with $x(0) = x_0$. Clearly $x \in AC(I, R^n)$ is a solution of the system (3.7.85). This verifies that if the the multifunction $\hat{F}$ has a fixed point then the differential inclusion has a solution. The reader can easily verify that the converse is also true. Thus it suffices to prove the existence of a fixed point of $\hat{F}$. Using Banach fixed point theorem for multi valued maps, we prove that $\hat{F}$ has a nonempty set of fixed points. For this we must show that $\hat{F}$ is nonempty with closed values in $2^{L_1(I,R^n)} \setminus \emptyset$ and that it is Lipschitz with respect to Hausdorff metric on $c(L_1(I, R^n))$ with Lipschitz constant less than 1. Let $g \in L_1(I, R^n)$ and define $x = \mathcal{R}g$. Clearly $x \in C(I, R^n)$. Since $(t, e) \longrightarrow F(t, e)$ is measurable in t on I and usc in e on R^n, it is clear that $t \longrightarrow F(t, x(t))$ is a nonempty measurable set valued function on I. Thus by Kuratowski-Ryll Nardzewski selection theorem (Appendix, Theorem A.7.9) the multifunction $t \longrightarrow F(t, x(t))$ has measurable selections. We must show that it has $L_1(I, R^n)$ selections. Since $x \in C(I, R^n)$ and $I \equiv [0, T]$ is a closed bounded interval, it is clear that there exists a finite positive number r such $x(t) \in B_r$ for all $t \in I$. Then by virtue of assumption (F2), it follows from a theorem on the existence of $L_1(I, R^n)$ selections ([48] Lemma 3.2, p175) that the multifunction $t \longrightarrow F(t, x(t))$ has a nonempty set of L_1 selections. Thus for each $g \in L_1(I, R^n)$, $\hat{F}(g) \neq \emptyset$. Further, since F is closed convex valued valued, $\hat{F}$ is also closed convex valued, that is, $\hat{F}(g) \in cc(L_1(I, R^n))$ for each $g \in L_1(I, R^n)$. For application of Banach fixed point theorem, we must now show that $\hat{F}$ is Lipschitz with Lipschitz constant less than 1. Define the function

$$\gamma(t) \equiv \int_0^t \ell(s) ds, t \in I.$$

Set $X_0 \equiv L_1(I, R^n)$ and, for $\delta > 0$, define the vector space X_δ by

$$X_\delta \equiv \left\{ z \in L_0(I, R^n) : \int_I \parallel z(t) \parallel e^{-2\delta\gamma(s)} ds < \infty \right\},$$

where $L_0(I, R^n)$ denotes the vector space of measurable functions with values in R^n. Furnished with the norm topology,

$$\parallel z \parallel_\delta \;\; \equiv \int_I \parallel z(s) \parallel e^{-2\delta\gamma(s)} ds,$$

X_δ is a Banach space. Let $g_1, g_2 \in L_1(I, R^n)$ and define $x_1 = \mathcal{R}(g_1), x_2 = \mathcal{R}(g_2)$. Clearly

$$\parallel x_1(t) - x_2(t) \parallel \leq \int_0^t \parallel g_1(s) - g_2(s) \parallel ds, t \in I.$$

Let $\varepsilon > 0$ and $z_1 \in \hat{F}(g_1)$, that is, $z_1(t) \in F(t, x_1(t)), t \in I$. Since F is Hausdorff Lipschitz with respect to its second argument, there exists a $z_2 \in \hat{F}(g_2)$ such that

$$\parallel z_1(t) - z_2(t) \parallel \leq \ell(t) \parallel x_1(t) - x_2(t) \parallel + \varepsilon, \; t \in I.$$

Define

$$h(t) \equiv \int_0^t \parallel g_1(s) - g_2(s) \parallel ds, t \in I.$$

Then it follows from the above inequality that

$$\parallel z_1(t) - z_2(t) \parallel \leq \ell(t) h(t) + \varepsilon, \;\; a.e \; t \in I.$$

Multiplying the above inequality on both sides by $e^{-2\delta\gamma(t)}$ and integrating over the interval I, we obtain

$$\int_I \| z_1(t) - z_2(t) \| e^{-2\delta\gamma(t)} dt \leq \int_I h(t) e^{-2\delta\gamma(t)} d\gamma(t) + \varepsilon T.$$

Integrating by parts the right hand expression of the above inequality, we obtain

$$\int_I \| z_1(t) - z_2(t) \| e^{-2\delta\gamma(t)} dt \leq (1/2\delta) \int_I \| g_1(t) - g_2(t) \| e^{-2\delta\gamma(t)} dt + \varepsilon T.$$

Thus we have

$$\| z_1 - z_2 \|_\delta \leq (1/2\delta) \| g_1 - g_2 \|_\delta + \varepsilon T.$$

Letting $D_{H,\delta}$ denote the Hausdorff distance on $c(X_\delta)$, the class of nonempty closed subsets of the metric space X_δ, it follows from the above inequality that

$$D_{H,\delta}(\hat{F}(g_1), \hat{F}(g_2)) \leq (1/2\delta) \| g_1 - g_2 \|_\delta + \varepsilon T.$$

Since $\varepsilon > 0$ is arbitrary and the above inequality holds for all such ε, we conclude that the multi function $\hat{F}$ is Lipschitz with respect to the Hausdorff metric $D_{H,\delta}$ on X_δ. Thus, for $\delta > (1/2)$, $\hat{F}$ is a multi valued contraction and hence by the generalized Banach fixed point theorem for multivalued maps (see Theorem 1.6.6), $\hat{F}$ has at least one fixed point $f^* \in X_\delta$. The reader can easily verify that

$$\| z \|_\delta \leq \| z \|_0 \leq c \| z \|_\delta$$

where $c = e^{\delta\gamma(T)}$. Thus the Banach spaces X_δ and X_0 are equivalent and hence f^* is also a fixed point of $\hat{F}$ in the Banach space X_0. This completes the proof. □

Under certain additional assumptions one can prove that the solution set $\mathcal{X}$ of the system (3.7.85) is closed. More precisely we have the following result.

Corollary 3.7.5. Suppose the assumptions of Theorem 3.7.4 hold and further there exists a $K \in L_1^+(I)$ such that the multifunction F satisfy the following growth condition

$$\sup\{\| z \|, z \in F(t, x)\} \leq K(t)(1 + \| x \|), x \in R^n.$$

Then the solution set $\mathcal{X}$ is a compact subset of $C(I, R^n)$.

Proof. Denote by

$$Fix\hat{F} \equiv \{f \in L_1(I, R^n) : f \in \hat{F}(f)\},$$

the set of fixed points of the multifunction $\hat{F}$. The set $Fix\hat{F}$ is nonempty by Theorem 3.7.4. By virtue of the growth assumption, it is a bounded subset of $L_1(I, R^n)$. Indeed, for any $f \in Fix\hat{F}$ it follows from the growth assumption that

$$\begin{aligned} \| f(t) \| &\leq K(t)(1+ \| \mathcal{R}_t(f) \|) \\ &\leq K(t)\Big(1+ \| x_0 \| + \int_0^t \| f(s) \| ds\Big). \end{aligned}$$

From this inequality one can easily verify that

$$\varphi(t) \leq C + \int_0^t K(s)\varphi(s)ds, t \in I,$$

where

$$\varphi(t) = \int_0^t \| f(s) \| ds$$

and

$$C = (1+ \| x_0 \|) \| K \|_{L_1(I)} \cdot$$

Thus it follows from Gronwall inequality that $\varphi(T) \leq C \exp\{\int_I K(s)ds\}$. Since C is independent of $f \in Fix(\hat{F})$, it follows from this that $Fix(\hat{F})$ is a bounded subset of $L_1(I, R^n)$. Let b denote this bound. Now we show that $Fix(\hat{F})$ is uniformly integrable. Again it follows from the growth assumption that every $f \in Fix(\hat{F})$ must satisfy the following inequality

$$\begin{aligned} \| f(t) \| &\leq K(t)\Big(1+ \| x_0 \| + \int_0^t \| f(s) \| ds\Big) \\ &\leq (1+ \| x_0 \| +b)K(t) \equiv \tilde{b}K(t), t \in I. \end{aligned}$$

Clearly it follows from this that

$$\lim_{\lambda(\sigma)\to 0} \int_\sigma \| f(t) \| dt \leq \tilde{b} \lim_{\lambda(\sigma)\to 0} \int_\sigma K(t)dt = 0$$

uniformly with respect to $f \in Fix(\hat{F})$. Hence by Dunford-Pettis theorem (Appendix, A.7.12), $Fix(\hat{F})$ is a relatively weakly sequentially compact subset of $L_1(I, R^n)$. Using the above results and upper semi continuity (usc) property of F, one can verify that $Fix(\hat{F})$ is weakly sequentially closed. Thus the set $Fix(\hat{F})$ is weakly sequentially compact. Since the solution set $\mathcal{X}$ is given by

$$\mathcal{X} \equiv \{\mathcal{R}(f), f \in Fix(\hat{F})\}$$

and $\mathcal{R}$ is an affine continuous map, one can easily verify that it is a compact subset of $C(I, R^n)$. This completes the proof. □

The reader may like to verify the last statement. Use Ascoli-Arzela theorem to verify that the set $\mathcal{X} \equiv \mathcal{R}(Fix(\hat{F}))$ is a compact subset of $C(I, R^n)$. Note that the continuous image of a weakly compact set need not be compact.

3.8 Exercises

(Q1) State the regularity of $f = f(t,x)$ required for C^2 solutions of equation $\dot{x} = f(t,x)$.
(Q2) Let $\mathcal{F}$ denote the class of Borel measurable vector fields $\{f : I \times R^n \longrightarrow R^n\}$ satisfying the following properties:

$$(P1): \quad \sup_{x \in R^n} \frac{\| f(t,x) \|}{1+ \| x \|} \leq L(t), \quad L \in L_1^+(I), \tag{3.8.86}$$

$$(P2): \quad \sup_{x,y \in B_r, x \neq y} \frac{\| f(t,x) - f(t,y) \|}{\| x - y \|} \leq K_r(t), \quad K_r \in L_1^+(I), \tag{3.8.87}$$

for any $r \geq 0$ and for all $f \in \mathcal{F}$ and consider the family of systems $\{\dot{x} = f(t,x), x(0) = x_0, f \in \mathcal{F}\}$. Show that the map $f \longrightarrow x$ from $\mathcal{F}$ to $C(I, R^n)$ is sequentially continuous.
(Q3): Consider the semilinear measure driven system

$$dx = f(t,x)dt + B(t)\mu(dt), x(0) = x_0 \in R^n, t \in I$$

where f is measurable in $t \in I$ and locally Lipschitz having linear growth in $x \in R^n$, B is continuous with values in $M(n \times d)$ and $\mu \in \mathcal{M}_c(I, R^d)$ the space of countably additive bounded vector measures having bounded total variation. Let x^n, x^o denote the solutions corresponding to the measures μ^n, μ^o respectively. Show that $x^n(t) \to x^o(t)$ for each $t \in I$ whenever $\mu^n \to \mu^o$ weakly.
(Q4): Show that the set M_L as described in the Remark at the end of Section 3.6 is weakly closed.
(Q5): Consider the example (E2) of Section 3.7 and justify that the variational inequality is equivalent to the differential inclusion given there.
(Q6): Refer to Corollary 3.7.5. Use Ascoli-Arzela theorem to verify that the set $\mathcal{X} \equiv \mathcal{R}(Fix(\hat{F}))$ is a compact subset of $C(I, R^n)$.
(Q7): Consider the control system $\dot{x} = f(x,u)$ subject to the constraint $u \in G(x)$ where G is a suitable multifunction. Clearly under this constraint,

the system is better described by the differential inclusion $\dot{x} \in F(x)$ where $F(\xi) \equiv f(\xi, G(\xi))$. Let $\mathcal{X}$ denote the family of solutions of the differential inclusion. State sufficient conditions for the function f and the multifunction G that guarantee compactness of the solution set $\mathcal{X} \subset C(I, R^n)$.

3.9 Bibliographical Notes

In general, the subject of nonlinear systems is vast and almost all physical problems, including engineering, involve nonlinear dynamics. Here we have considered only systems governed by nonlinear ordinary differential equations. Even this class is so advanced that one can write a separate book on this topic. For qualitative study of nonlinear differential equations the classical books due to Nemytski and Stepanov [70], Bellman and Cooke [17], Ames [8], La Salle and Lefschetz [58], Driver [32] and Hale [40] are all very well known. See also Ahmed [2] which includes a wide variety of applications to systems and control theory. Systems governed by differential equations on infinite dimensional spaces covering partial differential equations, the books by Fattorini [35], Pavel [75], Barbu [14], Zeidler [100], Deimling [27], Hu and Papgeorgiou [48], Sell and You [82], Tanabe [88], Ahmed and Teo [4], Ahmed [5, 6] cover a very large area of nonlinear functional analysis, nonlinear systems theory and their applications to systems and control. For chaotic dynamic systems see Thompson and Stewart [90]. For many other interesting engineering applications see Skowronski [85], Udwadia-Weber-Leitmann [91]. For impulsive dynamic systems on finite dimensional spaces, the book by Lakshmikantham, Bainov and Simeonov is classical [56], and the more recent book by Martynyuk [67] covers questions of stability for such systems. In recent years intense interest has been shown in the general area of impulsive dynamic systems both in finite and infinite dimensional spaces using vector valued measures. See the recent papers by Silva and Vinter [196], Maso and Rampazzo [188], and Ahmed [105, 106, 107, 112, 113, 114, 129, 132]. The method used by Maso and Rampazzo, and Silva and Vinter is based on graph completion technique while that due to Ahmed is direct and based on explicit dependence of jumps on the immediate past state. The later is compatible with the classical impulsive systems treated in the broad literature and applies to both finite and infinite dimensional systems. The graph completion technique provides robust solutions and is certainly powerful for finite dimensional systems. It is not certain if this technique can be extended to infinite dimensional problems involving generators of C_0 semigroups and vector measures taking values in general Banach spaces. For detailed discussion see [151]. Another popular topic in recent years is differential inclusions. There is also a vast literature covering questions of existence of solutions for finite and infinite dimensional systems. For finite dimensional problems see Aubin and Cellina and infinite dimensional problems see [12], Deimling [28]. For optimal control of such systems see Cesari [24], Ahmed [101, 112, 118, 121, 128, 131, 132, 136]. For application to uncertain systems concerning stability and control see [46].

Chapter 4

Basic Stability Theory

4.1 Introduction

The concept of stability of systems, natural or man made, is very important. If a system built by engineers fails to perform its function satisfactorily or shut down or burn out in the presence of slight disturbance we may say the system is unstable. For example, if a moderate storm hits a bridge for a short time and the bridge continues to vibrate and takes long time to stop, or, in the worst case, never stops the structure is unstable. In fact sustained vibration may lead to disastrous failure. In contrast if the vibration stops in short time after the disturbance has passed, the system is stable. We must illustrate this further mathematically. Consider the system

$$\dot{x} = f(x) \tag{4.1.1}$$

where $f(x) = Ax$ if the system is linear. Define the set of rest states or equilibrium states as

$$\mathcal{Z} \equiv \{\xi \in R^n : f(\xi) = 0\}.$$

It is possible for the system to rest in any one of these states. However if it is perturbed, the system may exhibit any one of the following behaviors.
(1): It may run away from the equilibrium and never return. In this case we say the system is unstable with respect to the particular equilibrium or the rest state.
(2): The system runs away from the rest state but stays nearby and executes motion around it. The system is said to be stable in the Lyapunov sense. Mathematically the system is stable, though, from engineering point of view, this may not be considered satisfactory or even desirable for some applications.

(3): The system is kicked out of the rest state (possibly by some external force) but eventually returns to it. In this case we say the system is asymptotically stable.
(4): The system is kicked out of the rest state but returns to it exponentially fast. In this case the system is exponentially stable. For many engineering problems this is a very desirable property and some time the system may be controlled to acquire this property.

Many natural systems, like the solar system, ecological systems, and even man made systems like the power system, orbiting artificial satellites, electrical oscillators (used in communication), all exhibit periodic motions or solutions. For such problems equation (4.1.1) has a periodic solution which we may denote by $x_p(t), t \in R$. One may ask the question of stability of this periodic motion. For example, if a giant star passes by the solar system causing perturbation of the orbiting planets, will the system return to its previous (prior to onset of disturbance) periodic orbit. If a power system experiences a heavy load for a very short time, due possibly to a short circuit, will it return to its previous state eventually? These are the type of questions which are answered by stability theory.

4.2 Stability of the Equilibrium

Mathematically the question of stability of the equilibrium can all be reduced to the study of stability of the origin of the state space or equivalently stability of the zero state. Let $z \in \mathcal{Z}$ be any rest state and suppose, due to some disturbance at time $t = 0$, the system exhibits the motion $x(t), t \geq 0$. Define $y(t), t \geq 0$, so that

$$x(t) \equiv y(t) + z, t \geq 0, \tag{4.2.2}$$

and let $\tilde{f}(\xi) \equiv f(\xi + z)$. Clearly $\tilde{f}(0) = 0$. In other words 0 is an equilibrium state of the modified system

$$\dot{y}(t) = \tilde{f}(y(t)), t \geq 0, \tag{4.2.3}$$

which is equivalent to the system (4.1.1) around the rest state z. Thus we have demonstrated that the study of stability of any equilibrium state of the original system is equivalent to the study of stability of the origin (zero state) of an equivalent system.

Similarly let us consider the system (4.1.1) having a periodic motion denoted by $x_p(t), t \geq 0$, and let $x(t), t \geq 0$, denote the perturbed motion.

Define $y(t), t \geq 0$, so that

$$x(t) = x_p(t) + y(t), t \geq 0. \tag{4.2.4}$$

Introduce the function

$$\tilde{f}(t, \xi) \equiv f(x_p(t) + \xi) - f(x_p(t)), t \geq 0, \xi \in R^n. \tag{4.2.5}$$

Clearly $\tilde{f}(t, 0) = 0$. Using the equations (4.2.4) and (4.2.5), we observe that the motion y is governed by the differential equation

$$\dot{y}(t) = \tilde{f}(t, y(t)), t \geq 0. \tag{4.2.6}$$

Thus again we have reduced the question of stability of periodic motion to the question of stability of the zero state.

In fact the preceding conclusions hold for any motion, not just for a fixed rest state or a periodic motion. Let $x^*(t), t \geq 0$, be any solution of an arbitrary time varying system

$$\dot{x}^*(t) = f(t, x^*(t)), t \geq 0, \tag{4.2.7}$$

and let $x(t), t \geq 0$, denote the motion due to certain perturbation. As before, define y so that $x(t) = x^*(t) + y(t), t \geq 0$ and

$$\tilde{f}(t, \xi) \equiv f(t, x^*(t) + \xi) - f(t, x^*(t)).$$

Clearly $\tilde{f}(t, 0) = 0$, and y satisfies the differential equation

$$\dot{y}(t) = \tilde{f}(t, y(t)), t \geq 0. \tag{4.2.8}$$

In view of the above observations we can always formulate any problem of stability in terms of the stability of the origin or the zero state and take f so that $f(t, 0) = 0$.

For simplicity of presentation, from now on we consider only time invariant systems like the system (4.1.1) and to begin with we assume that $f(0) = 0$. In other words we consider the system

$$(S) : \dot{x}(t) = f(x(t)), x(0) = x_0, \quad \text{with } f(0) = 0. \tag{4.2.9}$$

Define the solution operator T by

$$x(t, x_0) \equiv T_t(x_0). \tag{4.2.10}$$

In other words, $T_t, t \geq 0$, denotes the transition operator, nonlinear in the case of nonlinear systems, that maps the initial state x_0 into the current state $x(t)$. For any $\alpha > 0$, let B_α denote the open ball of radius α around the origin. That is

$$B_\alpha \equiv \{\zeta \in R^n : \| \zeta \| < \alpha\}.$$

Definition 4.2.1. (Lyapunov Stability) The system (S) is said to be stable in the Lyapunov sense (with respect to the zero state), if for every $R > 0$, there exists an $r = r(R) \leq R$, possibly dependent on R, so that

$$T_t(B_r) \subset B_R, \quad \forall t \geq 0. \tag{4.2.11}$$

In other words zero state is a stable equilibrium point for the system (4.2.9)

An Example. Consider the (undamped) system

$$\ddot{\xi} + \xi = 0, \xi(0) = \xi_0, \dot{\xi}(0) = \xi_1.$$

Defining the state vector $x = \begin{pmatrix} x_1 \\ x_2 \end{pmatrix} = \begin{pmatrix} \xi \\ \dot{\xi} \end{pmatrix}$ we can write the system in the canonical form

$$\dot{x}_1 = x_2 \;\; \dot{x}_2 = -x_1. \tag{4.2.12}$$

Define

$$E(t) \equiv V(x_1(t), x_2(t)) \equiv (1/2)x_1^2(t) + (1/2)x_2^2(t).$$

This represents the total energy of the system given by the sum of kinetic and elastic potential energies. Taking the time derivative, we have

$$\begin{aligned} \dot{E}(t) &= x_1(t)\dot{x}_1(t) + x_2(t)\dot{x}_2(t) \\ &= x_1(t)x_2(t) - x_2(t)x_1(t) = 0. \end{aligned}$$

This shows that the system is conservative. Thus

$$E(t) = E(0) = (1/2)\xi_0^2 + (1/2)\xi_1^2$$

and for any $R > 0$, $T_t(x_0) \in B_R$ whenever $x_0 \in B_r$ for $r \leq R$. Thus the zero state is stable in the Lyapunov sense. However, it is not stable in the sense of BIBO as considered in Chapter 2. Here the impulse response is $h(t) = \sin t$ and hence $\int_0^\infty |h(t)| dt = \infty$.

Similarly the reader may verify that for the system

$$\dot{x}_1 = x_2 - x_1 \;\; \dot{x}_2 = x_1 - x_2, \tag{4.2.13}$$

we have $E(t) \leq E(0)$ for all $t \geq 0$. Hence the system is stable with respect to the origin. In fact it is stable with respect to the set (diagonal) $D \equiv \{\xi \in R^2 : \xi_1 = \xi_2\}$. If at time $t = 0$, the system is at the origin, and then kicked off from this rest state by some external force, it will return to the set D and rest. Thus D is stable.

Definition 4.2.2. (Instability) The system (S) is said to be unstable with respect to the origin or simply the origin is unstable, if, for any given $R > 0$, there exists no positive number $r(> 0)$ for which $T_t(B_r) \subset B_R$, for all $t \geq 0$.

Example. Consider the system

$$\dot{x}_1 = x_1, \dot{x}_2 = x_2 \quad \text{with } x_1(0) = x_{10}, x_2(0) = x_{20}.$$

Define

$$E(t) \equiv V(x_1(t), x_2(t)) = (1/2)x_1^2(t) + (1/2)x_2^2(t).$$

Clearly

$$\dot{E}(t) = x_1^2(t) + x_2^2(t) = 2E(t).$$

Solving for E we find that $E(t) = E(0)e^{2t}$. This shows that energy grows exponentially with time and the system is unstable. A similar example is given by

$$\dot{x}_1 = x_1 - x_2, \dot{x}_2 = x_2 - x_1, \text{with } x_1(0) = x_{10}, x_2(0) = x_{20}.$$

Again for the same energy functional, we have

$$\dot{E}(t) = (x_1(t) - x_2(t))^2.$$

Clearly $E(t) \geq E(0)$ and it grows until the trajectory hits the diagonal D, that is, $x(t) \in D \equiv \{x \in R^2 : x_1 = x_2\}$. Note D is also a set of equilibrium, but one can easily check that it is unstable.

Definition 4.2.3. (Asymptotic Stability) The system (S) is said to be asymptotically stable with respect to the zero state, if, for any $x_0 \in R^n$,

$$\lim_{t \to \infty} T_t(x_0) = 0.$$

Example. Consider the system

$$\dot{x}_1 = x_2, \dot{x}_2 = -x_1 - x_2 \quad \text{with } x_1(0) = x_{10}, x_2(0) = x_{20}.$$

We show that the zero state is asymptotically stable. Take

$$E \equiv (1/2)x_1^2 + (1/2)x_2^2$$

and note that

$$\dot{E}(t) = -x_2^2(t) \le 0.$$

This only tells us that $E(t) \le E(0)$. In fact $\lim_{t\to\infty} E(t) = 0$. Suppose $x_2(t) = 0$ for all $t \ge t_0$, but $x_1(t)$ is not. This implies that $\dot{E}(t) = 0$ and $x_1(t) = c$ for some constant $c \ne 0$ for all $t \ge t_0$. But by the second equation, this means $\dot{x}_2(t) = -c$ and hence $x_2(t) \ne 0$ for $t \ge t_0$ contradicting the initial hypothesis. Thus the zero state is asymptotically stable.

Definition 4.2.4. (Exponential Stability) The system (S) is said to be exponentially stable with respect to the zero state, or simply the origin is exponentially stable, if there exist numbers $M > 0, \alpha > 0$ such that

$$\| T_t(x_0) \| \le Me^{-\alpha t}, t \ge 0.$$

Here M may depend on the intensity of perturbation $\| x_0 \|$.

Example. Consider the system

$$\dot{x}_1 = x_2 - x_1, \dot{x}_2 = -x_1 - x_2 \text{ with } x_1(0) = x_{10}, x_2(0) = x_{20}.$$

Define $E(t) \equiv V(x_1(t), x_2(t)) = (1/2)x_1^2(t) + (1/2)x_2^2(t)$. Clearly

$$\dot{E}(t) = -x_1^2(t) - x_2^2(t) = -2E(t).$$

Solving for E, we have $E(t) = E(0)e^{-2t}$. This shows that the energy decays exponentially with time and the system is exponentially stable. Here the constants M and α are given by $M = E(0) = (1/2) \| x_0 \|^2$ and $\alpha = 2$. Similarly, one can verify for the system,

$$\dot{x}_1 = x_2 - 4x_1, \dot{x}_2 = -x_1 - 4x_2 \text{ with } x_1(0) = x_{10}, x_2(0) = x_{20},$$

that the energy function E as defined above decays exponentially as

$$E(t) \le E(0)e^{-6t}.$$

4.3 Lyapunov Stability Theory

All the notions of stability introduced in the preceding section can be characterized and evaluated by use of a special function called the Lyapunov function named after its author A.M.Lyapunov who introduced the basic theory of stability around 1892. This is a real valued function defined on the whole state space or part of it. For many of the engineering problems, this function can be taken as the energy function given by the sum of kinetic and potential energies. However often this is found to be insufficient. One may consider the Lyapunov function as the generalized energy function. For simplicity let us consider the time invariant system (4.2.9).

A function $V : R^n \longrightarrow R$ is said to be a Lyapunov function for the system (S) given by equation (4.2.9), if it satisfies the following basic properties:

(L1): $V : R^n \longrightarrow \bar{R}_0 \equiv [0, \infty]$,
(L2): $V \in C^1$, that is, V is once continuously differentiable,
(L3): $V(x) > 0$ for $x \neq 0$ and it has its minimum at $x = 0$,
(L4): $(d/dt)V(x(t)) = (DV, f)(x(t)) \equiv (V_x(x(t)), f(x(t))) \leq 0$, along any solution trajectory $x(t) \equiv T_t(x_0) \equiv x(t, x_0), x_0 \in R^n$.

The basic theorems of Lyapunov can be stated as follows.

Theorem 4.3.1. (Lyapunov Theory) The system given by equation

$$\dot{x} = f(x), f(0) = 0, x(0) = x_0,$$

is stable with respect to the origin in the Lyapunov sense (or Lyapunov stable) if there exists a Lyapunov function V, (satisfying the properties (L1) - (L4)). Further more, the system is asymptotically stable if it is Lyapunov stable and also

$$(d/dt)V(T_t(x_0)) = (V_x(T_t(x_0)), f(T_t(x_0))) < 0, \quad t \geq 0,$$

that is, strict inequality holds for every $x_0 \in R^n$.

Proof. The proof is quite simple. Since this proof is similar to the proof of its local version given below in Theorem 4.3.2, we omit this. Readers interested in the proof may also consult the book [2]. □

For asymptotic stability we note that along any trajectory $T_t(x_0)$, the function $V(T_t(x_0))$ must be a strictly decreasing function of t.

Thus given a system, if we can find a Lyapunov function for it then we may conclude that the system is stable. Hence, in order to determine the stability of a system, it is necessary to find or construct a Lyapunov function. Unfortunately there is no systematic method to determine if a Lyapunov function exists for a given system. Usually if one has sufficient knowledge about the energy of the system, one may try the energy function and check if it is a Lyapunov function. It must be noted that, failure to find a Lyapunov function does not necessarily mean that the system is unstable.

Examples of Global Stability: Here we present several examples to illustrate this celebrated result of Lyapunov. Note that in all the examples given in the introduction, we took the energy function.

(E1): Consider the (undamped) system

$$m\ddot{\xi} + k\xi = 0, \xi(0) = x_{10}, \dot{\xi}(0) = x_{20} \tag{4.3.14}$$

Let us write this in the standard form. Define state $x \equiv \begin{pmatrix} x_1 \\ x_2 \end{pmatrix} \equiv \begin{pmatrix} \xi \\ \dot{\xi} \end{pmatrix}$.

Then we have the system

$$\begin{aligned} \dot{x}_1 &= x_2, \quad x_1(0) = x_{10} \\ \dot{x}_2 &= -(k/m)x_1, \quad x_2(0) = x_{20}. \end{aligned}$$

This is a mechanical system and the total energy is given by sum of the kinetic energy $(1/2)mx_2^2$ and the elastic potential energy $(1/2)kx_1^2$. We may try the energy function as the Lyapunov function

$$V(x) \equiv V(x_1, x_2) \equiv (1/2)mx_2^2 + (1/2)kx_1^2.$$

Taking the derivative of this function along the solution trajectory we find

$$\begin{aligned} \dot{V}(x(t)) &= mx_2\dot{x}_2 + kx_1\dot{x}_1 \\ &= -kx_1x_2 + kx_1x_2 = 0. \end{aligned}$$

Since here V represents energy, it follows from this identity that energy is conserved, that is, $V(x(t)) = V(x_0), t \geq 0$. In other words there is no dissipation of energy. Such a system is called conservative. In fact this is a mechanical oscillator and V is a Lyapunov function and the system is stable in the Lyapunov sense.

(E2): In contrast, consider the (damped) system given by

$$m\ddot{\xi} + b\dot{\xi} + k\xi = 0, \xi(0) = x_{10}, \dot{\xi}(0) = x_{20} \tag{4.3.15}$$

where $b > 0$ is the damping coefficient or coefficient of friction. The corresponding system is given by

$$\begin{aligned} \dot{x}_1 &= x_2, \quad x_1(0) = x_{10} \\ \dot{x}_2 &= -(k/m)x_1 - (b/m)x_2, \quad x_2(0) = x_{20}. \end{aligned}$$

Using the energy function of the previous example as the Lyapunov function we find that

$$\dot{V} = -bx_2^2 \leq 0.$$

Thus again the system is Lyapunov stable. In fact we can show that it is actually asymptotically stable with respect to the zero state. For this it suffices to show that $\dot{V}$ is strictly negative in $R^2 \setminus \{0\}$. This will follow if $x_2 \neq 0$ whenever $x_1 \neq 0$. But this follows immediately from the system equations. Indeed if $x_1 = c \neq 0$ and $x_2 = 0$, the ratio

$$\frac{dx_2}{dx_1}|_{x_1=c, x_2=0} = (+\infty / -\infty).$$

This shows that the system can not settle down any where but the origin. The energy is dissipated through friction or the damper.

(E3): (Disarmament a Reality) Let us consider two nations N1 and N2 engaged in arms race, an example being the last(!) cold war. Let S_1 denote the stockpile of N1 and S_2 that of N2. Suppose the U.N is able to achieve a treaty agreement between the two nations accepting the following program for arms reduction

$$\begin{aligned} \dot{S}_1 &= -\alpha S_1 + \beta(S_2 - S_1) \\ \dot{S}_2 &= -\gamma S_2 + \delta(S_1 - S_2) \end{aligned}$$

where the parameters $\{\alpha, \gamma > 0\}$ denote the natural depreciation rates of weapons and war materials of each nation respectively, and the parameters $\{\beta, \delta > 0\}$ are the rates by which each nation is allowed to augment or reduce it's stockpile in proportion to the disparity.

We show that the zero state is asymptotically stable leading to complete disarmament. Define

$$V = (1/2\beta)S_1^2 + (1/2\delta)S_2^2.$$

One can easily verify that

$$\dot{V} = 2S_1S_2 - (1 + (\alpha/\beta))S_1^2 - (1 + \gamma/\delta)S_2^2 \leq -(S_1 - S_2)^2.$$

Thus the arms race continues till the two nations have reached parity, that is, $S_1 = S_2$. At this point the race stops and the process slides down the line $\{(S_1, S_2) \in R^2 : S_1 = S_2\}$ leading to total disarmament. Note that if the signs of the parameters $\{\alpha, \gamma\}$ are reversed, the arms race can continue for ever leading to instability. This can happen if the two nations secretly compensate depreciation by replacement.

(E4): Here we give an example of a Lyapunov function which is not a quadratic function (so not an energy function) of the state. Consider the nonlinear system

$$\dot{x}_1 = x_2, \quad x_1(0) = x_{10}$$
$$\dot{x}_2 = -(1/m)g(x_1) - (b/m)x_2, \quad x_2(0) = x_{20}.$$

For example g could be any real valued function that lies strictly in the first and the third quadrant of the plane; in particular functions of the form $g(\zeta) = a\zeta^{2n+1}, a > 0$ do satisfy this condition. In this case we take

$$V(x_1, x_2) \equiv (1/2)mx_2^2 + \int_0^{x_1} g(\xi)d\xi.$$

Hence

$$\dot{V} = -mx_2\{(g(x_1)/m) + (b/m)x_2\} + g(x_1)x_2 = -bx_2^2.$$

Again by similar arguments as in example (E2), $\dot{V}$ is strictly negative definite and hence by Lyapunov theorem the zero state is asymptotically stable.

(E5): (Attitude Dynamics of a Satellite) The dynamics of a satellite in geosynchronous orbit (approximately 23000 miles from the center of the earth) is given by

$$I_x\dot{p} + (I_z - I_y)qr = T_x$$
$$I_y\dot{q} + (I_x - I_z)pr = T_y$$
$$I_z\dot{r} + (I_y - I_x)pq = T_z$$

where $\{I_x, I_y, I_z\}$ are the moment of inertia of the satellite and p, q, r are the spin rates all about the axes x, y, z respectively. The functions, $\{T_x, T_y, T_z\}$ are the applied torques. For the uncontrolled system, we show that the kinetic energy

$$V(p, q, r) \equiv (1/2)(I_xp^2 + I_yq^2 + I_zr^2)$$

is a Lyapunov function. Indeed taking the time derivative of this function we find that

$$\dot{V}(p(t), q(t), r(t)) = 0$$

implying that the system is conservative and $V(p(t), q(t), r(t)) = V(p_0, q_0, r_0)$ for all $t \geq 0$. This means that once the system is perturbed from the zero state, it remains in motion ever after without any attenuation and the satellite wobbles around. Note that there is no damping: aerodynamic or otherwise. At the geosynchronous orbit there is no atmosphere and hence no source of dissipation of energy. The amplitude of motion is dictated by the energy imparted to the system to begin with. This system is stable in the

Lyapunov sense but not asymptotically stable. Clearly this is unacceptable for a communication satellite with antennas directed towards the earth or other satellites. To stabilize this satellite one may choose to use feedback control of the form

$$T_x = -K_x p, T_y = -K_y q, T_z = -K_z r$$

or any control of the general form

$$T_x = -u_1(p), T_y = -u_2(q), T_z = -u_3(r)$$

where $u_i(\xi)$ is any odd function or more generally a function whose graph lies in the first and third quadrant of the plane. In this general case

$$\dot{V} = -\{u_1(p)p + u_2(q)q + u_3(r)r\}.$$

By appropriate choice of the feedback controls u_i, one can obtain not only asymptotic stability of the zero state but also exponential stability.

(E6): (Nonlinear Circuit) A network consists of a voltage source e, feeding into an electrical circuit consisting of a linear resistor R, connected in series to a parallel branch of linear inductor L, and a nonlinear capacitor with charge-voltage characteristics given by $q = f(v)$ where f is an increasing function of voltage v with $f(0) = 0$. One can verify that the correct state variable is $col(q, i)$ and the state equation is given by

$$(d/dt)q = (1/R)(e - g(q)) - i$$
$$(d/dt)i = (1/L)g(q),$$

where q is the charge on the capacitor and i is the current in the inductor and $g(q) = f^{-1}(q) = v$. The Lyapunov function V is given by

$$V \equiv (1/2)Li^2 + \int_0^q g(\xi)d\xi.$$

The time derivative of V along any solution is given by

$$\dot{V} = -(1/R)(g(q))^2 = -(1/R)v^2$$

where v is the voltage across the capacitor. In fact, from this one can show that the system is asymptotically stable.

So far we have discussed global stability in the sense that the system is defined on R^n and the Lyapunov function is also defined on all of R^n.

However this is not at all essential. In fact we will have examples where the system has positive solutions only, for example, population dynamics. Even more important is the fact that the system may be stable only locally around the equilibrium. For this, the Lyapunov theorem given above requires only minor modification. Since the stability of an equilibrium is equivalent to the stability of the origin, we can state the stability theorem with reference to the origin.

Theorem 4.3.2. (Local Stability) The system given by equation

$$\dot{x} = f(x), f(0) = 0, x(0) = x_0 \tag{4.3.16}$$

is stable with respect to the zero state in the Lyapunov sense (or Lyapunov stable) if there exists an open neighborhood Ω of the origin and a Lyapunov function V on Ω. Further more, the zero state is asymptotically stable if it is Lyapunov stable and also $\dot{V} \equiv (V_x, f) < 0$ on $\Omega \setminus \{0\}$.

Proof. Let Ω be the set as defined in the statement of the theorem. Let B_R denote the open ball of radius R around the origin. Clearly there exists an $R > 0$ so that $B_R \subset \Omega$. For any $c > 0$, define the set

$$E_c \equiv \{x \in \Omega : V(x) < c\}.$$

Since $V(0) = 0$, $0 \in E_c$. It follows from continuity of V that the set E_c, as defined above, is a connected open neighborhood of the origin. Thus there exists a positive number $c^* < c$ such that $E_{c^*} \cup \partial E_{c^*} \subset B_R$. Since E_{c^*} is an open neighborhood of the origin, there exists an $r > 0$ such that the open ball B_r, of radius r around the origin, is contained in E_{c^*}. Now take any point $x_0 \in B_r$ and let $x(t, x_0) \equiv T_t(x_0)$ denote the solution of system (4.3.16). Now examine the expression given by

$$V(T_t(x_0)) = V(x_0) + \int_0^t \dot{V}(T_s(x_0))ds, t \geq 0.$$

Since V is a C^1 function $\dot{V}$ is continuous. Thus by continuity of the solution trajectories, the integrand considered as a function of s is continuous and nonpositive. Hence $V(T_t(x_0)) \leq V(x_0)$. Since $x_0 \in B_r \subset E_{c^*}$, we have $V(T_t(x_0)) < c^*$ for all $t \geq 0$ and hence $T_t(x_0) \in B_R$. Thus we have shown that for every open ball B_R, contained in Ω, there exists an open ball $B_r \subset B_R$ such that $T_t(x_0) \in B_R$ for all $t \geq 0$ whenever $x_0 \in B_r$. This proves the first part of the theorem. The proof of the second part follows from the strict negativity of $\dot{V}$ and the fact that V is positive off the origin. In fact, this implies that the function $t \longrightarrow V(T_t(x_0))$ is monotone decreasing and hence it has a limit and the limit is zero. This also implies the convergence of $\dot{V}(T_t(x_0))$ to zero as $t \to \infty$. This ends the proof. □

Examples for Local Stability: We present some examples for illustration.

(E7): Consider the nonlinear system

$$\begin{aligned} \dot{x}_1 &= x_2 \\ \dot{x}_2 &= -x_1 - g(x_2). \end{aligned}$$

where $g(s) = \sin s$. This system is locally stable. Here $\Omega = \{(x_1, x_2) \in R^2 : |x_2| < \pi\}$. For the Lyapunov function we take $V = (1/2)x_2^2 + (1/2)x_1^2$. The Lyapunov derivative of V is given by

$$\dot{V} = -x_2 g(x_2)$$

and this is strictly negative if $|x_2| < \pi$ and the system is locally asymptotically stable near the origin.

(E8): (Simple Pendulum) The motion of a simple pendulum of mass m attached to an inextensible string of length ℓ in a vertical plane is described by the differential equation

$$m\ell\ddot{\theta} + m\ell g \sin\theta = \tau$$

where τ is the torque applied on the body. The angular position denoted by θ is measured with respect to the downward vertical axis. In contrast, if it is measured with respect to the upward vertical axis the equation changes to

$$m\ell\ddot{\varphi} - m\ell g \sin\varphi = \tau$$

which is the motion of the inverted pendulum. For the stability of the zero state, we set $\tau = 0$ giving

$$\ddot{\theta} + g\sin\theta = 0$$

and for the inverted pendulum the plus sign changes to the minus sign giving

$$\ddot{\varphi} - g\sin\varphi = 0.$$

In any case, for the Lyapunov function one may choose

$$V = (1/2)\dot{\theta}^2 + g\int_0^\theta \sin r dr.$$

For the simple pendulum, it is easy to verify that $\dot{V} = 0$ implying stability in the Lyapunov sense near the zero state. On the other hand for the inverted pendulum we have

$$\dot{V} = 2g\dot{\varphi}\sin\varphi$$

and the system is unstable near the zero state.

(E9): Consider the nonlinear prey-predator (rabbit-fox) model given by

$$\begin{aligned} \dot{R} &= \alpha R - \beta RF \\ \dot{F} &= -\gamma F + \delta RF, \end{aligned} \tag{4.3.17}$$

where all the coefficients are positive. The system has two equilibrium states given by the solution of the algebraic equation

$$\begin{aligned} 0 &= \alpha R - \beta RF \\ 0 &= -\gamma F + \delta RF. \end{aligned} \tag{4.3.18}$$

Solving this equation, we have the non zero equilibrium population given by

$$(R_e, F_e) = (\gamma/\delta, \alpha/\beta). \tag{4.3.19}$$

Linearizing any nonlinear system around the equilibrium solution one obtains, in general, the following linear system

$$\dot{\tilde{x}} = f_x(x^*)\tilde{x}, \tag{4.3.20}$$

where f_x stands for the matrix of gradients of the vector function f and $\tilde{x} \equiv x - x^*$ denotes the fluctuation around the equilibrium x^*. Using this linearization technique applied to the model (4.3.17) one obtains the following linear system

$$(d/dt)\begin{pmatrix} \tilde{R} \\ \tilde{F} \end{pmatrix} = \begin{pmatrix} 0 & -(\beta\gamma/\delta) \\ (\delta\alpha/\beta) & 0 \end{pmatrix}\begin{pmatrix} \tilde{R} \\ \tilde{F} \end{pmatrix}. \tag{4.3.21}$$

The characteristic roots for this system is determined from the expression,

$$0 = det(\lambda I - A) = (\lambda + i\sqrt{\alpha\gamma})(\lambda - i\sqrt{\alpha\gamma}).$$

Thus the characteristic equation of the linearized system has purely imaginary roots. From this we conclude that the population level fluctuates around the equilibrium and the system is Lyapunov stable. In fact using the equations for fluctuation one can easily verify that

$$(d\tilde{R}/d\tilde{F}) = -k^2(\tilde{F}/\tilde{R}),$$

where

$$k^2 = (\beta/\delta)^2(\gamma/\alpha).$$

Integrating this we obtain

$$\tilde{R}^2 + k^2\tilde{F}^2 = \tilde{R}_0^2 + k^2\tilde{F}_0^2 = C^2,$$

where the constant C, as indicated, depends on the initial state. This means that the population fluctuates around the equilibrium describing an elliptical orbit around it. Such an equilibrium point is called focus. In fact the solutions of equation (4.3.21) are periodic in time with frequency given by $\omega = \sqrt{\alpha\gamma}$. For better appreciation, the reader may draw several of these orbits and note that they are all in the first quadrant of the plane around the equilibrium. One may also draw the time periodic solutions for both $(\tilde{R}, \tilde{F})$ and note how one population follows the rise and decline of the other.

4.4 Lyapunov First Method

For a general nonlinear system it is very difficult to construct a Lyapunov function. For weakly nonlinear systems described by an equation of the form

$$\dot{x} = Ax + g(x), \tag{4.4.22}$$

where the linear part is dominant and g is weakly nonlinear in the sense that $\| g(x) \|$ is very small whenever x is in the neighborhood of the origin and vanishes at $x = 0$, the stability of the origin is predominantly determined by the stability of the linear part. Thus if all the eigen values of the matrix A have negative real parts the origin may be locally stable. In engineering applications, often a nonlinear system may be linearized around the rest state (or steady state) and its stability determined by looking at the stability of the linear part, here the equation $\dot{x} = Ax$. This is the essence of the Lyapunov first method. Thus our task now is to study the stability of the linear system

$$\dot{x} = Ax. \tag{4.4.23}$$

Stability of this system is determined by the eigen values of the matrix A which are given by its characteristic roots, that is the roots of the polynomial

$$P(s) \equiv det(sI - A).$$

This is a polynomial of degree n and has n roots counting the multiplicity. Complex roots appear in conjugate pairs. In general the solution of equation

(4.4.23) is given by

$$x(t) = e^{tA}x_0 = \sum_{k=1}^{r} e^{\lambda_k t} G_k(t) x_0$$

where $\{\lambda_k\}$ are the distinct roots of P or the eigen values of A of multiplicity m_k with $\sum_{k=1}^{r} m_k = n$ and $\{G_k\}$ are certain polynomials in t of degree $m_k - 1$ with coefficients which are certain matrices $C_{k,s}, 1 \leq s \leq m_k - 1, 1 \leq k \leq r$. That is

$$G_k(t) = \sum_{s=1}^{m_k-1} C_{ks} t^s.$$

For the purpose of our discussion here, it is not necessary to know what exactly these matrices are. From the general expression for the solution, it is clear that if all the roots $\{\lambda_k, 1 \leq k \leq r\}$ have negative real parts then the linear system is asymptotically stable (even more; exponentially stable) which is also expressed by saying that A is stable. This very property can be determined by solving certain algebraic equations known as the Lyapunov equation which is relatively much easier. For this purpose we need some basic notations.

Definition 4.4.1. A matrix K is said to be real if all its entries are real, it is symmetric if it equals its transpose, that is, $K = K'$, and it is said to be positive if the quadratic form

$$q(\xi) \equiv (K\xi, \xi) \geq 0 \ \forall \xi \in R^n.$$

The matrix K is said to be positive definite if $q(\xi) > 0 \ \forall \xi \neq 0$. In this case there exists a positive number k such that

$$q(\xi) = (K\xi, \xi) \geq k \parallel \xi \parallel^2 \ \ \forall \xi \in R^n.$$

Let us denote by M_s^+ the set of all real symmetric positive definite matrices of dimension n. Lyapunov theory tells that the question of stability of the matrix A is equivalent to the question of existence of solutions of certain algebraic equations. This is stated in the following theorem.

Theorem 4.4.2. The system (4.4.23) is asymptotically stable or equivalently the matrix A is stable if, and only if, the following algebraic equation

$$A'Q + QA = -\Gamma \tag{4.4.24}$$

has a solution $Q \in M_s^+$ for every $\Gamma \in M_s^+$.

Proof. First we prove the sufficient condition. Suppose that for every $\Gamma \in M_s^+$ equation (4.4.24) has a solution $Q \in M_s^+$. Define the quadratic function $V(t) \equiv (Qx(t), x(t))$ where $x = x(t, x_0)$ is any solution of the differential system (4.4.23) corresponding to the initial state $x_0 \in R^n$. Differentiating this with respect to t we obtain

$$\begin{aligned}\dot{V}(t) &= (QAx(t), x(t)) + (A'Qx(t), x(t)) \\ &= -(\Gamma x(t), x(t)). \end{aligned} \tag{4.4.25}$$

Integrating this over the interval $[0, T]$ we have

$$\int_0^T (\Gamma x(t), x(t))dt + V(T) = V(0).$$

Positivity of the matrix Q implies that $V(T) \geq 0$ and hence

$$\int_0^T (\Gamma x(t), x(t))dt \leq V(0), \quad \text{for all } \; T \geq 0. \tag{4.4.26}$$

Since Γ is also a positive definite matrix, there exists a number $\gamma > 0$ such that

$$(\Gamma \xi, \xi) \geq \gamma \parallel \xi \parallel^2 \quad \text{for all } \; \xi \in R^n.$$

Using this fact in the inequality (4.4.26) we obtain

$$\int_0^\infty \parallel x(t) \parallel^2 dt \leq V(0)/\gamma. \tag{4.4.27}$$

Since $V(0) < \infty$ it follows from this expression that $x \in L_2([0, \infty), R^n)$ and at the same time $x \in C([0, \infty), R^n)$. This implies that

$$\lim_{t \to \infty} x(t) = 0.$$

Thus the system is asymptotically stable. So, existence of a solution of the Lyapunov equation, satisfying the conditions stated in the theorem, is a sufficient condition for stability of the matrix A.

Now we show that this is also a necessary condition. Suppose now that A is a stability matrix or equivalently the system (4.4.23) is asymptotically stable. Then it follows from the general expression for the solution that for any $\zeta \in R^n$, $\lim_{t \to \infty} e^{tA}\zeta = 0$. Consider the matrix differential equation

$$\begin{aligned}\dot{Z}(t) &= A'Z(t) + Z(t)A, t \geq 0, \\ Z(0) &= \Gamma. \end{aligned} \tag{4.4.28}$$

This equation has a unique solution given by

$$Z(t) = e^{tA'}\Gamma e^{tA}. \tag{4.4.29}$$

This is easily verified by differentiating Z with respect to time and recalling that A commutes with any function of A. The uniqueness follows from Gronwall inequality applied to

$$Z(t) = Z_0 + \int_0^t A'Z(s)ds + \int_0^t Z(s)Ads,$$

and setting $Z_0 = 0$. Integrating equation (4.4.28) over the time interval $[0, T]$ we have

$$\begin{aligned} Z(T) &= \Gamma + \int_0^T A'Z(t)dt + \int_0^T Z(t)Adt \\ &= \Gamma + A'\left(\int_0^T Z(t)dt\right) + \left(\int_0^T Z(t)dt\right)A \end{aligned} \tag{4.4.30}$$

Since A is a stability matrix, Z is integrable on $[0, \infty)$ and $\lim_{T\to\infty} Z(T) = 0$. Define

$$Q \equiv \int_0^\infty Z(t)dt. \tag{4.4.31}$$

Clearly this integral is well defined. Hence, letting $T \longrightarrow \infty$, it follows from equation (4.4.30) that

$$A'Q + QA = -\Gamma. \tag{4.4.32}$$

Thus Q, given by the expression (4.4.31), is a solution of the Lyapunov equation. We must now show that $Q \in M_s^+$. Clearly the elements of the matrix Q are real. Since $Z(t)$ is symmetric for all t, its integral is also symmetric. Thus Q is a real symmetric matrix. For positivity note that

$$(Q\xi, \xi) = \int_0^\infty (Z(t)\xi, \xi)dt = \int_0^\infty (\Gamma e^{tA}\xi, e^{tA}\xi)dt \geq \gamma \int_0^\infty \| e^{tA}\xi \|^2 \, dt.$$

Since e^{tA} is always nonsingular, $\int_0^\infty \| e^{tA}\xi \|^2 \, dt > 0$, for every $\xi \neq 0$. Thus Q is positive definite. In fact non singularity of the transition operator e^{tA} implies the existence of a positive number β such that

$$\int_0^\infty \| e^{tA}\xi \|^2 \, dt \geq \beta \| \xi \|^2$$

for all $\xi \in R^n$. This completes the proof. □

E1: For illustration consider the second order system

$$m\ddot{\xi} + b\dot{\xi} + k\xi = 0$$

and define $x_1 = \xi$ and $x_2 = \dot{\xi}$. Written in the canonical form given by (4.4.23), we have

$$(d/dt)\begin{pmatrix} x_1 \\ x_2 \end{pmatrix} = \begin{pmatrix} 0 & 1 \\ -(k/m) & -(b/m) \end{pmatrix}\begin{pmatrix} x_1 \\ x_2 \end{pmatrix}. \qquad (4.4.33)$$

The Lyapunov equation $A'Q + QA = -I$ with $\Gamma = I$ is given by

$$\begin{pmatrix} 0 & -(k/m) \\ 1 & -(b/m) \end{pmatrix}\begin{pmatrix} q_{11} & q_{12} \\ q_{21} & q_{22} \end{pmatrix} + \begin{pmatrix} q_{11} & q_{12} \\ q_{21} & q_{22} \end{pmatrix}\begin{pmatrix} 0 & 1 \\ -(k/m) & -(b/m) \end{pmatrix} = \begin{pmatrix} -1 & 0 \\ 0 & -1 \end{pmatrix}.$$

From this we obtain the following set of algebraic equations

$$\begin{aligned} -(k/m)(q_{12} + q_{21}) &= -1 \\ -(k/m)q_{22} + q_{11} - (b/m)q_{12} &= 0 \\ q_{11} - (b/m)q_{21} - (k/m)q_{22} &= 0 \\ q_{12} + q_{21} - 2(b/m)q_{22} &= -1. \end{aligned}$$

From the middle two equations it follows that $q_{12} = q_{21}$ and hence the matrix Q is symmetric and

$$q_{12} = q_{21} = (m/b)q_{11} - (k/b)q_{22}. \qquad (4.4.34)$$

Clearly from the first equation and the symmetry of Q it follows that

$$q_{12} = q_{21} = (m/2k). \qquad (4.4.35)$$

Using this fact in the fourth equation we obtain

$$q_{22} = m(k+m)/2bk \qquad (4.4.36)$$

$$q_{11} = (b/m)q_{12} + (k/m)q_{22} = (b^2 + k(k+m))/2bk. \qquad (4.4.37)$$

Since all the parameters are real, Q is real. So Q is a real symmetric matrix. We show that it is positive definite. To prove positivity, it suffices to prove that the determinants of all the minors are positive. Clearly both q_{11} and q_{22} are positive if $\{m, k, b\}$ are positive. Thus it suffices to show that the determinant of Q is positive. The determinant is given by

$$detQ = (m/4k)(1 + (k+m)^2/b^2). \qquad (4.4.38)$$

Clearly for $b > 0$ it follows from the expressions (4.4.36), (4.4.37), (4.4.38) that Q is a positive definite matrix.

One can also demonstrate the positivity of Q by direct computation as follows:

$$\begin{aligned} V \equiv (Qx, x) &= ([b^2 + k(k+m)]/2bk)x_1^2 + (m/k)x_1x_2 + (m(m+k)/2bk)x_2^2 \\ &= (b/2k)(x_1 + (m/b)x_2)^2 + ((k+m)/2b)x_1^2 + (m/2bk)x_2^2 \\ &\geq ((k+m)/2b)x_1^2 + (m/2bk)x_2^2 \geq \nu \parallel x \parallel^2 \end{aligned} \quad (4.4.39)$$

where $\nu \equiv \min\{(k+m)/2b), (m/2bk)\} > 0$ for positive parameters. For $m = k = b = 1$ we have

$$\begin{aligned} (Qx, x) &= (3/2)x_1^2 + x_1x_2 + x_2^2 \\ &\geq (x_1^2 + (1/2)x_2^2) + (1/2)(x_1 + x_2)^2 \geq (1/2) \parallel x \parallel^2 \end{aligned}$$

which is positive. Thus if we take the Lyapunov function as $V \equiv (Qx, x)$ as given above we have $\dot{V} = - \parallel x \parallel^2$ and we can directly conclude that the system is asymptotically stable. This was not possible with the simple energy function

$$E = (1/2)(kx_1^2 + mx_2^2).$$

First Method Applied to Nonlinear Systems: The first method of Lyapunov can be applied to a class of weakly nonlinear problems of the form

$$\dot{x} = Ax + g(x) \quad (4.4.40)$$

where A is a constant matrix and $g : R^n \longrightarrow R^n$ is a nonlinear function. This is given formally in the following corollary.

Corollary 4.4.2. Suppose the linear part of the system (4.4.40), $(g = 0)$, is asymptotically stable in the sense of Theorem 4.4.1 and let Q denote the unique solution of the Lyapunov equation

$$A'Q + QA = -\Gamma$$

corresponding to any $\Gamma \in M_s^+$. Suppose g is continuous with $g(0) = 0$ and there exists a positive number δ such that the set

$$G_\delta \equiv \{\xi \in R^n : (\Gamma\xi, \xi) - 2(Q\xi, g(\xi)) > \delta \parallel \xi \parallel^2\}$$

is a nonempty open subset of R^n. Then the zero state is locally asymptotically stable.

Proof. Under the given assumptions, it is clear that 0 is an interior point of the set G_δ and hence there exists an $\varepsilon > 0$ such that $N_\varepsilon(0) \subset G_\delta$. Let $x_0 \in N_\varepsilon(0)$ and $x = x(t, x_0)$ any solution of the nonlinear equation (4.4.40). Define $V \equiv (Qx, x)$ where x is the solution defined above. Taking the time derivative along this trajectory we have

$$\dot{V} = -(\Gamma x, x) + 2(Qx, g(x)).$$

Since $x_0 \in N_\varepsilon(0)$ and the solution is a continuous function of time, there exists a positive number T_ε possibly dependent on ε so that $x(t, x_0) \in N_\varepsilon(0)$ for all $t \in [0, T_\varepsilon)$. Integrating $\dot{V}$ over this time interval, we obtain

$$V(x(T_\varepsilon)) = V(x_0) - \int_0^{T_\varepsilon} \{(\Gamma x(t), x(t)) - 2(Qx(t), g(x(t))\} dt.$$

Since by our assumption on the nonlinearity, $N_\varepsilon \subset G_\delta$, it follows from the previous equation that

$$V(x(T_\varepsilon)) \leq V(x_0) - \delta \int_0^{T_\varepsilon} \| x(t) \|^2 \, dt.$$

Hence $V(x(T_\varepsilon)) < V(x_0)$. Since Q is positive definite, this implies that V is monotonically decreasing along the trajectory. Hence the solution trajectory is moving closer to the zero state with increasing time. Thus the origin is locally asymptotically stable. □

E2: For illustration we consider the nonlinear system

$$m\ddot{\xi} + b\dot{\xi} + k\xi + c\xi\dot{\xi} = 0,$$

with all the parameters positive. We can write the system in the form given by (4.4.40) and using the corollary verify that the zero state is locally stable. Without any loss of generality, we may take $m = b = k = 1$, and in this case the system is given by

$$(d/dt)\begin{pmatrix} x_1 \\ x_2 \end{pmatrix} = \begin{pmatrix} 0 & 1 \\ -1 & -1 \end{pmatrix}\begin{pmatrix} x_1 \\ x_2 \end{pmatrix} + \begin{pmatrix} 0 \\ -cx_1x_2 \end{pmatrix}, \tag{4.4.41}$$

with the nonlinear term g, given by, $g_1 = 0, g_2 = -cx_1x_2$. As seen in example (E1), A is a stability matrix. For $\Gamma = I$, the solution of the Lyapunov equation, $A'Q + QA = -I$, is given by

$$Q = \begin{pmatrix} 3/2 & 1/2 \\ 1/2 & 1 \end{pmatrix}.$$

Using these matrices we have

$$\begin{aligned}(\Gamma x, x) - 2(Qx, g(x)) &= x_1^2 + x_2^2 + 2cx_1x_2((1/2)x_1 + x_2) \\ &= x_1^2(1 + cx_2) + x_2^2(1 + 2cx_1).\end{aligned}$$

Since $c > 0$, it is evident that, for $0 < \delta < 1$, the set

$$E_\delta \equiv \{x \in R^2 : x_1 > (1/2c)(\delta - 1), x_2 > (1/c)(\delta - 1)\}$$

is a nonempty open subset of R^2 containing the origin in it's interior. Define the set G_δ as

$$G_\delta \equiv \{x \in R^2 : (\Gamma x, x) - 2(Qx, g(x)) > \delta \parallel x \parallel^2\}.$$

The reader can easily check that whenever $x \in E_\delta$, it is also in G_δ. In other words $E_\delta \subset G_\delta$ and hence the set G_δ is nonempty containing the origin in it's interior. Thus the system is locally stable with respect to the zero state.

It is interesting to note that as c decreases, the set E_δ contained in G_δ expands and eventually converges to R^2 signifying global stability.

Remark. We will return to stability once again in Chapter 6 when we deal with the questions of stabilizability of unstable systems by use of state feed-back controls.

4.5 Stability of Elementary Impulsive Systems

Consider the linear impulsive system given by

$$dx(t) = Ax(t)dt + G(t)\nu(dt), x(0) = x_0. \tag{4.5.42}$$

Theorem 4.5.1. Consider the system (4.5.42) and suppose $A \in M(n \times n)$ is a stability matrix and G is a $M(n \times d)$ valued function, locally integrable with respect to the positive measure $|\nu|(dt)$ induced by the variation of the vector measure $\nu \in \mathcal{M}_c(R_0, R^d)$ which is locally of bounded variation. Suppose the function y given by

$$y(t) \equiv \int_0^t e^{-\theta A} G(\theta)\nu(d\theta), t \geq 0, \tag{4.5.43}$$

is bounded for all $t \geq 0$. Then the system is asymptotically stable with respect to the zero state.

Proof. The proof follows immediately from the expression

$$\begin{aligned} x(t) &= e^{tA}\left\{x_0 + \int_0^t e^{-sA}G(s)\nu(ds)\right\} \\ &= e^{tA}\{x_0 + y(t)\}. \end{aligned}$$

Since $\sup\{\| y(t) \|, t \geq 0\} = b < \infty$ and A is a stability matrix implying that all its eigen values have negative real parts, the assertion of the theorem follows. □

Remarks. Any of the following conditions are sufficient guaranteeing uniform boundedness of the function y. (C1): G is continuous, ν is purely atomic and there are only finite number of atoms on $R_0 \equiv [0, \infty)$. Indeed let

$$\nu(dt) = \sum_{i=1}^{m} \alpha_i \delta_{t_i}(dt)$$

where for each i, $\alpha_i \in R^d$ and $t_i \in R_0$. Note that in this case

$$\sup_{t\geq 0} \| y(t) \| \;\; \leq \;\; \sum_{i=1}^{m} \| e^{-t_i A} G(t_i) \| \| \alpha_i \| \equiv b.$$

Clearly this is finite. (C2): G is bounded measurable and for every $\varepsilon > 0$, there exists a finite $T_\varepsilon > 0$ such that

$$\int_{T_\varepsilon}^{\infty} \| e^{-sA} \| \, |\nu|(ds) < \varepsilon.$$

With this condition,

$$\sup_{t\geq 0} \| y(t) \| \;\; \leq \;\; \left(\sup_{0\leq s\leq T_\varepsilon} \| e^{-sA} G(s) \| \| \nu \|_{[0,T_\varepsilon)} \right) + \gamma_0 \varepsilon < \infty,$$

where $\gamma_0 \equiv \sup_{t\geq 0} \| G(t) \|$.

Similar results hold for the semilinear problem

$$dx(t) = Ax(t)dt + G(x(t-))\nu(dt), x(0) = x_0. \qquad (4.5.44)$$

Corollary 4.5.2. Consider the system (4.5.44) and suppose $A \in M(n \times n)$ is a stability matrix and

$$G : R^n \longrightarrow M(n \times d)$$

is continuous and uniformly bounded on R^n with $G(0) = 0$ and, for any $\varepsilon > 0$, there exists a finite positive number T_ε so that

$$\int_{T_\varepsilon}^{\infty} \| e^{-sA} \| \, |\nu|(ds) < \varepsilon.$$

Then the semilinear system is asymptotically stable with respect to the zero state.

Proof. Clearly the solution of the system (4.5.44) is given by the solution of the integral equation,

$$x(t) = e^{tA}x_0 + \int_0^t e^{(t-s)A} G(x(s-))\nu(ds), t \geq 0.$$

Define

$$y(t,x) \equiv x_0 + \int_0^t e^{-sA} G(x(s-))\nu(ds), t \geq 0.$$

Let $\gamma_0 \equiv \sup\{\| G(\xi) \|, \xi \in R^n\}$. By assumption this is a finite number. Hence for any $t > T_\varepsilon$, we have

$$\| y(t,x) \| \leq \left(\| x_0 + \int_0^{T_\varepsilon} e^{-sA} G(x(s-))\nu(ds) \| \right) + \gamma_0 \int_{T_\varepsilon}^t \| e^{-sA} \| \, |\nu|(ds).$$

Since $T_\varepsilon < \infty$, and the measure ν has bounded variation on bounded sets and e^{-tA} has bounded norms on bounded intervals, the first term is finite. By assumption

$$\int_{T_\varepsilon}^{\infty} \| e^{-sA} \| \, |\nu|(ds) \leq \varepsilon.$$

Thus the second term is less than $\gamma_0 \, \varepsilon$. This proves that along any solution of the system equation (4.5.44), $y(t,x), t \geq 0$, is uniformly bounded. Since A is a stability matrix, this implies asymptotic stability of the system (4.5.44). □

Now we consider the following nonlinear system

$$\dot{x}(t) = f(x(t)), t \in R_0 \setminus D \tag{4.5.45}$$

$$x(t_i+) \equiv x(t_i) = x(t_i-) + g_i(x(t_i-)), t_i \in D \equiv \{t_i, i \in N_0, t_0 = 0\}, \tag{4.5.46}$$

where $g_i(\xi)$ is a bounded continuous map from R^n to R^n denoting the jump size at time t_i. Define the following operators

$$(\mathcal{A}\psi)(\xi) \equiv < D\psi(\xi), f(\xi) >, \xi \in R^n$$

$$(\mathcal{B}_i\psi)(\xi) \equiv \int_0^1 < D\psi(\xi + \theta g_i(\xi)), g_i(\xi) > \, d\theta.$$

Note that the operator $\mathcal{A}$ denotes the classical Lyapunov derivative in the direction of the vector field f and the operators $\mathcal{B}_i$ denote the jump operators associated with jump intensity g_i. This is obtained by use of Lagrange formula given by

$$\psi(y) = \psi(x) + \int_0^1 dr < D\psi(x + r(y - x)), y - x >, x, y \in R^n$$

for any $\psi \in C^1(R^n)$. We may call $\mathcal{B}_i V$ the impulsive derivative of the Lyapunov function V at the jump instant t_i.

We have the following result based on the Lyapunov function of the unperturbed system.

Theorem 4.5.3. Consider the system (4.5.45), (4.5.46) and suppose $f(0) = 0, g_i(0) = 0$ for all $i \in N_0$. Suppose the nonlinear system

$$\dot{x}(t) = f(x(t)), t \in R_0 \tag{4.5.47}$$

is asymptotically stable with respect to the origin and let V be its Lyapunov function. Then the impulsive system (4.5.45), (4.5.46) is also asymptotically stable with respect to the origin if for every $\varepsilon > 0$, there exists $n_\varepsilon \in N_0$ such that for all $z \in R^n$,

$$\left| \left\{ \sum_{i=k+1}^{\infty} (\mathcal{B}_i V)(z) \right\} \right| \leq \varepsilon, \forall\, k \geq n_\varepsilon \tag{4.5.48}$$

Proof. Consider the system (4.5.45), (4.5.46) starting from any initial state $x_0 \in R^n$ and let $x(t) \equiv x(t, x_0)$ denote the corresponding solution. It is not difficult to verify that for any $t \in [t_k, t_{k+1})$

$$V(x(t)) = V(x_0) + \int_0^t (\mathcal{A}V)(x(s))ds + \sum_{i=1}^{k} (\mathcal{B}_i V)(x(t_i-)). \tag{4.5.49}$$

Thus given $x(t_k) \equiv x(t_k+)$ and $t \in [t_n, t_{n+1})$ with $k < n$ this can be written as

$$V(x(t)) = V(x(t_k)) + \int_{t_k}^t \mathcal{A}V(x(s))ds + \sum_{i=k+1}^{n} \mathcal{B}_i V(x(t_i-)).$$

Hence, for k sufficiently large, it follows from our assumption (4.5.48) that

$$V(x(t)) \leq V(x(t_k)) + \int_{t_k}^t \mathcal{A}V(x(s))ds + \varepsilon, t \geq t_k, k \geq n_\varepsilon.$$

Since the unperturbed system $\dot{x} = f(x), t \in R_0$, is assumed to be asymptotically stable, and $\varepsilon > 0$ is arbitrary, it follows from this inequality that under the given assumptions the perturbed system is also asymptotically stable. □

Similar result can be proved for the system

$$d\xi(t) = f(t, \xi(t))dt + G(t, \xi(t-))\nu(dt), t \geq 0, \tag{4.5.50}$$

where $f : R_0 \times R^n \longrightarrow R^n$ and $G : R_0 \times R^n \longrightarrow M(n \times d)$ measurable in the first argument and continuous in the second, and $\nu \in \mathcal{M}_c(R_0, R^d)$. For this purpose define the operators

$$(\mathcal{A}(t)\varphi)(t, x) = \partial\varphi(t, x)/\partial t + (f(t, x), D\varphi(t, x))$$
$$(\mathcal{B}(t)\varphi)(t, x) \equiv \int_0^1 d\theta\ G^*(t, x)D\varphi(t, x + \theta G(t, x)\nu(\{t\})), (t, x) \in I \times R^n,$$

where G^* denotes the adjoint of the operator G or the transpose of the matrix $G(t, x)$. Clearly these operators are well defined for every $\varphi \in C^{1,1}(R_0 \times R^n)$.

Theorem 4.5.4. Consider the system (4.5.50), and suppose the following assumptions hold: (a1): $f(t, 0) = 0, G(t, 0) = 0$, (a2): the regular system

$$\dot{x} = f(t, x), t \geq 0, \tag{4.5.51}$$

is exponentially stable with respect to the origin having decay rate $K > 0$, with $V \in C^{1,1}(R_0 \times R^n)$ being the corresponding Lyapunov function, (a3): there exist finite positive numbers $\{M, T\}$ and $\gamma \in [0, K)$ such that

$$\left| \int_T^t e^{Ks} < (\mathcal{B}(s)V)(s, \xi(s-)), \nu(ds) > \right| \leq Me^{\gamma(t-T)} \tag{4.5.52}$$

for all $t \geq T$. Then the impulsive system (4.5.50), is also exponentially stable with respect to the origin.

Proof. Let $\xi \in B^{\ell oc}(R_0, R^n)$ be any solution of equation (4.5.50). Computing V along this solution trajectory we obtain

$$dV = \mathcal{A}(t)Vdt + < \mathcal{B}(t)V, \nu(dt) > . \tag{4.5.53}$$

Since V is a Lyapunov function for the regular system (4.5.51) with decay rate K, it satisfies the (partial) differential inequality

$$\mathcal{A}(t)V + KV \leq 0,\ (t, x) \in R_0 \times R^n. \tag{4.5.54}$$

Substituting this in equation (4.5.53) and integrating by parts we arrive at the following expression for $t \geq T$,

$$\begin{aligned} V(t,\xi(t)) &\leq V(T,\xi(T))e^{-K(t-T)} \\ &\quad + e^{-Kt}\int_T^t e^{Ks} < (\mathcal{B}(s)V)(s,\xi(s-)),\nu(ds) > . \end{aligned} \tag{4.5.55}$$

By assumption (a3), we can choose T finite such that

$$\left| \int_T^t e^{Ks} < (\mathcal{B}(s)V)(s,\xi(s-),\nu(ds) > \right| \leq Me^{\gamma(t-T)} \tag{4.5.56}$$

for all $t \geq T$. Hence it follows from the inequality 4.5.55 that for all $t \geq T$ we have

$$\begin{aligned} V(t,\xi(t)) &\leq V(T,\xi(T))e^{-K(t-T)} + Me^{-\gamma T}e^{-(K-\gamma)t} \\ &\leq \left(V(T,\xi(T))e^{(KT-\gamma t)} + Me^{-\gamma T}\right)e^{-(K-\gamma)t} \\ &\leq \left(V(T,\xi(T))e^{-\gamma(t-T)} + Me^{-KT}\right)e^{-(K-\gamma)(t-T)}. \end{aligned} \tag{4.5.57}$$

Since $\xi \in B^{\ell oc}(R_0, R^n)$ and $T < \infty$, it is clear that $\| \xi(T) \| < \infty$ and so $V(T,\xi(T)) < \infty$ and hence it follows from the above inequality that the system (4.5.50) is exponentially stable. □

Remark. Clearly, the conclusion of the above theorem holds under the stronger assumption

$$\sup_{t\geq T}\left| \int_T^t e^{Ks} < (\mathcal{B}(s)V)(s,\xi(s-)),\nu(ds) > \right| \leq M. \tag{4.5.58}$$

An Example. For illustration of the above results we present an example from aerospace. The attitude dynamics of a geosynchronous satellite is given by the following equations

$$\begin{aligned} \dot{x}_1 &= ((I_2 - I_3)/I_1)x_2x_3 \\ \dot{x}_2 &= ((I_3 - I_1)/I_2)x_1x_3 \\ \dot{x}_3 &= ((I_1 - I_2)/I_3)x_1x_2 \end{aligned} \tag{4.5.59}$$

where $\{x_1, x_2, x_3\}$ denotes the angular velocities (spin rates) along the principal axes and $\{I_1, I_2, I_3\}$ the corresponding moments of inertia respectively. Taking the kinetic energy of the system as the Lyapunov function

$$V(x) \equiv (1/2)I_1x_1^2 + (1/2)I_2x_2^2 + (1/2)I_3x_3^2 \tag{4.5.60}$$

it is easily verified that $\dot{V}(x(t)) = 0$ along any solution of the system (4.5.59) and hence the system is conservative and stable in the Lyapunov sense. For communication satellites it is absolutely necessary that the system is asymptotically stable with respect to the rest state $x = 0$. This can be easily realized by use of negative state feedback as follows

$$\begin{aligned}\dot{x}_1 &= ((I_2 - I_3)/I_1)x_2x_3 - (K_1/I_1)x_1 \equiv f_1(x_1, x_2, x_3) \\ \dot{x}_2 &= ((I_3 - I_1)/I_2)x_1x_3 - (K_2/I_2)x_2 \equiv f_2(x_1, x_2, x_3) \\ \dot{x}_3 &= ((I_1 - I_2)/I_3)x_1x_2 - (K_3/I_3)x_3 \equiv f_3(x_1, x_2, x_3)\end{aligned} \tag{4.5.61}$$

where $\{K_1, K_2, K_3\}$ are suitable positive numbers representing the gain factors. It is easily verified that V, as given above, is also a Lyapunov function for the feedback system and

$$\dot{V}(x(t)) = -(K_1x_1^2 + K_2x_2^2 + K_3x_3^2). \tag{4.5.62}$$

Defining

$$K \equiv 2\min\{(K_1/I_1), (K_2/I_2), (K_3/I_3)\}$$

we find that

$$\dot{V}(x(t)) \leq -KV(x(t)), t \geq 0. \tag{4.5.63}$$

Thus the system with negative state feedback is asymptotically stable with respect to the rest (zero) state. Note that K can be chosen as large as necessary. In addition to forces caused by solar pressure and geomagnetic activities, occasionally the system may be subject to bombardment by micro meteorites in outer space displacing the satellite from its rest state. Including the torques produced by micro meteorites, the system (4.5.61) is described by the following measure driven equations

$$\begin{aligned}dx_1 &= f_1(x_1, x_2, x_3)dt + C_1x_1\ \nu_1(dt) \\ dx_2 &= f_2(x_1, x_2, x_3)dt + C_2x_2\ \nu_2(dt) \\ dx_3 &= f_3(x_1, x_2, x_3)dt + C_3x_3\ \nu_3(dt)\end{aligned} \tag{4.5.64}$$

Denoting by $f = col\{f_1, f_2, f_3\}$ and the matrix G by

$$\begin{pmatrix} C_1x_1 & 0 & 0 \\ 0 & C_2x_2 & 0 \\ 0 & 0 & C_3x_3 \end{pmatrix}$$

and $\nu = col\{\nu_1, \nu_2, \nu_3\}$, we can rewrite (4.5.64) as the system

$$dx(t) = f(x(t))dt + G(x(t-))\ \nu(dt), t \geq 0. \tag{4.5.65}$$

The maximum size of the meteorites determines the range of the vector measure ν. We may assume that this range is a compact set $\Gamma \subset R^3$ containing the origin. At times one may expect an intense shower of micro meteorites hitting the space craft and displacing it from its rest state $x = col(x_1, x_2, x_3) = col(0, 0, 0)$. If the shower persists long enough, the space craft may be destabilized without any possibility of recovery unless external forces are used. We note that this system satisfies the assumptions of theorem 4.5.4 provided the meteoritic shower activities are intermittent. For this example, the (differential) operators $\mathcal{A}$ and $\mathcal{B}$ are given by

$$\mathcal{A}V(x) \equiv -K_1x_1^2 - K_2x_2^2 - K_3x_3^2, x \in R^3, \tag{4.5.66}$$

and

$$\mathcal{B}(t)V(x) \equiv \int_0^1 d\theta \Big\{ G^*DV(x + \theta G(x)\nu(\{t\}) \Big\}, x \in R^3. \tag{4.5.67}$$

Performing the integration, we find that

$$\mathcal{B}(t)V(x) = \begin{pmatrix} I_1C_1x_1^2[1 + (1/2)C_1\nu_1(\{t\})] \\ I_2C_2x_2^2[1 + (1/2)C_2\nu_2(\{t\})] \\ I_3C_3x_3^2[1 + (1/2)C_3\nu_3(\{t\})] \end{pmatrix}. \tag{4.5.68}$$

Since the range of the vector measure ν is a compact set Γ, there exists a finite positive number β such that

$$\| \mathcal{B}(t)V(x) \| \leq \beta V(x), x \in R^3.$$

For negative state feedback $K_i > 0, i = 1, 2, 3$, there exists a positive number K such that

$$\mathcal{A}V(x) \leq -KV(x), x \in R^3.$$

Hence along any solution path of equation (4.5.65) we have

$$V(x(t)) \leq V(x_0) - K \int_0^t V(x(s))ds + \beta \int_0^t V(x(s-))|\nu|(ds), t \geq 0, \tag{4.5.69}$$

where $|\nu|(dt) \equiv \sup\{|\nu_i|(dt), i = 1, 2, 3\}$. Since intense meteoritic shower activities may be considered to be an intermittent phenomenon with low frequency of occurrence (may cease for sufficiently long time before the next strike), for any finite positive number b there exists a finite number $T_b > 0$ so that

$$\beta \int_{T_b}^{\infty} e^{Ks}V(x(s-))|\nu|(ds) \leq b.$$

So for $t \geq T_b$, we have

$$V(x(t)) \leq \left(V(x(T_b)) + be^{-KT_b}\right)e^{-K(t-T_b)}.$$

This shows that for a suitable choice of negative state feedback the system is exponentially stable.

Remark. If the meteoritic shower activities are intermittent and persist only for short time intervals compared to the intermittency, the system is exponentially stable. On the other hand if it is persistent without sufficient time gaps (pause), increase of gain will neither improve stability nor prevent instability. In contrast, if ν is absolutely continuous with respect to the Lebegue measure, suitable gain increase will prevent instability.

4.6 Exercises

(P1): Suppose there is a positive number $\alpha > 0$ and real symmetric positive semidefinite matrix Γ so that the Lyapunov equation

$$Y\left(A+\frac{\alpha}{2}\right)+\left(A+\frac{\alpha}{2}\right)'Y=-\Gamma$$

has a real symmetric positive definite solution Y. Prove that the system

$$\dot{x} = Ax, t \geq 0,$$

is exponentially stable.

(P2): Consider the system

$$\dot{x}_1 = \alpha x_2 - \beta x_1, \ \dot{x}_2 = \gamma x_1 - \delta x_2.$$

Show that if $2\beta, 2\delta > (\alpha+\gamma)$, then the system is exponentially stable,

$$E(t) \leq E(0)e^{-\nu t}$$

where $E \equiv (1/2)(x_1^2 + x_2^2)$ and $\nu = \min\{2\beta - (\alpha+\gamma), 2\delta - (\alpha+\gamma)\}$.

(P3): Consider the geosynchronous satellite whose attitude dynamics is given by the following system of equations

$$I_x\dot{p} + (I_z - I_y)r(q-\omega) = T_x$$
$$I_y\dot{q} + (I_x - I_z)pr = T_y$$
$$I_z\dot{r} + (I_y - I_x)p(q-\omega) = T_z,$$

where $\{I_x, I_y, I_z\}$ are the moments of inertia and $\{T_x, T_y, T_z\}$ are the torques (controls) with reference to the $\{x, y, z\}$ axes, and ω is the spin rate of the earth.

(a) Give a Lyapunov function for the system and show that for $T_x = T_y = T_z \equiv 0$ the system is conservative.

(b) Choose gains $\{K_x, K_y, K_z\}$ so that the feedback control law,

$$T_x = -K_x \; p, T_y = -K_y \; (q - \omega), T_z = -K_z \; r,$$

makes the system exponentially stable.

(P4): A network consists of a voltage source e, feeding into an electrical circuit consisting of a linear capacitor C, connected in series to a parallel branch of linear resistor R, and a nonlinear inductor with flux-current characteristics given by $\varphi = \ell(i)$ where ℓ is an increasing function of current i with $\ell(0) = 0$.
(a) Justify that the correct state variable is $col(\varphi, v)$ and the state equation is given by

$$\begin{aligned}\dot{\varphi} &= e - v \\ \dot{v} &= (1/C)g(\varphi) + (1/RC)(e - v),\end{aligned}$$

where v is the voltage across the capacitor and φ is the magnetic flux induced in the inductor and $g(\varphi) = \ell^{-1}(\varphi) = i$.
(b) Verify that the function V given by

$$V \equiv (1/2)Cv^2 + \int_0^{\varphi} g(\xi) d\xi,$$

is a Lyapunov function for the system and that the zero state is asymptotically stable.

(P5): A network consists of a current source I, feeding into a parallel branch of linear inductor L, and a nonlinear capacitor with charge-voltage characteristics given by $q = f(v)$ where f is an increasing function of voltage v with $f(0) = 0$.
(a) Justify that the correct state variable is $col(q, i)$ and the state equation is given by

$$\begin{aligned}(d/dt)q &= I - i \\ (d/dt)i &= (1/L)g(q),\end{aligned}$$

where q is the charge on the capacitor and i is the current in the inductor and $g(q) = f^{-1}(q) = v$.
(b) Verify that the function V given by

$$V \equiv (1/2)Li^2 + \int_0^q g(\xi)d\xi,$$

is a Lyapunov function for the system and that the system is conservative.

(P6): Use the Lyapunov second method to verify that the nonlinear system

$$\ddot{\xi} + (1 + d\dot{\xi})\dot{\xi} + (1 + c\dot{\xi})\xi = 0,$$

for $d, c > 0$, is locally stable with respect to the origin. (Hint: follow the procedure as in example E2 of Section 4)

4.7 Bibliographical Notes

The father of stability theory is Lyapunov (1892). His fundamental memoir published in Russian in 1892 and translated into French in 1907 under the title: Probbleme general de la stabilite de mouvement: planted the seed of stability theory. Again this is a very matured field and certainly requires a complete book to cover reasonable segment of available materials in the area. In this chapter only very basic stability theory is covered. An excellent introduction to the field is the book by La Salle and Lefschetz [58]. Many more interesting materials can be found in [17] and [2]. For functional differential equations and their stability see the books by Kolmanovskii [53], Driver [32], Hale [40]. For stochastic differential equations the book by Hasminskii [42] is classical. For application of Lyapunov theory to mechanical systems see Skowronski [85], Udwadia [91]. For stability of systems governed by partial differential equations, in particular space station and space crafts see Biswas and Ahmed [160, 161] and Lim and Ahmed [181]. For stability of suspension bridges see Ahmed and Harbi [143, 144, 145], Ahmed [116] and for stochastic versions see Ahmed [146]. A significant amount of materials on stability of classical impulsive systems can be found in Martynyuk [67] and several papers due to Liu, Kaul, Shen, Vatsala [182, 183, 184, 177, 178] and Akhmet [158]. The results presented in section 4.6 are nothing more than simple extension of Lyapunov theory. Stability theory for general impulsive systems driven by vector measures is not very well developed. This presents an attractive area for further research.

Chapter 5

Observability and Identification

5.1 Introduction

The concept of observability or identifiability of systems is very important in many applications. As we shall see soon, this is a class of inverse problems where the outcome or the consequence is known to the observer and he seeks to find the cause. For example, an aircraft went down; one wants to know what may have been the state of the system just prior to the crash. One observes sudden decline of a given population in a natural habitat where many different species of animals including prey and predators cohabit. From this information one may seek to identify the cause leading to the observed outcome. In this chapter we study these problems under the assumptions that there is a dynamic model that governs the system. Another class of inverse problems requires identification of system parameters from the knowledge of input and output. The later class of problems can be formulated as optimal control problems or optimization problems. Hence this is treated in Section 8.5 of Chapter 8.

5.2 Linear Time Invariant Systems

First we consider linear time invariant systems of the form

$$\dot{x} = Ax + Bu, \tag{5.2.1}$$

$$y = Hx, \tag{5.2.2}$$

where x denotes the state trajectory, and u the control policy taking values from R^d and y the output trajectory taking values from R^m. Clearly the matrices must have compatible dimension, for example, A is a $n \times n$ matrix, B is a $n \times d$ matrix and H is a $m \times n$ matrix. For identification of state we introduce the following definition.

Definition 5.2.1. The system $\mathcal{S}$ given by equations (5.2.1),(5.2.2) is said to be observable at time T if from the knowledge of the input u over the time interval $I_T \equiv [0, T]$ and the output y over the same time interval I_T, the state trajectory x can be determined (uniquely).

First note that if one can determine the initial state x_0 from the given data, the state trajectory x is uniquely determined by the expression

$$x(t) = e^{tA}x_0 + \int_0^t e^{(t-s)A}Bu(s)ds, t \in I_T. \tag{5.2.3}$$

So mathematically the problem of observability of the system is equivalent to the problem of identifiability of the initial state x_0. Thus the equivalent definition for observability can be formulated as follows:

Definition 5.2.2. The system $\mathcal{S}$ is observable if, from the given input and output data, the initial state is uniquely identifiable.

Thus we can formulate the observability problem as follows. Corresponding to an unknown initial state $x_0 \in R^n$, the system output y is given by

$$y(t, x_0) = He^{tA}x_0 + \int_0^t He^{(t-s)A}Bu(s)ds, t \in I_T. \tag{5.2.4}$$

Since y and u are given we may define

$$\begin{aligned}\tilde{y}(t) \equiv \tilde{y}(t, x_0) \equiv y(t) &- \int_0^t He^{(t-s)A}Bu(s)ds \\ &= He^{tA}x_0.\end{aligned} \tag{5.2.5}$$

Clearly, in order that the initial state be (uniquely) identifiable it is necessary that

$$\tilde{y}(t, \xi) \not\equiv \tilde{y}(t, \eta) \text{ on } I_T \text{ for } \xi \neq \eta. \tag{5.2.6}$$

Since He^{tA} is a rectangular matrix of dimension $m \times n$, it is obvious that this matrix has no inverse. Usually the dimension of the output is less than the dimension of the internal state of the system which is not directly accessible (*i.e* $m < n$). So we must look for some other indirect route. Towards this goal, we introduce the matrix

$$Q_T \equiv \int_0^T e^{tA'}H'He^{tA}dt. \tag{5.2.7}$$

This matrix is called the observability matrix for the system $\mathcal{S}$. Now we can state and prove the following result.

Theorem 5.2.3. The system $\mathcal{S}$ is observable if, and only if, the observability matrix Q_T is positive definite.

Proof. First we prove the sufficient condition, that is, the "if" part (Observability $\Leftarrow Q_T > 0$). We prove this by establishing a contradiction. Suppose $Q_T > 0$ but the system is not observable. This means that there are distinct (initial states) vectors $\xi \neq \zeta \in R^n$ such that $\tilde{y}(t,\xi) = \tilde{y}(t,\zeta)$, for all $t \in I_T$. This implies that

$$\int_0^T \| \tilde{y}(t,\xi) - \tilde{y}(t,\zeta) \|^2 \, dt = 0. \tag{5.2.8}$$

But

$$\begin{aligned}\int_0^T \| \tilde{y}(t,\xi) - \tilde{y}(t,\zeta) \|^2 \, dt &= \int_0^T \| He^{tA}(\xi - \zeta) \|^2 \, dt \\ &= (Q_T(\xi - \zeta), \xi - \zeta).\end{aligned} \tag{5.2.9}$$

Hence $(Q_T(\xi - \zeta), (\xi - \zeta)) = 0$. But Q_T is positive definite and therefore this identity is satisfied if and only if $\xi = \zeta$. This contradicts our hypothesis that $\xi \neq \zeta$ and hence the assertion follows. Now we prove the converse (Observability $\Rightarrow Q_T > 0$). Suppose the system is observable. Then for every pair of distinct vectors $\xi \neq \zeta \in R^n$, as initial states, $\tilde{y}(t,\xi) \not\equiv \tilde{y}(t,\zeta)$. This implies that

$$\int_0^T \| \tilde{y}(t,\xi) - \tilde{y}(t,\zeta) \|^2 \, dt > 0 \; \forall \xi \neq \zeta. \tag{5.2.10}$$

Thus

$$\int_0^T \| \tilde{y}(t,\xi - \zeta) \|^2 \, dt > 0 \; \forall \xi \neq \zeta. \tag{5.2.11}$$

This implies that

$$\begin{aligned}\int_0^T \| \tilde{y}(t,\xi - \zeta) \|^2 \, dt &= \int_0^T \| He^{tA}(\xi - \zeta) \|^2 \, dt \\ &= \int_0^T (He^{tA}(\xi - \zeta), He^{tA}(\xi - \zeta)) dt \\ &= (Q_T(\xi - \zeta), (\xi - \zeta)) > 0 \; \forall \xi \neq \zeta.\end{aligned} \tag{5.2.12}$$

This proves that the observability matrix Q_T is positive definite. This completes the proof of the theorem. □

Now the problem is: given that the system is observable, how do we identify the initial state and hence the entire state trajectory from the given data $\{y(t), u(t), t \in I_T\}$. Since y and u are given we define as before the process $\tilde{y}$ as

$$\tilde{y}(t) = y(t) - \int_0^T He^{(t-s)A}Bu(s)ds, t \in I_T.$$

In view of this we can prove the following result.

Corollary 5.2.4. If the system $\mathcal{S}$ is observable then the initial state x_0 and the state trajectory x are identified as follows

$$(a) : x_0 = Q_T^{-1} \int_0^T e^{tA'} H' \tilde{y}(t)dt, \tag{5.2.13}$$

$$(b) : x(t) = e^{tA}x_0 + \int_0^T e^{(t-s)A}Bu(s)ds, t \in I_T. \tag{5.2.14}$$

Proof. Let x_0 denote the initial state that we want to determine from the given data. Clearly

$$\tilde{y}(t) = He^{tA}x_0.$$

Multiply this expression by the matrix valued function $e^{tA'}H'$ on the left of each of the terms and integrate over the time interval I_T. This gives

$$\begin{aligned} \int_0^T e^{tA'} H' \tilde{y}(t)dt &= \int_0^T e^{tA'} H' He^{tA}x_0 dt \\ &= Q_T \, x_0. \end{aligned} \tag{5.2.15}$$

Since the system is observable, by the previous theorem we have Q_T positive definite and hence nonsingular and consequently invertible. Thus it follows from the above expression that

$$x_0 = Q_T^{-1} \int_0^T e^{tA'} H' \tilde{y}(t)dt. \tag{5.2.16}$$

This proves (a). Once the initial state x_0 is known and the control data is given, the state trajectory is given by (b). This completes the proof. □

If one wishes to compute the observability matrix numerically from given numerical data for A and H, it is easier to solve the following differential equation

$$\begin{aligned} \dot{M}_T(t) &= -A' M_T(t) - M_T(t)A - H' H, t \in I_T, \\ M_T(T) &= 0. \end{aligned} \tag{5.2.17}$$

This is solved backward in time starting from $M_T(T) = 0$ as indicated. The observability matrix Q_T is then given by

$$Q_T = M_T(0). \tag{5.2.18}$$

Verification of this is left as an exercise for the reader.

Intuitively one may feel that if a time invariant system is observable at time $T > 0$, then it may be observable at any time $\tau > 0$. Indeed this is true as presented in the following theorem.

Theorem 5.2.5. For a time invariant system, if it is observable at time $T > 0$, then it is observable at any time $\tau > 0$, smaller or greater than T.

Proof. By theorem 5.2.3, observability of the system is equivalent to positive definiteness of the associated observability matrix Q_T. Thus it suffices to demonstrate that if for some T, $Q_T > 0$, then it is positive definite for any and every $\tau > 0$. First define the matrix valued functions

$$F(t) \equiv e^{tA'} H' H e^{tA}, t \geq 0, \tag{5.2.19}$$

$$Q_t \equiv \int_0^t F(s) ds. \tag{5.2.20}$$

Since by our assumption the system is observable at time T, $Q_T > 0$. Take $\tau \geq T$ and note that

$$Q_\tau = Q_T + \int_T^\tau F(s) ds = Q_T + P_{T,\tau}, \tag{5.2.21}$$

where we have defined

$$P_{T,\tau} = \int_T^\tau F(s) ds.$$

Note that the matrix $P_{T,\tau}$ is at least positive semidefinite. Indeed

$$(P_{T,\tau}\xi, \xi) = \int_T^\tau (F(s)\xi, \xi) ds = \int_T^\tau \parallel H e^{sA} \xi \parallel^2 ds \geq 0.$$

Since Q_T is strictly positive, it follows from positive semidefiniteness of $P_{T,\tau}$ and (5.2.21) that Q_τ is strictly positive that is positive definite. Now let $\tau < T$. We must show that $Q_\tau > 0$ given that Q_T is. Suppose Q_τ is not positive definite. Then there exists a nonzero vector $\eta \in R^n$ so that $(Q_\tau \eta, \eta) = 0$. This means that

$$(Q_\tau \eta, \eta) = \int_0^\tau \parallel H e^{sA} \eta \parallel^2 ds = 0. \tag{5.2.22}$$

Thus

$$He^{sA}\eta = 0, \forall s \in (0,\tau).$$

Let $z \in R^m$ be any nontrivial vector and consider the function defined by

$$f(t) \equiv (He^{tA}\eta, z).$$

Clearly this function is well defined on $I_T \equiv [0,T]$ and it is an analytic function which vanishes on the open interval $(0,\tau)$. It is well known that if an analytic function vanishes on an open interval, it vanishes every where. This follows from Taylor's expansion. Indeed, for any $t \in [\tau, T]$, and any $s \in (0,\tau)$ we can write

$$f(t) = \sum_{k=0}^{\infty}(f^{(k)}(s)/k!)(t-s)^k, \tag{5.2.23}$$

where $f^{(k)}$ denotes the $k-th$ derivative of f. Since f is analytic and $f(s) \equiv 0$ on $(0,\tau)$, all its derivatives must also vanish on this interval. Hence it follows from the Taylor's expansion that

$$f(t) \equiv 0, \ \forall t \in I_T = [0,T],$$

and hence we have

$$f(t) \equiv (He^{tA}\eta, z) \equiv 0, \ \forall t \in [0,T].$$

Since z is arbitrary, this implies that

$$He^{tA}\eta \equiv 0, \forall t \in I_T.$$

Hence we have

$$\begin{aligned}(Q_T\eta,\eta) &= (Q_\tau\eta,\eta) + (P_{\tau,T}\eta,\eta),\\ &= (Q_\tau\eta,\eta) + \int_\tau^T \| He^{sA}\eta \|^2\\ &= 0.\end{aligned}$$

Since we are given that Q_T is positive definite, this implies that $\eta = 0$ which contradicts the hypothesis that there exists a nontrivial vector $\eta \in R^n$ for which $(Q_\tau\eta,\eta) = 0$. The contradiction completes the proof. □

Remark. Note that from the proof of the previous theorem, one can realize that for a time invariant system, if it is observable over any period of time of positive length, then it is also observable over any other period of time of positive length. The reader is encouraged to justify this.

To determine if a given system is observable or not, according to the previous results, it is necessary to determine if the associated observability matrix is positive definite or not. This is not always easy. It involves integration of matrix valued functions and verification of its positivity. There are other simpler alternatives. This is stated in the following result.

Theorem 5.2.6. The necessary and sufficient condition for observability of the system $\mathcal{S}$ is that the rank of the matrix $\{H', A'H', (A')^2H', \cdots, (A')^{n-1} H'\}$ is full (= n). Compactly one may state this as follows

$$\text{Observability} \Leftrightarrow \text{Rank}\{H', A'H', (A')^2H', \cdots, (A')^{n-1}H'\} = n.$$

Proof. By theorem 5.2.3, observability is equivalent to positivity of the observability matrix Q_T. Thus it suffices to show that

$$Q_T > 0 \Leftrightarrow \text{Rank}\{H', A'H', (A')^2H', \cdots, (A')^{n-1}H'\} = n.$$

First we prove that if $Q_T > 0$, then the rank condition holds. Since $Q_T > 0$,

$$(Q_T\xi, \xi) = \int_0^T \| He^{tA}\xi \|^2 \, dt > 0 \; \forall \xi \neq 0. \tag{5.2.24}$$

For any matrix Γ we use the notation,

$$Ker\Gamma \equiv N(\Gamma) \equiv \{\xi \in R^n : \Gamma\xi = 0\},$$

to denote its null space. Thus it follows from the expression (5.2.24), that

$$Ker\{He^{tA}, t \in I_T\} = \{0\}.$$

In other words,

$$\{He^{tA}\xi = 0 \; \forall t \in I_T\} \Leftrightarrow \xi = 0. \tag{5.2.25}$$

This is equivalent to the statement

$$\{(He^{tA}\xi, z) = 0 \; \forall t \in I_T \text{ and } \forall z \in R^m\} \Leftrightarrow \xi = 0, \tag{5.2.26}$$

which, in turn, is equivalent to the statement

$$\{(\xi, e^{tA'}H'z) = 0 \; \forall t \in I_T \text{ and } \forall z \in R^m\} \Leftrightarrow \xi = 0. \tag{5.2.27}$$

Define

$$f(t) \equiv (\xi, e^{tA'} H' z), t \in I_T.$$

Clearly this function is analytic and we can use Taylor's expansion to write

$$f(t) = \sum_{k=0}^{\infty} f^{(k)}(0)(t^k/k!),$$

where $f^{(k)}$ denotes the $k-th$ derivative of the function f. This function can be identically zero if and only if all its derivatives at $t = 0$ vanish. That is, $f^{(k)}(0) = 0$ for all k from the set of nonnegative integers N_0. This means that

$$(\xi, (A')^k H' z) = 0, \ \forall \ k \in N_0.$$

Thus the statement (5.2.27) is equivalent to

$$\{(\xi, (A')^k H' z) = 0 \ \forall k \in N_0, \ \forall z \in R^m\} \Leftrightarrow \xi = 0. \tag{5.2.28}$$

Now note that k runs through all of N_0 and hence this is an infinite sequence. By use of Caley-Hamilton theorem, this can be reduced to an equivalent finite sequence. According to Caley-Hamilton theorem, any square matrix A of dimension n satisfies its characteristic equation

$$P(\lambda) \equiv det(\lambda I - A) = c_0 I + c_1\lambda + c_2\lambda^2 + \cdots, +c_{n-1}\lambda^{n-1} + \lambda^n = 0. \tag{5.2.29}$$

In other words

$$P(A) = c_0 I + c_1 A + c_2 A^2 + \cdots, +c_{n-1}A^{n-1} + A^n = 0.$$

This means that any integer power (equal to n and beyond) of the matrix A can always be written as a function of the powers of A up to $k = n - 1$. In other words $A^k, k \geq n$ do not carry any information which is not contained in $\{A^k, k = 0, 1, \cdots, n-1\}$. Thus the expression (5.2.28) can be truncated at $k = n - 1$ giving

$$\{(\xi, (A')^k H' z) = 0 \ \forall k \in \{0, 1, 2, \cdots, n-1\}, \ \forall z \in R^m\} \Leftrightarrow \xi = 0. \tag{5.2.30}$$

Clearly this means that $\bigcup_{k=0}^{n-1}(A')^k H'(R^m)$ is all of R^n. In other words

$$\text{Range}\{H', A'H', (A')^2 H', \cdots, (A')^{n-1}H'\} = R^n.$$

Equivalently the column vectors of the matrix within the parenthesis, span the whole space R^n. Hence by definition,

$$\text{Rank}\{H', A'H', (A')^2 H', \cdots, (A')^{n-1}H'\} = n.$$

For the proof of the converse, suppose the rank condition holds. Since observability is equivalent to the positive definiteness of Q_T, we must verify that $Q_T > 0$. We prove this by contradiction. Suppose Q_T is not positive definite, then there exists a vector $\xi(\neq 0) \in R^n$ such that $(Q_T\xi, \xi) = 0$. But we have seen before in Theorem 5.2.5 that, this implies $He^{tA}\xi \equiv 0$ for all $t \in I_T$. Hence

$$(He^{tA}\xi, \eta) = 0, \ \forall t \in I_T, \ \text{and} \ \forall \eta \in R^m. \tag{5.2.31}$$

Following similar steps as in the proof of the first part we arrive at the following expression

$$\{(\xi, (A')^k H' \eta) = 0 \ \forall k \in \{0, 1, 2, \cdots, n-1\}, \ \forall \eta \in R^m\}. \tag{5.2.32}$$

But since

$$\text{Rank}\{H', A'H', (A')^2 H', \cdots, (A')^{n-1} H'\} = n,$$

it follows from this that $\xi = 0$. This is a contradiction. Thus $Q_T > 0$. This completes the proof. □

5.3 Some Examples

In this section we present some examples from well known physical systems.

(E1): An electrical D.C. motor is governed by the following second order differential equation.

$$\begin{aligned} J\ddot{\Theta} + b\dot{\Theta} &= T \equiv u, \\ y &= \dot{\Theta}, \end{aligned} \tag{5.3.33}$$

where J denotes the moment of inertia, b denotes the friction coefficient which may combine the bearing friction as well as the aerodynamic resistance. $T = u$ is the control torque and y denotes the output. Using the rank condition we show that the system is not observable. Writing $x_1 = \Theta, x_2 = \dot{\Theta}$ we obtain the canonical form

$$(d/dt)\begin{pmatrix} x_1 \\ x_2 \end{pmatrix} = \begin{pmatrix} 0 & 1 \\ 0 & -(b/J) \end{pmatrix}\begin{pmatrix} x_1 \\ x_2 \end{pmatrix} + \begin{pmatrix} 0 \\ (1/J) \end{pmatrix} u, \tag{5.3.34}$$

with the output given by

$$y = \begin{pmatrix} 0 & 1 \end{pmatrix}\begin{pmatrix} x_1 \\ x_2 \end{pmatrix}. \tag{5.3.35}$$

Computing the rank we have

$$\text{Rank}\{H', A'H'\} = \text{Rank}\left\{\begin{matrix} 0 & 0 \\ 1 & -(b/J) \end{matrix}\right\} = 1.$$

Thus the system is not observable.

On the other hand if the output is taken as $y = \Theta$ the matrix H changes to $\begin{pmatrix} 1 & 0 \end{pmatrix}$ and this makes a big difference and we have

$$\text{Rank}\{H', A'H'\} = \text{Rank}\left\{\begin{matrix} 1 & 0 \\ 0 & 1 \end{matrix}\right\} = 2$$

and hence the system is observable.

Remark. From engineering intuition, it is obvious that by simply measuring the speed of a rotating machine one can not identify the initial state.

(E2): Let us consider the model

$$\begin{aligned} J\ddot{\Theta} + b\dot{\Theta} + k\Theta &= T \equiv u, \\ y &= \dot{\Theta}. \end{aligned} \tag{5.3.36}$$

This represents a cylindrical beam, one end of which is rigidly attached to a vertical wall so that the beam is horizontal. J denotes the moment of inertia about the central axis, b represents friction and k denotes the stiffness of the beam. Strictly speaking this represents a particular mode. Here Θ measures the angular displacement of the free end due to applied torque. The reader may easily verify that in this case the rank is two, unlike the example 1. Thus by simply measuring the displacement rate one can identify the initial state $\{\Theta(0).\dot{\Theta}(0)\}$. This was not possible for a rotating machine.

(E3): Consider an electrical circuit which consists of a parallel LC branch one end of which is connected to one terminal of a voltage source while the other end is connected in series with a resistor (R) which terminates at the other terminal of the source. Let $u(t), t \geq 0$, denote the source voltage and i the current in the inductor and v the voltage across the capacitor. The system equation is given by

$$\begin{aligned} R(C(d/dt)v + i) + v &= u \\ L(d/dt)i &= v. \end{aligned}$$

Defining the state as $x_1 = v$ and $x_2 = i$, the state space model is given by

$$(d/dt)\begin{pmatrix} x_1 \\ x_2 \end{pmatrix} = \begin{pmatrix} -(1/RC) & -(1/C) \\ (1/L) & 0 \end{pmatrix}\begin{pmatrix} x_1 \\ x_2 \end{pmatrix} + \begin{pmatrix} (1/RC) \\ 0 \end{pmatrix} u. \tag{5.3.37}$$

If we observe the voltage across the capacitor we have

$$y = v = Hx = \begin{pmatrix} 1 & 0 \end{pmatrix} \begin{pmatrix} x_1 \\ x_2 \end{pmatrix}. \tag{5.3.38}$$

Computing H' and $A'H'$ one can easily check that

$$\text{Rank}\{H', A'H'\} = \text{Rank} \begin{pmatrix} 1 & -(1/RC) \\ 0 & -(1/C) \end{pmatrix} = 2.$$

Hence, given the generator voltage, from the knowledge of the voltage across the capacitor one can identify the initial state and hence the entire state trajectory. The reader may verify that the same conclusion is valid if one observes only the current in the inductor. One can make other choices for observation like the voltage across the series resistor and find that the system is observable. The reader is encouraged to do this problem.

(E4): Consider the Rabbit-Fox population

$$\begin{aligned} \dot{R} &= \alpha R - \beta F \\ \dot{F} &= -\gamma F + \delta R - u, \end{aligned} \tag{5.3.39}$$

where u is the control which may be zero or a positive function of time determining the removal rate of foxes. The conservationist wants to preserve the rabbit population by controlled removal of the predator population, in this case fox. In the absence of any control action he wants to determine the fox population in the habitat. The foxes are rather foxy and it is difficult to enumerate them since they can hide. The rabbits are gentle and it is easier to find them and hence enumerate them. The question is: by observing the rabbit population is it possible to determine the fox population? Clearly according to our theory, it suffices to determine the initial population and then determine the fox population from the state equation. This is an observability question. So here the output is given by

$$y = R = Hx = \begin{pmatrix} 1 & 0 \end{pmatrix} \begin{pmatrix} x_1 \\ x_2 \end{pmatrix} = \begin{pmatrix} 1 & 0 \end{pmatrix} \begin{pmatrix} R \\ F \end{pmatrix}.$$

For the rank we have

$$\text{Rank}\{H', A'H'\} = \text{Rank}\begin{Bmatrix} 1 & \alpha \\ 0 & -\beta \end{Bmatrix} = 2.$$

Thus the system is observable and hence from the count of the rabbit population we can exactly determine the fox population given the control.

(E5): Population count is sometimes very difficult and costly. On the other hand, the population growth rate may be easier to determine from the observation of the growth rate in a particular region of the habitat. The underlying assumption here is that the growth rate in a particular region is the same as for the entire habitat. This is reasonably true if the opportunities and the climate are uniform throughout the habitat. Here we take the output as

$$y = \dot{R} = \alpha R - \beta F = Hx = \begin{pmatrix} \alpha & -\beta \end{pmatrix} \begin{pmatrix} x_1 \\ x_2 \end{pmatrix} = \begin{pmatrix} \alpha & -\beta \end{pmatrix} \begin{pmatrix} R \\ F \end{pmatrix}.$$

In this case, if $\alpha\gamma \neq \beta\delta$, we have

$$\text{Rank}\{H', A'H'\} = \text{Rank}\begin{Bmatrix} \alpha & \alpha^2 - \beta\delta \\ -\beta & \beta(\gamma - \alpha) \end{Bmatrix} = 2$$

and therefore, the system is observable. However, if $\alpha\gamma = \beta\delta$, then the rank is one and hence the system is not observable.

Remark. These examples clearly indicate that the observability of a system is dictated by the combined interaction of the system matrix A and the measurement matrix H. Given the matrix A, the system may be observable for one choice of measurement strategy, H, and not for another. Thus one may consider the following interesting problem. Let $M(m \times n)$ denote the space of $m \times n$ real matrices and define

$$(A|H) \equiv (H', A'H', (A')^2 H', \cdots, (A')^{n-1} H')$$

and the set

$$\mathcal{H} \equiv \{H \in M(m \times n) : \text{Rank}(A|H) = n\}.$$

Clearly this is a proper subset of the vector space $M(m \times n)$ and if this is empty, there exists no (linear) measurement strategy in this class that makes the system observable.

5.4 Linear Time Varying Systems

In this section we present the necessary and sufficient conditions for observability of linear time varying systems given by

$$\dot{x}(t) = A(t)x(t) + B(t)u(t), s \le t \le T, \tag{5.4.40}$$

$$y(t) = H(t)x(t), t \in I_{s,T} \equiv [s, T]. \tag{5.4.41}$$

In fact some of these conditions are very similar to those of the time invariant system.

Definition 5.4.1. The system (5.4.40), (5.4.41) is said to be observable over the time period $I_{s,T}$ if, from the input and the output data $u(t), t \in I_{s,T}, y(t), t \in I_{s,T}$, respectively, the initial state (and hence the entire state trajectory) is identifiable.

From chapter 2 we know that the solution of equation (5.4.40), corresponding to any initial state $x(s) = \xi$, is given by

$$x(t) = \Phi(t,s)\xi + \int_s^t \Phi(t,\theta)B(\theta)u(\theta)d\theta, t \in I_{s,T}. \tag{5.4.42}$$

Corresponding to an arbitrary $\xi \in R^n$, considered as the initial state, define

$$\begin{aligned} \tilde{y}(t,\xi) &= y(t,\xi) - \int_s^t H(t)\Phi(t,\theta)B(\theta)u(\theta)d\theta \\ &= H(t)\Phi(t,s)\xi, \end{aligned} \tag{5.4.43}$$

where $y(t,\xi)$ denotes the output corresponding to the initial state ξ and the fixed but arbitrary control input u.

As in the time invariant case, we introduce the observability matrix $Q_T(s)$ as follows:

$$Q_T(s) \equiv \int_s^T \Phi^{'}(t,s)H^{'}(t)H(t)\Phi(t,s)dt. \tag{5.4.44}$$

In the time varying case it is quite conceivable that a system may be observable over one period of time and not so on another period of time. This is because both the intrinsic characteristics of the system determined by the matrix $A(t)$ and the measurement strategy influenced by $H(t)$ are time varying. Hence the observability matrix is truly a function of the intervals.

Theorem 5.4.2. The system (5.4.40, 5.4.41) is observable over the time interval $I_{s,T}$ if and only if the observability matrix $Q_T(s)$ is positive definite.

Proof. First we prove that Observability $\Rightarrow Q_T(s) > 0$. Since the system is observable, two distinct initial states produce two distinct outputs. Thus for $\xi \neq \eta$, $\tilde{y}(t,\xi) \not\equiv y(t,\eta)$. Hence

$$\begin{aligned} &\int_s^T \| \tilde{y}(t,\xi) - y(t,\eta) \|^2 dt > 0, \forall \xi \neq \eta \\ \Rightarrow &\int_s^T \| H(t)\Phi(t,s)(\xi - \eta) \|^2 dt > 0, \forall \xi \neq \eta, \end{aligned}$$

$$\Rightarrow \int_s^T \{(\Phi'(t,s)H'(t)H(t)\Phi(t,s)(\xi-\eta),(\xi-\eta))\}dt > 0, \forall \xi \neq \eta,$$
$$\Rightarrow (Q_T(s)(\xi-\eta),(\xi-\eta)) > 0, \forall \xi \neq \eta. \tag{5.4.45}$$

This shows that the observability matrix is necessarily positive definite whenever the system is observable. For the converse, we prove that strict positivity of $Q_T(s)$ implies observability. We demonstrate this by contradiction. Suppose $Q_T(s)$ is positive definite but the system is not observable. This means that there exist two distinct initial states producing identical outputs. That is, $\tilde{y}(t,\xi) \equiv \tilde{y}(t,\eta), t \in I_{s,T}$. Retracing the steps in equation (5.4.45), it is clear that this implies

$$(Q_T(s)(\xi-\eta),(\xi-\eta)) = 0.$$

Since $Q_T(s)$ is positive definite this implies that $\xi = \eta$. This is a contradiction of our hypothesis on the distinctness of ξ and η. Hence the system is observable, given that the observability matrix is positive definite. This completes the proof. □

Corollary 5.4.3. The observability matrix $Q_T(t)$ considered as a function of the starting time $t \in I_{s,T}$ satisfies the following matrix differential equation.

$$\dot{Q}_T(t) = -A'(t)Q_T(t) - Q_T(t)A(t) - H'(t)H(t), t \in [s,T),$$
$$Q_T(T) = 0. \tag{5.4.46}$$

Proof. Clearly it follows from the expression (5.4.44) that

$$Q_T(t) \equiv \int_t^T \Phi'(\theta,t)H'(\theta)H(\theta)\Phi(\theta,t)d\theta. \tag{5.4.47}$$

Taking derivative of this function with respect to t and using the properties of the transition operator $\Phi(t,\tau), \tau \leq t$, one obtains the differential expression. The terminal condition follows simply from the integral defining the observability matrix. This completes the proof. □

Solving the equation (5.4.46) backward in time we obtain the original observability matrix $Q_T(s)$. Unfortunately, unlike the linear time invariant system, there is no rank condition for time varying systems. Thus one may have to use a numerical technique to determine observability. Here we wish to explore this possibility. Given that the initial time s is fixed, we may consider Q as a function of the terminal time, that is, $Q_t(s), t \geq s$. Our problem is to find if there exists a time $t = \tau > s$ such that $Q_\tau(s)$ is positive definite. Once we find such a time the problem is solved and the system is observable. Towards this goal,we introduce the function

$$g(t) \equiv \inf_{\|\xi\|=1} (Q_t(s)\xi, \xi). \tag{5.4.48}$$

As seen below, this is the smallest eigen value of the matrix $Q_t(s)$.

Corollary 5.4.4. The function $t \longrightarrow g(t), t \geq s$, is the smallest eigen value of the matrix $Q_t(s)$ and it is a real valued, nonnegative and nondecreasing function of $t \geq s$ and it has a limit $r \in [0, \infty]$, that is

$$\lim_{t \to \infty} g(t) = r.$$

Proof. That $g(t)$ is the smallest eigen value of the matrix $Q_t(s)$ follows from the definition of eigen values of nonnegative matrices. Since $Q_t(s)$ is a real symmetric matrix, all its eigen values must be real and nonnegative and consequently $g(t)$ is real and nonnegative. One can also prove this by using Lagrange multiplier technique and minimizing the function

$$F(\xi, \lambda) \equiv (Q_t(s)\xi, \xi) + \lambda(1 - \| \xi \|^2)$$

without constraints. Setting the derivatives to zero one finds that

$$Q_t(s)\xi = \lambda\, \xi \tag{5.4.49}$$

$$\| \xi \| = 1. \tag{5.4.50}$$

The set of pairs $\{\lambda, \xi\}$ that satisfy these equations are precisely the eigen values and the normalized eigen vectors. Scalar multiplying by ξ on either side of the first equation, one obtains

$$\lambda = \frac{(Q_t(s)\xi, \xi)}{\| \xi \|^2} = (Q_t(s)e, e), \| e \| = 1.$$

Hence $g(t) = \inf_{\|e\|=1}(Q_t(s)e, e)$ is the smallest λ that satisfies the eigen value problem. We now show that g is a nondecreasing function of t. Indeed by definition of g we have

$$g(t) = \inf_{\|\xi\|=1} \int_s^t \| H(\theta)\Phi(\theta, s)\xi \|^2 \, d\theta.$$

Taking the infimum under the integral sign we have

$$g(t) = \inf_{\|\xi\|=1} \int_s^t \| H(\theta)\Phi(\theta, s)\xi \|^2 \, d\theta \geq \int_s^t \inf_{\|\xi\|=1} \| H(\theta)\Phi(\theta, s)\xi \|^2 \, d\theta. \tag{5.4.51}$$

Since the integrand under the last integral sign is nonnegative, it follows from the above that $g(t)$ is a monotone nondecreasing function of time. Any such function must have a limit in the extended real number system. But since $g(s) = 0$ and g is nondecreasing for $t \geq s$, the limit r must belong to $\bar{R}_0 = [0, \infty]$. This completes the proof. □

It is interesting to note that if $r = 0$, the system is never observable after time s. If on the other hand $r > 0$, then it follows from continuity of g that there exists a finite time T such that the system is observable over the time horizon $[s, T]$.

Numerical Algorithms for Observability. Here we present two algorithms for determining the observability of time varying systems.

Algorithm A

Step 1: Solve the matrix differential equation (5.4.46) backward in time for $Q_T(s)$.
Step 2: Find $\inf_{\|\xi\|=1}(Q_T(s)\xi, \xi) \equiv g(T)$. For the minimization, one may use khun-Tucker technique.

Step 3: If $g(T) > 0$, stop; the system is observable over the time interval $[s, T]$. If not, set $T \longrightarrow T + \alpha$ for some α positive and repeat steps 1-3.

Algorithm B
Step 1: Solve the following equations

$$\dot{Y}(r) = -Y(r)A(r), Y(0) = I, 0 \leq r \leq s, \tag{5.4.52}$$

$$\dot{X}(r) = A(r)X(r), X(s) = Y^{-1}(s), s \leq r \leq t. \tag{5.4.53}$$

Step 2: Compute

$$\Phi(\theta, s) = X(\theta)Y(s), s \leq \theta \leq t, \tag{5.4.54}$$

$$Q_t(s) = \int_s^t \Phi'(\theta, s)H'(\theta)H(\theta)\Phi(\theta, s)d\theta. \tag{5.4.55}$$

Step 3: Minimize $(Q_t(s)\xi, \xi)$, over $\| \xi \| = 1$, giving $g(t)$.

Step 4: If $g(t) > 0$, stop, the system is observable over the period $[s, t]$. If not, repeat the steps 1-3 by incrementing t to $t + \alpha$ for any suitable positive α.

5.5 Input Identification

In Sections 2 and 4 we have considered identification of state from input and output data. Now we consider the problem of identifying or detecting the input u from the initial state and the output data y. This is the so called inverse problem. We consider the time invariant system. It follows from the system equations (5.2.1), (5.2.2) that

$$\tilde{y}(t,u) \equiv y(t) - He^{tA}x_0 = \int_0^t He^{(t-s)A}Bu(s)ds, t \in I_T \equiv [0,T].$$

Here the output y and the initial state x_0 are known; the unknown is the input u. The most natural space of control strategies is the linear space of essentially bounded measurable functions taking values from R^d. This is usually denoted by $L_\infty(I_T, R^d)$. Any two elements of this space that differ only on a set of measure zero are identified as one and the same or equivalent. This is reasonable not only theoretically but also practically. With a little reflection, the reader can see that any two inputs which are bounded measurable and differ only on a set of isolated points or a set of measure zero produce the same output. Thus some times one partitions this space into equivalence classes and ignores the original space. Hence one may keep using the same notation for the original space as well as the quotient space, that is the space of equivalence classes. We take $L_\infty(I_T, R^d)$ as the space of admissible controls. Another equally important class of controls is the class of finite energy controls denoted by the Hilbert space $L_2(I_T, R^d)$.

Definition 5.5.1. The system input is said to be detectable (identifiable), or the inverse problem has a solution, if the input $u \in L_\infty(I_T, R^d)$ can be uniquely determined (unique up to equivalence class) from the given data $\{x_0, y(t), t \in I_T\}$.

In fact a more relaxed definition for identifiability or detectability is obtained by only requiring that the input-output map is 1-1.

Theorem 5.5.2. Consider the system represented by the equations, (5.2.1) and (5.2.2). The input is detectable if and only if

$$\text{Rank}\{B'H', B'A'H', \cdots, B'(A')^{n-1}H'\} = d. \tag{5.5.56}$$

Proof. First we prove that if the rank condition holds then the system input is identifiable. We demonstrate this by contradiction. Suppose the rank condition holds but the system input is not identifiable. This means that there exist two distinct inputs, u^1 and u^2 which produce identical outputs. That is,

$$\tilde{y}(t, u^1 - u^2) \equiv \tilde{y}(t, w) \equiv 0, t \in I_T.$$

Thus

$$\tilde{y}(t, w) \equiv \int_0^t He^{(t-s)A} Bw(s) ds = 0, \ \forall \ t \in I_T.$$

This is possible for every $t \in I_T$ if and only if

$$He^{(t-s)A} Bw(s) = 0, \text{for almost all } s \in [0, t] \text{ and every } \ t \in I_T. \quad (5.5.57)$$

Since the rank condition given by (5.5.56) is equivalent to the condition

$$\text{Ker}\{He^{\tau A}B, \tau \in I_T\} = \{0\},$$

(for proof see the hints given in (P3) (5.6)) it follows from (5.5.57) that $w(s) = 0$, for almost all $s \in [0, t]$ for every $t \in I_T$. Hence $u^1 = u^2$. This contradiction proves the assertion. In other words, the system input is detectable or identifiable whenever the rank condition holds.

For the necessary condition, we prove that the rank condition holds given that the system input is detectable. Since the input is detectable,

$$\left\{ \int_0^t He^{(t-s)A} Bu(s) ds = 0, \forall t \in I_T \right\} \Rightarrow u \equiv 0. \quad (5.5.58)$$

This is equivalent to

$$\left\{ \int_0^t He^{sA} Bv(s) ds = 0, \forall t \in I_T \right\} \Rightarrow v \equiv 0, \quad (5.5.59)$$

which, in turn, is equivalent to

$$\left\{ \int_0^t (v(s), B' e^{sA'} H' \xi) ds = 0, \forall t \in I_T, \forall \xi \in R^m \right\} \Rightarrow v \equiv 0. \quad (5.5.60)$$

Again, this last statement is equivalent to

$$\{(v(s), B' e^{sA'} H' \xi) = 0, \text{for almost all } \ s \in I_T, \forall \xi \in R^m\} \Rightarrow v(s) = 0 \ a.e. \quad (5.5.61)$$

Hence (neglecting a set of measure zero) this is equivalent to

$$\{(\eta, B' e^{tA'} H' \xi) = 0, \text{for all } t \in I_T, \forall \xi \in R^m\} \Rightarrow \eta = 0. \quad (5.5.62)$$

This means that

$$\text{Range}\{B'e^{tA'}H', t \in I_T\} = R^d.$$

Using Caley-Hamilton theorem and following similar steps as in the proof of the rank condition for observability, it follows from this expression that

$$\text{Range}\{B'(A')^k H', k = 0, 1, 2, \cdots, n-1\} = R^d. \tag{5.5.63}$$

In other words, $Rank\{B'(A')^k H', k = 0, 1, 2, \cdots, n-1\} = d$. Thus the rank condition holds whenever the input is identifiable. This ends the proof. □

Using this result and an additional assumption we can show that one can identify the input from the initial state and the output data at the terminal time T. This is certainly a surprising result. Define

$$\tilde{y}(T) \equiv y(T) - He^{TA}x_0 \tag{5.5.64}$$

and the operator L from $L_2(I_T, R^d)$ to R^m by

$$Lu \equiv \int_0^T He^{(T-s)A}Bu(s)ds. \tag{5.5.65}$$

Here we may allow the Hilbert space $L_2(I_T, R^d)$ to be the space of admissible controls containing $L_\infty(I_T, R^d)$. By simple change of the variable of integration, we have an equivalent definition of the operator L given by

$$Lv \equiv \int_0^T He^{\tau A}Bv(\tau)d\tau, \quad v \in L_2(I_T, R^d). \tag{5.5.66}$$

Since $y(T)$ is the output, it is obvious that there is at least one input $u \in L_2(I_T, R^d)$ so that $\tilde{y}(T) = Lu$. Our problem is to find conditions under which this is unique. Clearly if u is the actual input,

$$v \longrightarrow q(v) \equiv \| \tilde{y}(T) - Lv \|_{R^m}^2$$

attains its minimum at $v = u$, and $q(u) = 0$. Writing the quadratic form q as

$$q(v) = (\tilde{y}(T) - Lv, \tilde{y}(T) - Lv) = \| \tilde{y}(T) \|^2 + (L^*Lv, v) - 2(L^*\tilde{y}(T), v),$$

and differentiating this (in the sense of Gateaux), it follows from the symmetry of the operator L^*L that u must be a solution of the equation

$$L^*\tilde{y}(T) = L^*Lv. \tag{5.5.67}$$

Note that L is a bounded linear operator from $L_2(I_T, R^d)$ to R^m and thus its adjoint, denoted by L^*, is a bounded linear map from R^m to $L_2(I_T, R^d)$ and is given by

$$(L^*\xi)(t) = B^{'}e^{tA^{'}}H^{'}\xi, \quad t \in I_T, \xi \in R^m. \tag{5.5.68}$$

Clearly the operator L^*L mapping $L_2(I_T, R^d)$ to $L_2(I_T, R^d)$ is given by

$$(L^*Lv)(t) = \int_0^T (B^{'}e^{tA^{'}}H^{'}He^{sA}B)v(s)ds. \tag{5.5.69}$$

Thus equation (5.5.67) can be written as

$$w(t) = \int_0^T R(t,s)v(s)ds, t \in I_T \tag{5.5.70}$$

where

$$w(t) \equiv (L^*\tilde{y}(T))(t), t \in I_T \tag{5.5.71}$$

$$R(t,s) \equiv (B^{'}e^{tA^{'}}H^{'}He^{sA}B), t, s \in I_T \times I_T, \tag{5.5.72}$$

are given and we seek for u that solves the equation (5.5.70). Clearly $w \in L_2(I_T, R^d)$ and $R \in L_2(I_T \times I_T, M(d \times d))$ where $M(d \times d)$ denotes the space of d-dimensional square matrices. Further note that the kernel is symmetric, that is,

$$R^{'}(t,s) = R(s,t), (t,s) \in I_T \times I_T.$$

In fact under our situation, both w and R are continuous. The equation (5.5.70) is the celebrated Fredholm integral equation of the first kind and our task is to invert the associated operator if it exists. This is stated in the following theorem.

Theorem 5.5.3. Suppose the rank condition of Theorem 5.5.2 holds. Then for every $w \in L_2(I_T, R^d)$ the integral equation (5.5.70) has a unique solution $u \in L_2(I_T, R^d)$. Further, the input u can be constructed in terms of the eigen values and eigen functions of the integral operator (5.5.70) as

$$u = \sum_{k=1}^{\infty}(w_k/\lambda_k)\phi_k$$

where $\{w_k\}$ are the Fourier coefficients of w with respect to the eigen functions $\{\phi_k\}$ of the kernel R with $\{\lambda_k\}$ being the corresponding eigen values.

Proof. Since $R(t,s) = R'(s,t)$ for $(s,t) \in I_T \times I_T$, it is clear that the operator $\mathcal{R}$ defined by

$$(\mathcal{R}v)(t) \equiv \int_{I_T} R(t,s)v(s)ds, t \in I_T,$$

is a self adjoint operator in the Hilbert space $E \equiv L_2(I_T, R^d)$. Since

$$(\mathcal{R}\phi, \phi) = \parallel \int_0^T He^{tA}B\phi(t)dt \parallel^2 \geq 0,$$

$\mathcal{R}$ is a positive self adjoint operator. It follows from the rank condition that this operator is actually positive definite. It is known that any positive self adjoint operator has a nonempty spectrum consisting of a countable set of eigen values. By virtue of self adjointness these eigen values are all real and due to positive definiteness, they are all strictly positive. Let us denote the eigen values by $\{\lambda_k\}$ and the corresponding eigen functions by $\{\phi_k\} \subset E$. These eigen functions are orthogonal and we may consider them normalized. Without loss of generality we may arrange these eigen values in the order of increasing magnitude setting

$$0 < \lambda_1 \leq \lambda_2 \leq \lambda_3, \cdots, \leq \lambda_k, \cdots,$$

and repeating them according to their multiplicity. By virtue of Hilbert-Schmidt theorem for symmetric L_2-kernels, the Kernel has the representation

$$R(t,s) = \sum_{k=1}^{\infty} \lambda_k \phi_k(t) \otimes \phi_k(s). \tag{5.5.73}$$

Further, any $w \in E$ given by an expression like (5.5.70) can be expressed using the eigen functions by the sum

$$w = \sum_{k=1}^{\infty} w_k \phi_k \tag{5.5.74}$$

where $\{w_k\}$ are the Fourier coefficients of w with respect to the family $\{\phi_k\}$. In case R is merely an L_2 kernel, the infinite series converges almost uniformly. In our case, the kernel R, as well as the function w, is continuous and hence both the series converge absolutely uniformly. Substituting these expressions in equation (5.5.70) one finds that for every integer k,

$$w_k = \lambda_k u_k$$

where $\{u_k = (u, \phi_k)\}$ are the Fourier coefficients of u with respect to the family $\{\phi_k\}$. For the existence of an L_2 solution, it is necessary and sufficient that

$$\sum_{k=1}^{\infty}(u_k)^2 = \sum_{k=1}^{\infty}(w_k/\lambda_k)^2 < \infty$$

and in this case the solution is given by

$$u = \sum_{k=1}^{\infty}(w_k/\lambda_k)\phi_k.$$

Since the smallest eigen value is strictly positive, we have

$$\sum_{k=1}^{\infty}(w_k/\lambda_k)^2 < (1/\lambda_1)^2\sum_{k=1}^{\infty}w_k^2 < \infty.$$

Hence we have an L_2 solution u and

$$\| u \|_E \leq (1/\lambda_1) \| w \|_E .$$

This result shows that

$$\| \mathcal{R}u \|_E \ \geq \lambda_1 \| u \|_E,$$

and this implies also uniqueness of solution. In other words the operator $\mathcal{R}$ has a bounded inverse in E and the solution $u = \mathcal{R}^{-1}w$. This ends the proof. □

Given the initial state, we see that by observing the output simply at the final time T, we are able to identify the input over the entire period $[0, T]$. This is what the previous theorem tells us. One obviously expects that if one can identify the input from observation of the output at one point (terminal time) only, one must be able to do so given the output data for the entire period. In this case the output y, the operator L, the adjoint operator L^* and the kernel R are given by

$$\tilde{y}(t) \equiv y(t) - He^{tA}x_0, t \in I_T, \tag{5.5.75}$$

$$(Lu)(t) \equiv \int_0^t He^{(t-s)A}Bu(s)ds, t \in I_T, \tag{5.5.76}$$

$$(L^*\eta)(t) \equiv \int_t^T B'e^{(s-t)A'}H'\eta(s)ds, t \in I_T \tag{5.5.77}$$

$$R(t,\tau) \equiv \int_{t\vee\tau}^T dsB'e^{(s-t)A'}H'He^{(s-\tau)A}B, (t,\tau) \in I_T \times I_T. \tag{5.5.78}$$

Following the same procedure we arrive at the Fredholm integral equation of the first kind as before

$$w(t) = \int_0^T R(t,s)u(s)ds,$$

where now w is given by

$$w(t) \equiv \int_t^T B'e^{(s-t)A'}H'\tilde{y}(s)ds.$$

Again, as before, one can verify that the operator $\mathcal{R}$ determined by the symmetric kernel R is a positive self adjoint operator in the Hilbert space $E \equiv L_2(I_T, R^d)$ and under the assumptions of Theorem 5.5.3, the Fredholm equation, written in the abstract form $w = \mathcal{R}u$, has a unique solution. In other words Theorem 5.5.3 remains valid also in this case.

Remark. For the time invariant case notice that $w \in C(I_T, R^d)$ and $R \in C(I_T \times I_T, M(d \times d))$. Thus one may expect a solution $u \in C(I_T, R^d)$.

Remarks on Time-varying Systems. For time-varying systems one can develop a similar result under assumptions which are not easily verifiable as in the time-invariant case where the rank condition has been used. In this case the kernel of the Fredholm operator is given by

$$R(t,s) \equiv B'(t)\Phi'(T,t)H'(T)H(T)\Phi(T,s)B(s), \quad t,s \in I_T \times I_T$$

where Φ is the transition operator corresponding to the system matrix $\{A(t), t \in I_T\}$. We have the following result.

Corollary 5.5.4. Suppose

$$Ker\{H(T)\Phi(T,t)B(t), t \in I_T\} = \{0\}.$$

Then, from the output $y(T)$ and the initial state x_0, the input u can be uniquely identified.

Proof. Note that, under the given assumption, the kernel R is real symmetric and positive definite and therefore the Fredholm operator $\mathcal{R}$ given by

$$(\mathcal{R}u)(t) \equiv \int_{I_T} R(t,s)u(s)ds$$

is a selfadjoint positive definite operator in the Hilbert space $E \equiv L_2(I_T, R^d)$. Hence by Theorem 5.5.3, the time-varying system is also identifiable in the input-output sense. □

Identification of Impulsive Inputs.

Time Invariant Case: Using the above results we can also identify impulsive inputs. Given the initial state x_0 and the output $y(T)$, we know $\tilde{y}(T)$ is given by (5.5.64). Let u denote the control that produces this output. Then using the operator L as given by (5.5.65), we have

$$\tilde{y}(T) = Lu = \int_0^T He^{(T-s)A}Bu(s)ds.$$

Since the kernel of this integral operator is continuous, impulsive inputs are admissible. Suppose the input u is given by a single impulsive force that acts at time $\tau \in (0,T)$ as described below

$$u(t) = z\delta(t-\tau),$$

for some $z \in R^d$. If the rank condition of Theorem 5.5.2 holds, then we can determine z as follows. Substituting this input we obtain

$$\tilde{y}(T) = He^{(T-\tau)A}Bz.$$

Premultiplying on either side of this expression by the matrix $B'e^{(T-\tau)A'}H'$ we obtain

$$B'e^{(T-\tau)A'}H'\tilde{y}(T) = (B'e^{(T-\tau)A'}H'He^{(T-\tau)A}B)z.$$

By virtue of the rank condition, the matrix

$$F \equiv (B'e^{(T-\tau)A'}H'He^{(T-\tau)A}B)$$

is nonsingular and so invertible. The result is

$$z = F^{-1}B'e^{(T-\tau)A'}H'\tilde{y}(T).$$

In fact one can also identify impulsive inputs of the form

$$u(t) = \sum_{k=1}^{\ell} z_k\delta(t-\tau_k), \tau_k \in (0,T), z_k \in R^d,$$

provided ℓ is a finite positive integer. Here we have tacitly assumed that the times of occurrence of the impulses are known. In some problems, this may also be an unknown variable to be identified. This is considered below.

Time Varying Case: Consider the linear system with impulsive input given by

$$dx(t) = A(t)x(t)dt + B(t)u(dt), t \in I$$

with output given by

$$y(t) = H(t)x(t), t \in I.$$

Suppose we observe the output y over the period of time $I \equiv [0, T]$. Then for $t \in I$ and $(s, \theta) \in I \times I$, we have

$$\tilde{y}(t) = y(t) - H(t)\Phi(t, 0)x_0, \tag{5.5.79}$$

$$w(t) \equiv (L^*\tilde{y})(t) = \int_t^T dr\{B'(t)\Phi'(r,t)H'(r)\tilde{y}(r)\}, \tag{5.5.80}$$

$$R(s, \theta) \equiv \int_{s \vee \theta}^T dr\{B'(s)\Phi'(r,s)H'(r)H(r)\Phi(r,\theta)B(\theta)\}. \tag{5.5.81}$$

Again for identification of impulsive inputs, we must solve the Fredholm equation of the first kind driven by vector valued measure

$$w(t) = \int_0^T R(t,s)u(ds), t \in I, \tag{5.5.82}$$

where the Kernel R and the function w are known. Clearly the question of detectability or identifiability of the input which is a vector measure in this case, is equivalent to unique solvability of equation(5.5.82). Using similar approach as in Theorem 5.5.3 we can prove the following result.

Theorem 5.5.5. Suppose the matrices B and H are continuous and the Fredholm operator $\mathcal{R}$ determined by the Kernel $R(t,s), (t,s) \in I \times I$ is self adjoint and positive definite. Then the input is identifiable.

Proof. Clearly under the given assumptions, it follows from the same procedure as in Theorem 5.5.3, that the input measure must satisfy the following relation

$$w_i/\lambda_i = \int_0^T < \phi_i(t), u(dt) >, \text{for all } i \in N, \tag{5.5.83}$$

where λ_i and ϕ_i are the eigen values and the eigen functions of the operator $\mathcal{R}$ corresponding to the kernel R. Since the kernel R is continuous the eigen functions are also continuous. This is very similar to an infinite system of moment equations. Here every thing is known except the vector measure u. We claim that this expression determines the input measure uniquely. We prove this by contradiction. Suppose there are two measures u and v that satisfy the same expression (5.5.83). Then we must have

$$0 = \int_0^T < \phi_i(t), u(dt) - v(dt) >, \ \forall\ i \in N. \tag{5.5.84}$$

Since any function $\psi \in C(I, R^d)$ can be uniformly approximated by a linear combination of the eigen functions $\{\phi_i\}$, it follows from the above identity that

$$0 = \int_0^T < \psi(t), u(dt) - v(dt) >, \ \forall\ \psi \in C(I, R^d). \tag{5.5.85}$$

This is impossible unless $u = v$.

Nonlinear systems: Identification or detection of inputs for nonlinear systems is rather difficult. Consider the impulsive system with output given by y,

$$dx(t) = f(t, x(t))dt + G(t, x(t-))u(dt), x(0) = x_0, t \in I, \tag{5.5.86}$$
$$y(t) = H(x(t)), t \in I, u \in \mathcal{M}(I, R^d). \tag{5.5.87}$$

Let $\mathcal{S}$ denote the solution map $u \longrightarrow x(u) \equiv \mathcal{S}(u)$ from $\mathcal{M}(I, R^d)$ to the space of trajectories $B(I, R^n)$ and $\mathcal{H}$ the Nemytski operator corresponding to the map H, that is $\mathcal{H}(x)(t) = H(x(t)), t \in I$. Recall that $B(I, R^n)$ denotes the space of bounded measurable functions on I with values in R^n. Define the composition map $F \equiv \mathcal{H}o\mathcal{S}$ which maps $\mathcal{M}(I, R^d)$ to $B(I, R^m)$. The objective is to determine u from the observation of the output over the interval I. This problem is also known as the inverse problem and generally it is ill posed. However if one looks for an admissible input having minimum norm, the problem may be well posed. Let $\mathcal{U}_{ad} \subset \mathcal{M}(I, R^d)$ denote the class of admissible controls and

$$Y_{ad} = F(\mathcal{U}_{ad}) \equiv \{y \in B(I, R^m) : y = F(u) \ \text{ for some } u \in \mathcal{U}_{ad}\}.$$

For each $y^* \in Y_{ad}$, the set (inverse map)

$$F^{-1}(y^*) \equiv \{u \in \mathcal{U}_{ad} : y^* = F(u)\} \tag{5.5.88}$$

is clearly a nonempty subset of the set of admissible controls. In general this is a multi valued map. Obviously if this turns out to be a single valued map the input is uniquely identifiable for any given output. Under certain assumptions we can prove that there exists an admissible control (input) of minimum norm corresponding to any given admissible output.

Theorem 5.5.6. Consider the system (5.5.86)-(5.5.87) and suppose f and G satisfy the assumptions of Corollary 3.6.7 and H is a continuous bounded map from R^n to R^m and the admissible inputs $\mathcal{U}_{ad}$ is a weakly compact subset of $\mathcal{M}(I, R^d)$. Then for every admissible output there exists an admissible control (input) of minimum (variation) norm.

Proof. We present a brief outline of the proof. Define the functional N given by the total variation norm, $N(u) \equiv \parallel u \parallel_V$ on $\mathcal{M}(I, R^d)$. Let $u^n \xrightarrow{w} u^o$. Since any norm is weakly lower semi continuous, we have

$$N(u^o) \leq \liminf_{n \to \infty} N(u^n).$$

We know that a weakly lower semi continuous functional on a weakly compact set attains its infimum. Thus it suffices to verify that the inverse map $F^{-1}(y)$ is weakly sequentially compact for any $y \in Y_{ad}$. Since $\mathcal{U}_{ad}$ is assumed to be weakly sequentially compact (see appendix, Theorem A.7.7) and $F^{-1}(y)$ is a subset of it, it suffices to verify that $F^{-1}(y)$ is weakly closed. Suppose $\{u^n\} \in F^{-1}(y)$ and that $u^n \xrightarrow{w} u^o$. Clearly $u^o \in \mathcal{U}_{ad}$, we must show that it is in $F^{-1}(y)$. Let $x^n \in B(I, R^n)$ denote the solution corresponding to u^n and x^o the solution corresponding to the control u^o. Clearly $y(t) = H(x^n(t)), t \in I$ for all positive integers n. Since H is continuous, it suffices to show that $x^n \xrightarrow{u} x^o$ in $B(I, R^n)$. Note that x^o, x^n satisfy the integral equations

$$x^o(t) = x_0 + \int_0^t f(s, x^o(s))ds + \int_0^t G(s, x_o(s-))u^o(ds), t \in I$$

and

$$x^n(t) = x_0 + \int_0^t f(s, x^n(s))ds + \int_0^t G(s, x_n(s-))u^n(ds), t \in I$$

respectively. Subtracting the second from the first equation we have

$$\begin{aligned} x^o(t) - x^n(t) = & \int_0^t [f(s, x^o(s)) - f(s, x^n(s))]ds \\ & + \int_0^t [G(s, x_o(s-)) - G(s, x^n(s-))]u^n(ds) \\ & + \int_0^t G(s, x^o(s-))(u^o(ds) - u^n(ds)), t \in I. \end{aligned}$$

Using this identity and generalized Gronwall inequality, one can verify that

$$\parallel x^o - x^n \parallel_{B(I,R^n)} \ \leq \ C \ \parallel \eta_n(T) \parallel$$

for all positive integers n where

$$\eta_n(T) \equiv \int_I G(s, x^o(s-))(u^o(ds) - u^n(ds)).$$

The constant C is given by

$$C = \exp\left\{\int_I K(s)ds + \ell \sup\{\| u^n \|_V\}\right\},$$

and it is positive and finite because the set $\mathcal{U}_{ad}$ is bounded. Since $x^o \in B(I, R^n)$ and $L(t), t \in I$, is uniformly bounded by ℓ, and u^n converges weakly to u^o, $\eta_n(T) \to 0$. Hence $x^n \xrightarrow{u} x^o$ and consequently $H(x^n(t)) \xrightarrow{u} H(x^o(t))$ on the interval I. Thus $u^o \in F^{-1}(y)$ proving that it is a weakly closed subset of a weakly compact set and so it is weakly compact. Thus N, being weakly lower semi continuous, attains its minimum on $F^{-1}(y)$. This completes the proof. □

Using similar approach we can prove the inverse problem (minimal-norm) for the system

$$\dot{x}(t) = f(t, x(t)) + G(t, x(t-))u(t), x(0) = x_0, t \in I, \quad (5.5.89)$$
$$y(t) = H(x(t)), t \in I, u \in \mathcal{U}_{ad} \subset L_\infty(I, R^d). \quad (5.5.90)$$

Here $\mathcal{U}_{ad}$ is assumed to be a weak star compact subset of $L_\infty(I, R^d)$. The reader is encouraged to carry out the details.

5.6 Exercises

(P1): Verify that the observability matrix Q_T can be computed by solving the matrix differential equation

$$\dot{M}_T(t) = -A' M_T(t) - M_T(t)A - H' H, t \in I_T,$$
$$M_T(T) = 0, \quad \text{and setting } Q_T = M_T(0). \quad (5.6.91)$$

(P2): Prove that the function g defined by equation (5.4.48) is continuous.
(P3): Prove that the rank condition

$$\text{Rank}\{B' H', B' A' H', \cdot, \cdot, B' (A')^{n-1} H'\} = d, \quad (5.6.92)$$

is equivalent to the condition

$$\text{Ker}\{He^{\tau A}B, \tau \in I_T\} = \{0\}.$$

Hints: The rank condition implies the range condition,

$$\bigcup_{k=0}^{n-1} B^{'}(A^{'})^k H^{'}(R^m) = R^d.$$

This is equivalent to

$$\{(\xi, HA^k B\eta) = 0, \ \forall \xi \in R^m, 0 \leq k \leq n-1\} \Rightarrow \eta = 0,$$

which, in turn, is equivalent to

$$\text{Ker}\{HA^k B, 0 \leq k \leq n-1\} = \{0\}.$$

By Caley-Hamilton theorem, this is equivalent to

$$\text{Ker}\{HA^k B, k \geq 0\} = \{0\}.$$

Then use the exponential series to express the function $t \longrightarrow He^{tA}B$ to obtain the result.

(P4): Construct the relevant system of algebraic equations whereby one can identify impulsive inputs occurring at a finite number of time instants on the open interval $(0, T)$.

(P5): Prove the inverse problem (minimal-norm) for the system

$$\dot{x}(t) = f(t, x(t)) + G(t, x(t-))u(t), x(0) = x_0, t \in I, \quad (5.6.93)$$

$$y(t) = H(x(t)), t \in I, u \in \mathcal{U}_{ad} \subset L_\infty(I, R^d). \quad (5.6.94)$$

Here $\mathcal{U}_{ad}$ is assumed to be a weak star compact subset of $L_\infty(I, R^d)$.

(P6): Consider the examples of Section 5.3 and suppose the initial states and the outputs are given. Using the rank condition of Theorem 5.5.2, verify if the inputs of the systems are identifiable.

5.7 Bibliographical Notes

Some of the materials presented in this chapter can be found in many books on control systems such as Ahmed [2], Brogan [23], Shinners [83], Kailath. Observability problem is a particular class of inverse problems or equivalently identification problems. Observability of the state given the input and the output is the well known classical result. Identifiability of the input from the knowledge of the initial state and the output appears to be a new addition here. Inverse problems for time varying systems and for impulsive systems are also new additions. In general inverse problems may not have unique solutions unless some additional constraints are imposed. By imposing minimum norm constraint one can identify input from the knowledge of initial state and the output. Another class of more interesting and difficult inverse problems involves identification of system matrices (or operators in case of infinite dimensional systems) from the knowledge of input and outputs. These kind of problems have been treated in the infinite dimensional context. Interested readers are referred to Ahmed [6, 154].

Chapter 6

Controllability and Stabilizability

6.1 Introduction

The concept of controllability is as important as observability. Given a system, the current state (initial state), the target state, and the admissible inputs, and the time horizon, the question is, can we find a control policy from the given class that will drive the system from the current state to the target or goal state. This kind of problem is encountered in many physical situations. For example, in government planning the control resources are known, the current state of the process (like: economy, health care, employment level etc.) is known, the problem is to find a control strategy that will achieve the government target within a specified period of time. This is the main problem we consider here. Another problem of significant interest is the question of stabilizability of unstable systems. This problem arises whenever a natural or a man made system is not stable to begin with and something can be done to stabilize it. We shall see that if the system is controllable then it is stabilizable by an appropriate state feed back control law. In case the full state is not accessible but an output is available, we shall see how compensators are designed based on the output feedback so as to achieve stability. These are the questions we discuss in this chapter.

6.2 Linear Time Invariant Systems

First we consider the linear time invariant system given by

$$\dot{x} = Ax + Bu, x(0) = x_0 \in R^n, t \in R_0 \equiv (0, \infty), \tag{6.2.1}$$

where x denotes the state trajectory corresponding to the initial state $x(0) = x_0$ and the control policy u taking values from R^d. Clearly the matrices must have compatible dimension, for example, A is a $n \times n$ matrix and B is a $n \times d$ matrix. We call this system $\mathcal{S}$ for convenience of reference. Let U be any suitable subset of R^d and define the class of admissible control strategies as

$$\mathcal{U}_{ad} \equiv \{u : u \text{ is measurable and } u(t) \in U, t \geq 0\}.$$

The basic problem is: given the initial (current) state $x_0 \in R^n$, the target state $x_1 \in R^n$ and possibly the time horizon $[0, T]$, is it possible to find a control policy $u \in \mathcal{U}_{ad}$ that takes the system from the current state to the goal state? This is precisely the controllability problem. The basic problem is the same also for nonlinear systems. For nonlinear systems the problem is more difficult. In addition to considering linear problems, here we present some interesting results also for nonlinear systems. The approach is based on Banach fixed point theorem.

Definition 6.2.1. The system $\mathcal{S}$ given by equation (6.2.1) and the admissible controls $\mathcal{U}_{ad}$, is said to be controllable at time T with respect to the pair $\{x_0, x_1\}$, if there exists a control strategy $u \in \mathcal{U}_{ad}$ over the time interval $I_T \equiv [0, T]$ so that the state attained at time T coincides with the given target state x_1. The system is said to be globally controllable over the period $[0, T]$ if it is controllable from any (every) initial state $x_0 \in R^n$ to any (every) target state $x_1 \in R^n$. It is said to be globally controllable if it is controllable for every triple $\{T, x_0, x_1\} \in R_0 \times R^n \times R^n$.

Controllability of a system can be viewed as it's capacity to reach out to desirable targets using available resources. This prompts us to introduce the concept of attainable or reachable sets.

Definition 6.2.2. For any time $\tau \geq 0$ and any $x_0 \in R^n$, the set $\mathcal{A}(\tau, x_0)$, given by

$$\mathcal{A}(\tau, x_0) \equiv \left\{\xi \in R^n : \xi = e^{\tau A} x_0 + \int_0^T e^{(\tau - s)A} Bu(s) ds, u \in \mathcal{U}_{ad}\right\}, \quad (6.2.2)$$

is called the attainable set at time $t = \tau$ starting from the state x_0 at time $t = 0$. These are the set of states that the system can attain at time τ. Equivalently one can define the reachable set as the set given by

$$\mathcal{R}(\tau) \equiv \left\{\xi \in R^n : \xi = \int_0^T e^{-sA} Bu(s) ds, u \in \mathcal{U}_{ad}\right\}. \quad (6.2.3)$$

The two sets are related by the following expression

$$e^{-\tau A} \mathcal{A}(\tau, x_0) = x_0 + \mathcal{R}(\tau), \tau \geq 0. \quad (6.2.4)$$

The only important difference between the two sets is that the reachable set $\mathcal{R}(\tau)$ is a nondecreasing set valued function of τ in the sense that

$$\mathcal{R}(\tau_1) \subset \mathcal{R}(\tau_2) \ \text{ for all } \tau_1 < \tau_2. \tag{6.2.5}$$

In case both the initial state x_0 and the final state x_1 are arbitrary one can define

$$z(t) \equiv z(t, x_0, x_1) \equiv x_1 - e^{tA}x_0, t \geq 0, \tag{6.2.6}$$

and pose the question if there is an admissible control u so that at some finite time, say $t = \tau$,

$$\int_0^\tau e^{(\tau-s)A}Bu(s)ds = z(\tau). \tag{6.2.7}$$

This problem is equivalent to the question of existence of a finite time τ so that $z(\tau) \in \mathcal{A}(\tau)$ where $\mathcal{A}(\tau)$ is given by

$$\mathcal{A}(\tau) \equiv \left\{z \in R^n : z = \int_0^\tau e^{(\tau-s)A}Bu(s)ds, u \in \mathcal{U}_{ad}\right\}. \tag{6.2.8}$$

These sets actually tell us how much the system can do given the admissible control resources. It is clear from the above expressions that in addition to the limitations imposed by control resources, there are intrinsic limitations due to the (algebraic) properties of the system matrix A and the control matrix B through which external forces are brought to bear upon the system. It is evident that if $\mathcal{A}(\tau) = R^n$, at some time τ, then the system is controllable from any initial state to any target state in time τ. In general, if

$$\bigcup_{t\geq 0} \mathcal{A}(t) = R^n$$

or equivalently

$$\bigcup_{t\geq 0} e^{tA}\mathcal{R}(t) = R^n,$$

the system is globally controllable (in finite time) otherwise not. Note that since $t \longrightarrow \mathcal{R}(t)$ is a monotone nondecreasing set valued function

$$\bigcup_{t\geq 0} \mathcal{R}(t) = \lim_{t\to\infty} \mathcal{R}(t).$$

Clearly if $\mathcal{A}(t) = R^n$, then $\mathcal{R}(t) = R^n$ also. If $\mathcal{A}(t) = R^n$ for every $t > 0$, the system is globally controllable (g.c). Temporarily, for global controllability we relax the constraint on the control set and take $U = R^d$, and the admissible control strategies $\mathcal{U}_{ad}$ as the space of bounded measurable and locally square integrable functions with values in R^d. Later in the sequel we study in details some useful topological and algebraic properties of this attainable set and its continuity with respect to time. Now we introduce the so called controllability matrix. Let $T > 0$, and define the matrix

$$C_T \equiv \int_0^T e^{(T-s)A} B B' e^{(T-s)A'} ds. \tag{6.2.9}$$

It is clear that the matrix C_T is symmetric and positive semidefinite. The following result relates controllability with the matrix C_T.

Theorem 6.2.3. The system is globally controllable (g.c) at time T, if and only if the controllability matrix C_T is positive definite. Compactly $g.c \Leftrightarrow C_T > 0$.

Proof. First we prove that positivity of the controllability matrix implies global controllability at T. That is, $C_T > 0 \Rightarrow g.c$. We prove this by actually constructing a control that does the job. Given the initial state $x_0 \in R^n$ and the target state $x_1 \in R^n$, define

$$z \equiv x_1 - e^{TA} x_0.$$

We show that under the given assumption we can construct a control u^* such that

$$z = \int_0^T e^{(T-s)A} B u^*(s) ds. \tag{6.2.10}$$

Indeed, since C_T is positive definite it is nonsingular and hence it has a bounded inverse. Take u^* given by

$$u^*(t) = B' e^{(T-t)A'} C_T^{-1} z, t \in I_T. \tag{6.2.11}$$

Thus the control u^*, as given, is well defined and obviously is in $L_\infty(I_T, R^d) \subset L_2(I_T, R^d)$. In fact this control is continuous and bounded. Substituting this in (6.2.10) we see that the identity holds. In other words, this control takes the system from the initial state x_0 to the target state x_1 at time T. This proves the first part. Now we show that if the system is globally controllable at time T, the controllability matrix C_T must be positive definite. That is, we prove $g.c \Rightarrow C_T > 0$. If the system is globally controllable at T, it is evident that both $\mathcal{A}(T) = R^n$ and $\mathcal{R}(T) = R^n$. Hence

$$(w, \xi) = 0 \text{ for all } w \in \mathcal{A}(T) \Rightarrow \xi = 0. \tag{6.2.12}$$

Since the transition matrix e^{tA} is always nonsingular, this is equivalent to

$$(e^{TA}w, \xi) = 0 \text{ for all } w \in \mathcal{R}(T) \Rightarrow \xi = 0. \tag{6.2.13}$$

It follows from the structure of the set $\mathcal{R}(T)$ that this is equivalent to the following statements

$$\left(\int_0^T e^{(T-s)A}Bu(s)ds, \xi\right) = 0 \text{ for all } u \in L_2(I_T, R^d) \Rightarrow \xi = 0, \tag{6.2.14}$$

and

$$\int_0^T (u(s), B' e^{(T-s)A'} \xi)ds = 0 \text{ for all } u \in L_2(I_T, R^d) \Rightarrow \xi = 0. \tag{6.2.15}$$

But if

$$\int_0^T (u(s), B' e^{(T-s)A'} \xi)ds = 0 \text{ for all } u \in L_2(I_T, R^d)$$

then

$$B' e^{(T-t)A'} \xi \equiv 0, \text{ for } a.e. \ t \in I_T.$$

Since $t \longrightarrow B' e^{(T-t)A'} \xi$ is continuous, this identity holds for all $t \in I_T$. In view of this the statement (6.2.15) is equivalent to

$$\{B' e^{(T-t)A'} \xi \equiv 0, \ \forall \ t \in I_T\} \Rightarrow \xi = 0. \tag{6.2.16}$$

Clearly this is equivalent to the following statement

$$\left\{\int_0^T \parallel B' e^{(T-t)A'} \xi \parallel^2 dt = 0\right\} \Rightarrow \xi = 0. \tag{6.2.17}$$

Thus it follows from the definition of the controllability matrix C_T that

$$\{(C_T\xi, \xi) = 0\} \Rightarrow \xi = 0. \tag{6.2.18}$$

Since C_T is always positive semidefinite, this shows that C_T is actually positive definite. This completes the proof. □

As a corollary of this we have the following interesting result.

Corollary 6.2.4. If the system is globally controllable at time $T \in R_0$ then it is also globally controllable for any $\tau \in R_0$.

Proof. By virtue of the previous theorem, we know that global controllability at time T is equivalent to the positive definiteness of the controllability matrix C_T. Thus it suffices to show that if $C_T > 0$ then $C_\tau > 0$ for any $\tau \in R_0$. If $\tau > T$, it is obvious that $C_\tau > 0$. So it remains to prove this for $\tau < T$. We prove this by contradiction. Suppose $C_T > 0$ but $C_\tau \not> 0$. Then there exists a $\xi(\neq 0) \in R^n$ such that $(C_\tau \xi, \xi) = 0$. This implies that

$$\int_0^\tau \| B' e^{(\tau - s)A'} \xi \|^2 \, ds = 0.$$

This means that

$$B' e^{sA'} \xi \equiv 0, \forall s \in I_\tau.$$

Hence for any $v \in R^d$, the function

$$f(s) \equiv (B' e^{sA'} \xi, v) = (\xi, e^{sA} Bv) \equiv 0, \forall s \in I_\tau.$$

Clearly f is an analytic function vanishing on I_τ. Hence it must vanish identically on R_0. This implies that $(C_T \xi, \xi) = 0$ also. Since $C_T > 0$, we must have $\xi = 0$ which contradicts our hypothesis. This completes the proof. □

Examining the expression for the controllability matrix given by (6.2.9) it is evident that computationally it is not at all trivial specially when the system is very high dimensional. Like in the case of observability problems, we can also obtain rank condition which is equivalent to global controllability. This is given in the following theorem.

Theorem 6.2.5. The system is globally controllable (g.c) if and only if the following rank condition holds

$$\text{Rank}(A|B) \equiv \text{Rank}\{B, AB, A^2B, \cdot, \cdot, A^{n-1}B\} = n. \tag{6.2.19}$$

Proof. Since by the previous theorem, $g.c \Leftrightarrow C_T > 0$, for any $T \in R_0$, it suffices to prove that $C_T > 0 \Leftrightarrow \text{Rank}(A|B) = n$. First we prove that $\{C_T > 0\} \Rightarrow \text{Rank}(A|B) = n$. We have seen in the course of the proof of the previous theorem, (see 6.2.11), that the positivity of the controllability matrix C_T is equivalent to the condition that

$$\text{Ker}\{B' e^{(T-t)A'}, t \in I_T\} = \{0\}.$$

But clearly this is equivalent to the following statement

$$\{(B' e^{(T-t)A'} \xi, v) = 0, \ \forall \, t \in I_T, \ \forall \, v \in R^d\} \Rightarrow \xi = 0. \tag{6.2.20}$$

Using Cayley-Hamilton theorem and the analyticity of the function $t \longrightarrow (B^{'}e^{(T-t)A^{'}}\xi, v)$, it follows from Taylor's expansion, precisely as we have seen in the proof of the rank condition for observability, that

$$\text{Range}\{B, AB, A^2B, \cdot, \cdot, A^{n-1}B\} = R^n.$$

This means that $\text{Rank}(A|B) = n$. Thus we have proved that positive definiteness of the controllability matrix implies that the rank must equal the dimension of the system. Now we prove the reverse implication $\{\text{Rank}(A|B) = n\} \Rightarrow C_T > 0$. We assume that the rank condition holds. Clearly this implies that

$$\text{Ker}\{B^{'}e^{(T-t)A^{'}}, t \in I_T\} = \{0\}.$$

Hence

$$\int_0^T \parallel B^{'}e^{(T-t)A^{'}}\xi \parallel^2 dt = 0,$$

implies that $\xi = 0$, which, in turn, implies positivity of C_T. This completes the proof. □

In fact it is not at all necessary to compute the controllability matrix using the basic definition given by (6.2.9). We show that it can be obtained by solving a matrix differential equation.

$$\dot{C}(t) = AC(t) + C(t)A^{'} + BB^{'}, t \in I_T, \tag{6.2.21}$$

$$C(0) = 0. \tag{6.2.22}$$

In fact this follows immediately from the time derivative of the function

$$C(t) \equiv \int_0^t e^{(t-s)A}BB^{'}e^{(t-s)A^{'}} ds, t \in I_T. \tag{6.2.23}$$

Thus all that is required is to solve this equation as an initial value problem and set $C_T \equiv C(T)$.

It is evident from the previous results, that if the rank condition, Rank $(A|B) = n$, holds the system is controllable to any moving target provided the target, as a function of t, is continuous and bounded on bounded intervals. The same conclusion remains valid also in the presence of persistent disturbance provided the disturbance is measurable. Let $\{\eta(t), t \geq 0\}$ denote the target with values $\eta(t) \in R^n$ and suppose it is continuous in $t \geq 0$; and let $\{D(t), t \geq 0\}$ denote the disturbance so that

$$\dot{x}(t) = Ax(t) + Bu(t) + D(t), t \geq 0.$$

If D is locally integrable and the target satisfies the mild condition as stated above, the system is controllable to the target at any finite time, say, $T > 0$. In this case we define

$$z = \eta(T) - e^{TA}x_0 - \int_0^T e^{(T-s)A}D(s)ds.$$

Assuming that the disturbance is known to the controller, the control driving the system to the target is given by

$$u^o(t) \equiv B'e^{(T-t)A'}C_T^{-1}z, t \in [0, T].$$

In many physical situations the control is not only limited in energy, it may also be limited in magnitudes. In other words in almost all engineering problems the control set U may be a compact (closed bounded) subset of R^d. The preceding results have been proved under the assumption that $U = R^d$. Here we wish to prove that similar results are valid under some additional condition, even when U is a compact subset of R^d.

For this purpose we define our admissible controls as

$$\mathcal{U}_{ad} \equiv \{u : u \text{ measurable}, u(t) \in U \text{ a.e } t \geq 0\}, \qquad (6.2.24)$$

where U is a compact subset of R^d. We prove the following result.

Theorem 6.2.6. Suppose the following assumptions hold:

(A1): Rank $(A|B) = n$,

(A2): $\int_0^\infty \| B'e^{tA'}\eta \| dt = \infty$ for any $\eta \neq 0$,

(A3): $U \subset R^d$ is compact with $0 \in$ Int (U).

Then the system is globally controllable in finite time.

Proof. Since $0 \in$ Int (U), there exists a number $r > 0$ such that the ball

$$B_r \equiv \{v \in R^d :\| v \|\leq r\}$$

is contained in the interior of U. For each $t \geq 0$, define the (constrained) attainable set $\mathcal{A}_r(t)$ by

$$\mathcal{A}_r(t) \equiv \left\{\xi \in R^n : \xi = \int_0^t e^{(t-s)A}Bu(s)ds, u \text{ measurable}, u(s) \in B_r, 0 \leq s \leq t\right\}. \qquad (6.2.25)$$

First note that $\mathcal{A}_r(t)$ is a convex set valued function of t. Convexity follows from linearity of the operator

$$(Lu)(t) \equiv \int_0^t e^{(t-s)A}Bu(s)ds, t \geq 0,$$

and the convexity of the ball B_r. For global finite time controllability it suffices to show that

$$\bigcup_{t\geq 0} \mathcal{A}_r(t) = R^n,$$

or equivalently $\mathcal{A}_r(t)$ is not contained in any proper subset of R^n for all finite $t \geq 0$. We prove this by contradiction. Suppose this is false. In other words the set

$$\mathcal{A}_r \equiv \bigcup_{t\geq 0} \mathcal{A}_r(t)$$

is a proper subset of R^n. But then it's closed convex hull denoted by

$$\hat{\mathcal{A}}_r \equiv \ell co \mathcal{A}_r$$

is also a proper subset of R^n. Thus there exists a $z(\neq 0) \in R^n \setminus \hat{\mathcal{A}}_r$. Since $\hat{\mathcal{A}}_r$ is a convex set, for every point z out side of $\hat{\mathcal{A}}_r$ or on its boundary, there exists a hyperplane, passing through z, that separates the set $\hat{\mathcal{A}}_r$ from the point z. In other words there exists a non zero vector $\eta \in R^n$, having unit norm, such that

$$(w, \eta) \leq (z, \eta), \text{ for all } w \in \mathcal{A}_r \subset \hat{\mathcal{A}}_r.$$

This means that

$$((Lu)(t), \eta) \leq (z, \eta), \text{ for all } t \geq 0, \text{ and } \forall\ u \text{ such that } u(t) \in B_r.$$

In other words, for all $t \geq 0$,

$$\int_0^t (u(s), B' e^{(t-s)A'} \eta) ds \leq (z, \eta), \ \forall\ u :\ u(t) \in B_r.$$

But this is equivalent to

$$\int_0^t (v(s), B' e^{sA'} \eta) ds \leq (z, \eta), \forall\ t \geq 0, \ \forall\ v : v(t) \in B_r. \tag{6.2.26}$$

We have seen before that the rank condition is equivalent to the condition

$$\text{Ker}\{B' e^{tA'}, t \geq 0\} = \{0\}.$$

Since $\eta \neq 0$, it follows from this that the set

$$J \equiv \left\{ t \geq 0 : \| B' e^{tA'} \eta \| = 0 \right\}$$

is a set of Lebesgue measure zero. Take $\alpha \in (0, r]$ and define

$$v(t) \equiv \begin{cases} 0, & \text{if } t \in J; \\ \alpha \frac{B' e^{tA'} \eta}{\| B' e^{tA'} \eta \|}, & \text{if } t \in R_0 \setminus J. \end{cases}$$

Clearly v, as defined, is an admissible control policy. Substituting this control in (6.2.26) we obtain

$$\alpha \int_0^t \| B' e^{sA'} \eta \| \, ds \leq (z, \eta) \leq \| z \| < \infty, \forall \, t \geq 0. \tag{6.2.27}$$

Under the assumption (A2), this leads to the following absurdity,

$$\infty = \int_0^\infty \| B' e^{tA'} \eta \| \, dt < \infty.$$

Thus $\mathcal{A}_r = R^n$ and the system is globally controllable in finite time. This completes the proof. □

Remark. Note that it is not necessary to have *zero* as an interior point of U as implied by assumption (A3). It suffices if U has an interior point v because in that case one can choose $U_v \equiv U - v \equiv \{\xi : \xi = u - v, u \in U\}$ instead of U.

An interesting fact about the control given by the expression (6.2.11) which is reproduced here,

$$u^*(t) \equiv B' e^{(T-t)A'} C_T^{-1} z, t \in I_T,$$

is that it has the minimum energy compared to all the controls that achieve the same goal over the same period of time.

Corollary 6.2.7. The control u^* given by

$$u^*(t) \equiv B' e^{(T-t)A'} C_T^{-1} z, t \in I_T,$$

sends the system from the state x_0 to the target state x_1 at time T spending minimum energy. Further, this control is unique in the sense that any other control that achieves the same goal at time T with minimum energy must equal u^* almost every where on I_T.

Proof. Suppose $u \in L_2(I_T, R^d)$ is any other control that does the same job. In other words

$$z \equiv (x_1 - e^{TA}x_0) = \int_0^T e^{(T-t)A}Bu(t)dt.$$

For convenience of notation, for the Hilbert space $L_2(I_T, R^d)$, we use

$$(u,v) \equiv \int_0^T (u(t), v(t))dt, \text{ and } \| u \| \equiv \sqrt{\int_0^T \| u(t) \|^2 \, dt}$$

for the scalar product and the norm respectively. Clearly

$$\| u \|^2 = \| u - u^* + u^* \|^2 = \| u - u^* \|^2 + \| u^* \|^2 + 2(u - u^*, u^*).$$

We show that $(u - u^*, u^*) = 0$. Indeed

$$\begin{aligned}(u - u^*, u^*) &= \int_0^T (u(t) - u^*(t), u^*(t))dt \\ &= \int_0^T (u(t) - u^*(t), B' e^{(T-t)A'} C_T^{-1} z)dt \\ &= \left(\int_0^T e^{(T-t)A} B(u(t) - u^*(t))dt, C_T^{-1} z \right) \\ &= 0.\end{aligned}$$

The last identity follows from the fact that both the control policies u and u^* take the system from x_0 to x_1 at the same time T. In other words u^* is orthogonal to $u - u^*$. Hence we conclude that

$$\| u \|^2 = \| u - u^* + u^* \|^2 = \| u - u^* \|^2 + \| u^* \|^2 \geq \| u^* \|^2 .$$

Since u is any control achieving the same goal, it follows from this that u^* does so with minimum energy. For the proof of uniqueness, let u^o be another control policy over the time horizon I_T that achieves the same goal with also minimum energy. Then

$$\begin{aligned}\| u^o - u^* \|^2 &= (u^o - u^*, u^o) - (u^o - u^*, u^*) \\ &= (u^o - u^*, u^o) = \| u^o \|^2 - (u^*, u^o) \\ &= \| u^o \|^2 - (u^*, u^o - u^* + u^*) \\ &= \| u^o \|^2 - (u^*, u^o - u^*) - \| u^* \|^2 \\ &= \| u^o \|^2 - \| u^* \|^2 = 0.\end{aligned}$$

In the middle two lines we have used the previous result that states that u^* is orthogonal to $u^o - u^*$. The last identity follows from the fact that both the controls have the same minimum energy. Thus u^o must equal u^* in the L_2 sense and hence almost every where on I_T. This completes the proof. □

Some Examples: Here we present few examples based on the rank condition.

(E1) Consider the third order system,

$$\xi^{(3)} + a_2\xi^{(2)} + a_1\xi^{(1)} + a_0\xi = u$$

with scalar control. Define the state vector as $x_1 = \xi, x_2 = \xi^{(1)} \equiv \dot{\xi}, x_3 = \xi^{(2)} \equiv \ddot{\xi}$. This system can be written in the canonical form $\dot{x} = Ax + bu$, with

$$A = \begin{pmatrix} 0 & 1 & 0 \\ 0 & 0 & 1 \\ -a_0 & -a_1 & -a_2 \end{pmatrix} \text{ and } b = \begin{pmatrix} 0 \\ 0 \\ 1 \end{pmatrix}.$$

The reader can check that

$$b = \begin{pmatrix} 0 \\ 0 \\ 1 \end{pmatrix}, Ab = \begin{pmatrix} 0 \\ 1 \\ -a_2 \end{pmatrix}, A^2b = \begin{pmatrix} 1 \\ -a_2 \\ -a_1 + a_2^2 \end{pmatrix}.$$

Since the rank of a square matrix is full (equal to it's dimension) if it's determinant is non zero, it suffices to verify that the $det(b, Ab, A^2b) \neq 0$. Indeed, $det\{b, Ab, A^2b\} = -1$, and hence the system is globally controllable. Note that the conclusion is independent of the values of the parameters $\{a_0, a_1, a_2\}$ and hence independent of stability or instability of the system. In fact any time invariant system given by an ordinary differential equation of order n, driven by a scalar control u is always globally controllable.

(E2) (Electrical Motor) Consider the universal motor

$$J\ddot{\theta} + D\dot{\theta} = T$$

where the input torque T is the control. This is a special case of example 1, and the reader can easily verify that $det(b, Ab) = -1 \neq 0$ and so globally controllable.

(E3) (Inverted Pendulum) An inverted pendulum is described by the differential equation

$$\ddot{\theta} - k \sin\theta = u, k > 0.$$

Near the zero state, this is given by $\ddot{\theta} - k\theta = u$. Define the state as $x_1 = \theta, x_2 = \dot{\theta}$. Again it follows from the first example, for $a_2 = 0, a_1 = 0, a_0 = -k$, that the system is globally controllable. Thus instability or stability does not affect controllability.

(E4) (Ecological System) Consider the rabbit-fox model

$$\dot{R} = 10R - F - \theta u$$
$$\dot{F} = -(1/2)F + 5R - 20u,$$

where u denotes the control (killing rate of foxes) and $\theta > 0$ is the collateral damage (accidental killing) done to the rabbit population. Clearly the matrices A and b are

$$A = \begin{pmatrix} 10 & -1 \\ 5 & -(1/2) \end{pmatrix}, \; b = \begin{pmatrix} -\theta \\ -20 \end{pmatrix}.$$

Again the matrix (b, Ab) is a square matrix and it's determinant is given by

$$det(b, Ab) = 5\theta^2 - 210\theta + 400 \equiv P(\theta).$$

Clearly for $\theta = 0$, the system is globally controllable. However, if there are real non zero roots of the polynomial P, and θ equals any one of these roots, the system is not globally controllable. For example, one can easily verify that $P(\theta) = 0$ for $\theta = \{2, 40\}$. Thus for these values of θ the system is not controllable.

(E5) (A Spherical Satellite) A spherically symmetric satellite is governed by the system of differential equations,

$$I\dot{p} = -\omega I r + \alpha_1 u_1 \;\; I\dot{q} = \alpha_2 u_2 \;\; I\dot{r} = I\omega p + \alpha_3 u_3,$$

where $\{p, q, r\}$ represent the spin velocities of the body around its principal axes, I the mass moment of inertia of the body, ω denotes the spin rate of the earth and $\{u_1, u_2, u_3\}$ are the control torques provided by thrusters. The parameters $\alpha_i = 1$ or $0, i = 1, 2, 3$ depending on whether or not the i-th thruster is in operating or failed state. The reader can easily verify that $Rank(B, AB, A^2B) = 3$ under any one of the following situations: (i): all the thrusters are active, (ii): thruster 1 or 3 but not both, failed. Thus under these situations, the system is globally controllable. On the other hand, if thruster 2 has failed, the rank is only 2 and hence the system is not controllable. This is also evident from the system equations.

Another interesting fact is that, mathematically controllability and observability are equivalent concepts and they are dual problems in the sense stated in the following corollary.

Corollary 6.2.8. If the pair (A, B) is controllable then the pair $(A^{'}, B^{'})$ is observable. Similarly if the pair (A, H) is observable, the pair $(A^{'}, H^{'})$ is controllable.

Proof. The proof is straight forward and follows from the Theorems 5.2.6 and 6.2.5. For instance, consider the first statement. We know from Theorem 5.2.6 that the pair $(A^{'}, B^{'})$ is observable if and only if the $Rank(A^{'}|B^{'}) = n$. Since

$$\begin{aligned} Rank(A^{'}|B^{'}) &= Rank\{(B^{'})^{'}, (A^{'})^{'}(B^{'})^{'}, \cdots, ((A^{'})^{'})^{n-1}(B^{'})^{'}\} \\ &= Rank(A|B) = n, \end{aligned}$$

the conclusion follows. The proof of the second statement is identical. □

6.3 Linear Time Varying Systems

In this section we consider controllability problems for linear time varying systems. The system is given by

$$\begin{aligned} \dot{x}(t) &= A(t)x(t) + B(t)u(t), t \in R_0, \\ x(s) &= x_0, u \in \mathcal{U}_{ad} \equiv L_2^{\ell oc}(R_0, R^d). \end{aligned} \tag{6.3.28}$$

The reason for choosing arbitrary s as the starting time is important in the case of time varying systems. Characteristic features of such systems may change with time and hence the system may behave differently at different times and different time horizons. Here also we define the controllability matrix as follows.

$$C_s(t) \equiv \int_s^t \Phi(t, \theta)B(\theta)B^{'}(\theta)\Phi^{'}(t, \theta)d\theta, s \leq t. \tag{6.3.29}$$

We prove the following result.

Theorem 6.3.1. The system given by (6.3.28) is globally controllable over the time horizon $I_{s,T} \equiv [s, T]$ if and only if $C_s(T) > 0$, compactly written as $g.c \Leftrightarrow C_s(T) > 0$.

Proof. The proof is quite similar to that of time invariant case. First we prove the implication $\{C_s(T) > 0\} \Rightarrow g.c.$ Let $x_0, x_1 \in R^n$ be any two elements and define $z \equiv x_1 - \Phi(T,s)x_0$. We choose a control that takes the system from the state x_0 at time s to the state x_1 at time T. Since $C_s(T)$ is nonsingular, the control policy given by

$$u^o(t) \equiv B^{'}(t)\Phi^{'}(t,s)C_s^{-1}(T)z \tag{6.3.30}$$

is well defined. If B is measurable, with all its elements locally square integrable, then the control u^o is measurable and also square integrable, that is $u^o \in L_2(I_{s,T}, R^d)$. If the elements of B are essentially bounded measurable functions then $u^o \in L_\infty(I_{s,T}, R^d)$. If B is only continuous and bounded, the control is also continuous and bounded. In any case substituting this control in the expression

$$x(T) = \Phi(T,s)x_0 + \int_s^T \Phi(T,\tau)B(\tau)u(\tau)d\tau$$

we find that $x(T) = \Phi(T,s)x_0 + z = x_1$. This shows that the control u^o does the job as required. Now we prove the reverse implication, $g.c \Rightarrow \{C_s(T) > 0\}$. Define, as in the time invariant case, the attainable set at time T by

$$\mathcal{A}_s(T) \equiv \left\{\xi \in R^n : \xi = \int_s^T \Phi(T,\theta)B(\theta)u(\theta)d\theta, u \in \mathcal{U}_{ad}\right\}. \tag{6.3.31}$$

Clearly the vector z defined by

$$z \equiv x_1 - \Phi(T,s)x_0$$

describes the whole space R^n as $\{x_0, x_1\}$ describe the space $R^n \times R^n$. Hence for global controllability (over the period $[s,T]$) it is obvious that we must have $\mathcal{A}_s(T) = R^n$. Since $g.c$ holds, we have

$$\{(a,\xi) = 0, \forall\ a \in \mathcal{A}_s(T)\} \Rightarrow \xi = 0.$$

In other words,

$$\left\{\int_s^T (u(\tau), B^{'}(\tau)\Phi^{'}(T,\tau)\xi)d\tau = 0, \forall\ u \in \mathcal{U}_{ad}\right\} \Rightarrow \xi = 0. \tag{6.3.32}$$

This is possible if and only if

$$B^{'}(t)\Phi^{'}(T,t)\xi = 0,\ \ a.e\ t \in I_{s,T}.$$

Thus the statement (6.3.32) is equivalent to the following one

$$\left\{\int_s^T \parallel B^{'}(\tau)\Phi^{'}(T,\tau)\xi \parallel^2 d\tau = 0, \forall\ u \in \mathcal{U}_{ad}\right\} \Rightarrow \xi = 0. \tag{6.3.33}$$

By definition of the controllability matrix $C_s(T)$, it follows from the above statement that

$$\{(C_s(T)\xi, \xi) = 0\} \Rightarrow \xi = 0.$$

Hence we have shown that $C_s(T) > 0$. This completes the proof. □

Remark. Unfortunately for time varying systems, there is no rank condition. So we do not have equivalent results. We discuss this further in the sequel.

Again it is not necessary to compute the controllability matrix from its definition (6.3.29). By simply differentiating this expression we can derive the following differential equation

$$\dot{C}_s(t) = A(t)C_s(t) + C_s(t)A'(t) + B(t)B'(t), t \geq s \tag{6.3.34}$$

$$C_s(s) = 0. \tag{6.3.35}$$

Solving this equation forward in time one can determine $C_s(T)$ for any $T > s$. Clearly $C_s(t)$ is a symmetric positive semidefinite matrix for all $t \geq s$. For fixed s, we have seen that for global controllability in finite time T, we must have $C_s(T)$ positive definite. Clearly if this matrix is never positive definite the system is not controllable. In order to find the first time this is positive definite one may, introduce the function

$$g_s(t) \equiv \inf\{(C_s(t)\xi, \xi), \parallel \xi \parallel = 1\}$$

as we did for observability problems. Clearly this is a nonnegative function of t on $[s, \infty)$ satisfying $g_s(s) = 0$. One can verify that if, for some $T > s$, $g_s(T) > 0$, then $g_s(\tau) > 0$ for all $\tau \geq T$. This means that if the system is controllable over the time horizon $[s, T]$ then it is also controllable over any time horizon $[s, \tau]$ for $\tau \geq T$. For fixed terminal time, say $\tau > 0$, the function $t \to g_t(\tau)$ on the interval $[s, \tau]$, is a decreasing, more precisely, nonincreasing function converging to $g_\tau(\tau) = 0$. Thus if $g_t(\tau) = 0$ for all $t \in [s, \tau]$, the system is not controllable over the time horizon $[s, \tau]$.

A more convenient indicator of controllability can be defined using the reachability matrix. This is related to the reachable set as defined below

$$R_s(t) \equiv \{\xi \in R^n : \xi = \int_s^t X^{-1}(\theta)B(\theta)u(\theta)d\theta, u \in \mathcal{U}_{ad}\} \tag{6.3.36}$$

which is very similar to the attainable set as introduced earlier. The reachability matrix can then be defined as

$$\tilde{C}_s(t) \equiv \int_s^t X^{-1}(\theta)B(\theta)B'(\theta)(X^{-1}(\theta))' d\theta, -\infty < s \leq t < \infty. \tag{6.3.37}$$

Note that the controllability and reachability matrices $C_s(t)$ and $\tilde{C}_s(t)$, given by (6.3.29) and (6.3.37), respectively are related by the following expression

$$C_s(t) = X(t)\tilde{C}_s(t)X^{'}(t), -\infty < s \leq t < \infty, \tag{6.3.38}$$

where $X(t), t \in R$, is the fundamental solution of the matrix differential equation $\dot{X} = A(t)X, X(0) = I$, as introduced in Chapter 2.

Corollary 6.3.2. The system given by (6.3.28) is globally controllable over the time horizon $I_{s,T} \equiv [s,T]$ if and only if the reachability matrix $\tilde{C}_s(T) > 0$, compactly written as $g.c \Leftrightarrow \tilde{C}_s(T) > 0$.

Proof. Since $g.c \Leftrightarrow C_s(T) > 0$, it suffices to show that $\{C_s(T) > 0\} \Leftrightarrow \{\tilde{C}_s(T) > 0\}$. But this is an immediate consequence of the fact that the fundamental matrix $X(t), t \in R$, is nonsingular. □

Using the reachability matrix, we may now introduce the function

$$\Pi_s(t) = \inf\{(\tilde{C}_s(t)\xi, \xi), \parallel \xi \parallel = 1\}, t \geq s, \tag{6.3.39}$$

The function $\Pi_s(t), t \geq s$, is nonnegative and one can easily verify that

$$\Pi_s(\tau) = \Pi_s(t) + \Pi_t(\tau) \ \geq \ \Pi_s(t), \text{ for } s \leq t \leq \tau,$$

and thus the function, $t \longrightarrow \Pi_s(t), t \geq s$, is a nondecreasing function of t with $\Pi_s(s) = 0$. Since it is a monotone non decreasing function of t, there exists a number $\alpha \in [0, \infty]$ so that

$$\lim_{t \to \infty} \Pi_s(t) = \alpha.$$

Clearly if $\alpha > 0$, the system is (globally) controllable in finite time, otherwise not. The first time the system is controllable is given by

$$T_1 \equiv \inf\{t \geq s : \Pi_s(t) > 0\}$$

The significance of the starting time s can be appreciated from the following simple example.

An Example. Consider the system

$$\dot{x}(t) = Ax(t) + B(t)u(t), t \geq s \geq 0, \tag{6.3.40}$$

$$x(s) = x_0, \tag{6.3.41}$$

where

$$B(t) = \begin{cases} B_0 & \text{if } t \in I_{s,T} \\ B_1 & \text{for } t > T. \end{cases}$$

Suppose $\text{Rank}(A|B_0) = n$ but $\text{Rank}(A|B_1) < n$. The reader can verify that this system is globally controllable for all starting times s satisfying $0 \leq s < T$, but it fails to be globally controllable if $s \geq T$. It is interesting to note that even if $B_1 = 0$, the system is globally controllable for all $\tau > T$ provided the starting time $s \in [0, T)$. The reader may verify that the appropriate control for this situation is given by

$$u^o(t) = \begin{cases} B_0' e^{(T-t)A'} C_s(T)^{-1} e^{-(\tau-T)A} z, & \text{for } t \in I_{s,T}, \tau \geq T \\ B_0' e^{(\tau-t)A'} C_s(\tau)^{-1} z & \text{for } t \in I_{s,T}, s < \tau < T, \end{cases}$$

where $z \equiv x_1 - e^{(\tau-s)A} x_0$. The same conclusion is valid if the fixed target $x_1 \in R^n$ is replaced by a moving target $\gamma(t) \in R^n$. In this case $z = \gamma(\tau) - e^{(\tau-s)A} x_0$. Of course all this is possible if the control is unlimited that is, the set $U = R^d$.

Before we complete this section, we may point out another interesting fact. Instead of considering Π as a function of the interval $[s, t]$, one could look at it as a measure on the Borel field B_R of subsets of the real line R. Assuming that the elements of $B(t), t \in R$, are locally square integrable, it is clear that for any $J \in B_R$,

$$\tilde{C}(J) \equiv \int_J X^{-1}(r) B(r) B'(r) (X^{-1})' dr$$

is a positive semi definite matrix and that

$$\nu(J) \equiv \inf\{(\tilde{C}(J)\xi, \xi), \| \xi \| = 1\}$$

is a positive measure. In fact it is a countably additive, sigma-finite positive measure. If for any $I \in B_R$, $\nu(I) > 0$, the system is globally controllable and since, for any $\Gamma \in B_R$ with $\Gamma \supset I$, $\nu(\Gamma) \geq \nu(I)$, it is also globally controllable on Γ.

Next we consider the question of controllability with control constraints. In case the controls are limited in magnitude, the situation is altogether different. However, under some additional assumptions, as we have seen in the time invariant case, we can achieve global controllability in finite time. This is presented in the following result.

Theorem 6.3.3. Suppose the following conditions hold:

(A1): $C_s(T) > 0$ for some $T > s$,

(A2): $\lim_{t \to \infty} \int_s^t \| B'(\theta) \Phi'(t, \theta) \eta \| d\theta = \infty$ for any $\eta \neq 0$.

(A3): U compact and $0 \in$ int (U).

Then the system is globally controllable in finite time.

Proof. The proof is quite similar to that of time invariant case. Since $0 \in$ Int (U), there exists a number $r > 0$ such that the ball $B_r \equiv \{v \in R^d : \| v \| \leq r\}$ is contained in the interior of U. For each $t \geq 0$, define the (constrained) attainable set $\mathcal{A}_r(t)$ by

$$\mathcal{A}_r(t) \equiv \left\{\xi \in R^n : \xi = \int_s^t \Phi(t,\theta)B(\theta)u(\theta)d\theta, u \text{ measurable}, u(\theta) \in B_r, s \leq \theta \leq t\right\}. \tag{6.3.42}$$

First note that $\mathcal{A}_r(t)$ is a convex set valued function of t. Again convexity follows from linearity of the operator

$$(Lu)(t) \equiv \int_s^t \Phi(t,\theta)B(\theta)u(\theta)d\theta, t \geq s,$$

and the convexity of the ball B_r. For global finite time controllability it suffices to show that

$$\bigcup_{t\geq 0} \mathcal{A}_r(t) = R^n,$$

or equivalently $\mathcal{A}_r(t)$ is not contained in any proper subset of R^n for all finite $t \geq s$. We prove this by contradiction. Suppose this is false. In other words the set

$$\mathcal{A}_r \equiv \bigcup_{t\geq s} \mathcal{A}_r(t)$$

is a proper subset of R^n. But then it's closed convex hull denoted by

$$\hat{\mathcal{A}}_r \equiv \mathcal{cl}co\mathcal{A}_r$$

is also a proper subset of R^n. Thus there exists a $z(\neq 0) \in R^n \setminus \hat{\mathcal{A}}_r$. Since $\hat{\mathcal{A}}_r$ is a convex set, for every point z out side of $\hat{\mathcal{A}}_r$ or on its boundary, there exists a hyperplane, passing through z, that separates the set $\hat{\mathcal{A}}_r$ from the point z. In other words there exists a non zero vector $\eta \in R^n$, dependent on z, having unit norm, such that

$$(w,\eta) \leq (z,\eta), \text{ for all } w \in \mathcal{A}_r \subset \hat{\mathcal{A}}_r.$$

This means that

$$((Lu)(t),\eta) \leq (z,\eta), \text{ for all } t \geq s, \text{ and } \forall\ u : u(t) \in B_r.$$

In other words, for all $t \geq s$,

$$\int_s^t (u(\theta), B^{'}(\theta)\Phi^{'}(t,\theta)\eta)d\theta \leq (z,\eta), \ \forall \ u \text{ such that } u(t) \in B_r.$$

Clearly, for any choice of $\tau > T$, we must have

$$\int_s^T (u(\theta), B^{'}(\theta)\Phi^{'}(\tau,\theta)\eta)d\theta \leq (z,\eta), \forall \ \tau \geq T, \ \forall \ u : u(t) \in B_r. \tag{6.3.43}$$

Since $C_s(T) > 0$, $C_s(\tau) > 0$ also for any $\tau \geq T$. But the positivity of the controllability matrix implies that

$$\text{Ker}\{B^{'}(\theta)\Phi^{'}(\tau,\theta), s \leq \theta \leq \tau\} = \{0\}.$$

Define the set

$$J_\tau \equiv \{s \leq t \leq \tau :\| B^{'}(t)\Phi^{'}(\tau,t)\eta \|= 0\}.$$

Since $C_s(\tau) > 0$, the Lebesgue measure of this set is strictly less than $(\tau - s)$. Take $\alpha \in (0, r]$ and define

$$u(t) \equiv \begin{cases} 0, & \text{if } t \in J_T; \\ \alpha \frac{B^{'}(t)\Phi^{'}(\tau,t)\eta}{\|B^{'}(t)\Phi^{'}(\tau.t)\eta\|}, & t \in I_{s,\tau} \setminus J_\tau. \end{cases}$$

Clearly u, as defined, is an admissible control policy. Substituting this control in (6.3.43) we obtain

$$\alpha \int_s^T \| B^{'}(t)\Phi^{'}(\tau,t)\eta \| \, dt \ \leq \ (z,\eta) \ \leq \ \| z \| \ < \infty \, \forall \tau \geq T. \tag{6.3.44}$$

Under the assumption (A2), this leads to the following absurdity,

$$\infty = \lim_{T \to \infty} \int_s^T \| B^{'}(t)\Phi^{'}(\tau,t)\eta \| \, dt < \infty.$$

Thus $\mathcal{A}_r = R^n$ and the system is globally controllable in finite time. This completes the proof. □

6.4 Perturbed (Linear/Nonlinear) Systems

In this section we study the question of controllability for perturbed linear and some nonlinear systems. We show that if the unperturbed system is globally controllable, then, under some restrictions, the perturbed system is also globally controllable.

First we consider the linear system

$$\dot{x}(t) = Ax(t) + (B + \tilde{B}(t))u(t), t \geq 0, \tag{6.4.45}$$
$$x(0) = x_0, \tag{6.4.46}$$

where the matrix valued function $\tilde{B}$ denotes the perturbation of the control matrix B. The question is, is the perturbed system globally controllable in finite time $T > 0$, given that the unperturbed system is. We show that this question is equivalent to the question of existence of a solution of a Fredholm integral equation of the second kind. We also present sufficient conditions for existence of a solution of the integral equation.

Theorem 6.4.1. Suppose the unperturbed system

$$\dot{x} = Ax + Bu, x(0) = x_0,$$

is globally controllable. Then the question of controllability of the perturbed system given by (6.4.45) is equivalent to the question of existence of a solution over some finite time interval $I_T \equiv [0, T]$ of the integral equation,

$$u(t) = g_T(t) + \int_0^T K_T(t, s)u(s)ds, t \in I_T, \tag{6.4.47}$$

where

$$g_T(t) \equiv B' e^{(T-t)A'} C_T^{-1}(x_1 - e^{TA}x_0), t \in I_T$$
$$K_T(t, s) \equiv -B' e^{(T-t)A'} C_T^{-1} e^{(T-s)A}\tilde{B}(s), (s, t) \in I_T \times I_T.$$

Further, if the perturbation $\tilde{B}$ is sufficiently small in norm, the integral equation has a unique solution for any given $T > 0$ and the system is globally controllable.

Proof. Given any $T > 0$, and the target state x_1, and control u, define

$$z_0 \equiv x_1 - e^{TA}x_0,$$
$$Z_T(u) \equiv z_0 - \int_0^T e^{(T-s)A}\tilde{B}(s)u(s)ds.$$

Notice that if $\tilde{B}$ is essentially bounded in norm, then, Z_T is a bounded linear map from $L_2(I_T, R^d)$ to R^n and $Z_T(u) \in R^n$ for every such control. Since the unperturbed system is globally controllable, considering $Z_T(u)$ as the target, it is clear that there exists a control $u^o \in L_2(I_T, R^d)$ so that

$$u^o(t) = B' e^{(T-t)A'} C_T^{-1} Z_T(u), t \in I_T. \tag{6.4.48}$$

Define the operator R_T by

$$R_T(u)(t) \equiv B' e^{(T-t)A'} C_T^{-1} Z_T(u), t \in I_T.$$

Thus, it follows from equation (6.4.48) that, for the existence of a control policy over the time horizon I_T that drives the system from the state x_0 to the target x_1, it is necessary that the operator R_T has a fixed point in $L_2(I_T, R^d)$, that is, there exist a $u^* \in L_2(I_T, R^d)$ so that

$$u^* = R_T u^*. \tag{6.4.49}$$

For the proof of sufficiency, note that if u^* is a fixed point of the operator R_T, then

$$u^*(t) = B' e^{(T-t)A'} C_T^{-1} Z_T(u^*), t \in I_T.$$

Substituting this control in the system equation (6.4.45) and using the expression for its solution

$$x(t) = e^{tA} x_0 + \int_0^t e^{(t-s)A}(B + \tilde{B}(s))u^*(s)ds$$

one can easily verify that $x(T) = x_1$. Now note that the operator equation (6.4.49) is precisely the integral equation (6.4.47). For the proof of existence of a solution of this integral equation, define

$$ess - sup\{\| \tilde{B}(t) \|, t \in I_T\} \equiv b.$$

Clearly if b is sufficiently small one can verify that the operator R_T is a contraction, that is

$$\| R_T u - R_T v \|_{L_2(I_T, R^d)} \leq \rho \| u - v \|_{L_2(I_T, R^d)},$$

with $\rho \in (0, 1)$. Hence for each such $\tilde{B}$, it follows from Banach fixed point theorem that the operator R_T has a unique fixed point and consequently the integral equation (6.4.47) has a unique solution $u^* \in L_2(I_T, R^d)$. Since $T > 0$ is arbitrary, this proves that the perturbed system (6.4.45) is globally controllable if the unperturbed system is.

Remarks. There are other sufficient conditions that guarantee the existence of solutions for the integral equation (6.4.47). We mention here two such conditions. Define

$$\alpha_T \equiv \sup_{0 \leq s,t \leq T} \| B' e^{(T-t)A'} C_T^{-1} e^{(T-s)A} \|$$

$$\| \tilde{B} \|_\infty \equiv ess - sup\{\| \tilde{B}(t) \|, t \geq 0\}, \text{ for } \tilde{B} \in L_\infty(R_0, M(n \times d))$$

$$\| \tilde{B} \|_2 \equiv \left(\int_0^T \| \tilde{B}(t) \|^2 dt \right)^{1/2}, \text{ for } \tilde{B} \in L_2^{\ell oc}(R_0, M(n \times d)).$$

It follows from the definition of the operator R_T, that, for the L_∞ case, we have

$$\| R_T u - R_T v \|_{L_2(I_T,R^d)} \;\le\; (T\ \alpha_T \parallel \tilde{B} \parallel_\infty) \parallel u - v \|_{L_2(I_T,R^d)}, \qquad (6.4.50)$$

and for the L_2 case we have

$$\| R_T u - R_T v \|_{L_2(I_T,R^d)} \;\le\; (\sqrt{T}\ \alpha_T \parallel \tilde{B} \parallel_2) \parallel u - v \|_{L_2(I_T,R^d)}\,. \qquad (6.4.51)$$

Thus, in order for R_T to be a contraction in L_2 space, it suffices if there exists a $\rho \in (0,1)$ such that

$$(T\ \alpha_T \parallel \tilde{B} \parallel_\infty) \;\le\; \rho \ \text{ for } \tilde{B} \in L_\infty \qquad (6.4.52)$$

and

$$(\sqrt{T}\ \alpha_T \parallel \tilde{B} \parallel_2) \;\le\; \rho \ \text{ for } \tilde{B} \in L_2. \qquad (6.4.53)$$

Under any one of these conditions, the integral equation (6.4.47) has a unique solution. Since the unperturbed system is globally controllable, the terminal time T can be freely chosen from the set $(0, \infty)$. Thus it appears from the expressions (6.4.52) and (6.4.53) that these inequalities are always satisfied for small enough $T > 0$. Unfortunately this is not generally true since $\lim_{T\downarrow 0} \alpha_T = \infty$. However, it is quite possible that one of the following conditions may hold

$$\lim_{T\downarrow 0} T\alpha_T = 0, \quad \lim_{T\downarrow 0} \sqrt{T}\alpha_T = 0.$$

In that case we can always find a T sufficiently small but positive so that R_T is a contraction in the Hilbert space $L_2(I_T, R^d)$ guaranteeing global controllability.

Next we consider nonlinear perturbation of the linear system governed by the following semilinear equation

$$\dot{x} = Ax + Bu + G(x), t \ge 0, \quad x(0) = x_0. \qquad (6.4.54)$$

Under standard assumptions on the nonlinear map G (see Chapter 3), this equation has a unique solution for each initial state $x_0 \in R^n$ and control $u \in L_2^{loc}$. For a fixed initial state, and any $t > 0$, we let $F_t(u)$ denote the control to solution map $u \longrightarrow x(t,u)$. Clearly this is a nonlinear non anticipative (or causal) map from the space of control policies to the state space.

Theorem 6.4.2. Suppose the unperturbed system, $\dot{x} = Ax + Bu, x(0) = x_0$, is globally controllable and the nonlinear map $G : R^n \longrightarrow R^n$ is Lipschitz with Lipschitz constant ℓ, and $\| G(0) \| < \infty$. Then the question of controllability of the perturbed system given by (6.4.54) is equivalent to the question of existence of a solution over some finite time interval $I_T \equiv [0,T]$ of the nonlinear functional equation,

$$u(t) = g_T(t) + \int_0^T L_T(t,s)G(F_s(u))ds, t \in I_T, \tag{6.4.55}$$

where

$$g_T(t) \equiv B' e^{(T-t)A'} C_T^{-1}(x_1 - e^{TA}x_0) \equiv B' e^{(T-t)A'} C_T^{-1} z_0, t \in I_T$$
$$L_T(t,s) \equiv -B' e^{(T-t)A'} C_T^{-1} e^{(T-s)A}, (s,t) \in I_T \times I_T.$$

Further, if the Lipschitz constant ℓ is sufficiently small, the integral equation has a unique solution for any given $T > 0$ and the system is globally controllable.

Proof. First we prove the equivalence of the controllability problem with the question of existence of solutions of the integral equation (6.4.55). Suppose $u^* \in L_2(I_T, R^d)$ is a solution of the integral equation and $x(\cdot, u^*)$ is the (unique) solution of the perturbed system (6.4.54) corresponding to the control u^*. We must show that $x(T, u^*) = x_1$, that is, any solution of the integral equation is a control that drives the system from the state x_0 to the desired target state x_1 at time T. This will mean that the existence of a solution of the integral equation implies the controllability of the perturbed system. Since u^* is a solution of the integral equation, we have

$$\begin{aligned} u^*(t) &= g_T(t) + \int_0^T L_T(t,s)G(F_s(u^*))ds, \ t \in I_T, \\ &= g_T(t) + \int_0^T L_T(t,s)G(x(s,u^*))ds, \ t \in I_T, \\ &= B' e^{(T-t)A'} C_T^{-1}\Big(z_0 - \int_0^T e^{(T-s)A}G(x(s,u^*))ds\Big), \\ &\equiv B' e^{(T-t)A'} C_T^{-1}\eta, \end{aligned} \tag{6.4.56}$$

where we have used η to denote the vector within the parenthesis. On the other hand, since $x(\cdot, u^*)$ is the unique solution of the perturbed system corresponding to the control policy u^*, the value of $x(\cdot, u^*)$ at time T is given by

$$x(T,u^*) = e^{TA}x_0 + \int_0^T e^{(T-s)A}Bu^*(s)ds + \int_0^T e^{(T-s)A}G(x(s,u^*))ds. \tag{6.4.57}$$

Substituting (6.4.56) into the second term of equation (6.4.57) we arrive at

$$x(T,u^*) = e^{TA}x_0 + \left\{\eta + \int_0^T e^{(T-s)A}G(x(s,u^*))ds\right\}$$
$$= e^{TA}x_0 + z_0 = e^{TA}x_0 + x_1 - e^{TA}x_0 = x_1.$$

Thus we have shown that if u^* is a solution of the integral equation (6.4.55), then it is also a solution of the controllability problem. Now we prove the converse. We show that any control, that drives the system from the state x_0 to the state x_1 at time T spending minimum energy, is a solution of the integral equation. Suppose $u^o \in L_2(I_T, R^d)$ is the control of minimum norm that takes the system from state x_0 to the target state x_1 at time T. Then we must have the following identity

$$\left(x_1 - e^{TA}x_0 - \int_0^T e^{(T-s)A}G(x(s,u^o))ds\right) = \int_0^T e^{(T-s)A}Bu^o(s)ds, \quad (6.4.58)$$

where $x(\cdot, u^o)$ is the solution of the perturbed equation (6.4.54) corresponding to the control u^o. For notational convenience, letting

$$\eta_T(u^o) \equiv \left(x_1 - e^{TA}x_0 - \int_0^T e^{(T-s)A}G(x(s,u^o))ds\right)$$
$$\equiv \left(x_1 - e^{TA}x_0 - \int_0^T e^{(T-s)A}G(F_s(u^o))ds\right) \quad (6.4.59)$$

the identity (6.4.58) reduces to

$$\eta_T(u^o) = \int_0^T e^{(T-s)A}Bu^o(s)ds. \quad (6.4.60)$$

Before we can proceed further, it is essential to verify that η_T given by (6.4.59) belongs to R^n with $\| \eta_T \| < \infty$. But this follows from the following arguments. By virtue of the assumptions on the nonlinear operator G, it follows from Gronwall inequality that, for every $u \in L_2(I_T, R^d)$, $x(\cdot, u) \in C(I_T, R^n)$ and $\sup\{\| x(t,u) \|, t \in I_T\} < \infty$. Using this result in (6.4.59) and the assumptions on G, just stated, one can easily check that $\eta_T(u^o) \in R^n$. Now consider the system

$$\dot{\zeta} = A\zeta + Bu, t \in I_T, \zeta(0) = 0. \quad (6.4.61)$$

The problem is to find a control $u \in L_2(I_T, R^d)$ that takes the system from the zero state to the state $\eta_T(u^o)$ at time T. Since this system is globally controllable, the unique control that takes the system from the zero state to the target state spending minimum energy, (see Corollary 6.2.6) is given by

$$\tilde{u} \equiv B' e^{(T-t)A'} C_T^{-1} \eta_T(u^o). \tag{6.4.62}$$

From equation (6.4.62) and uniqueness result of Corollary 6.2.7 it follows that u^o is $\tilde{u}$ and we arrive at the following identity

$$u^o \equiv B' e^{(T-t)A'} C_T^{-1} \eta_T(u^o).$$

This is precisely the integral equation (6.4.55). To complete the proof we must show that the functional equation (6.4.55) has a solution. We write this equation in the abstract form

$$u = R_T u \tag{6.4.63}$$

where R_T denotes the nonlinear operator given by the expression on the right hand side of equation (6.4.55). For the existence it suffices to show that R_T is a contraction in the Hilbert space $L_2(I_T, R^d)$. Define

$$\alpha_T \equiv \sup\{\| L_T(t,s) \|, 0 \leq s, t \leq T\}$$
$$a \equiv \| A \|, \quad b \equiv \| B \| .$$

Note that for finite $T > 0$, α_T is finite. Using Gronwall inequality, one can verify that

$$\| R_T u - R_T v \|_{L_2(I_T, R^d)} \leq \gamma \| u - v \|_{L_2(I_T, R^d)}$$

where

$$\gamma = \ell T \alpha_T \, [b/(a+\ell)] \; exp(a+\ell)T.$$

It is clear from this expression that if $\gamma < 1$, R_T is a contraction and hence by Banach fixed point theorem, it has a unique fixed point which is the unique solution of the integral equation (6.4.55). This completes the proof. □

Remark. Again we note that, for existence of a solution of the functional equation (6.4.63), it suffices if $\lim_{T \downarrow 0} (T\alpha_T) = 0$. This is due to the fact that the unperturbed system is globally controllable and hence $T > 0$ can be chosen as small as required for R_T to be a contraction. In that situation it is not necessary to impose any undue limitation on the Lipschitz constant ℓ.

An Example. Consider the inverted pendulum $\ddot{\theta} - k\sin\theta = u$. Defining the state variables as $x_1 = \theta, x_2 = \dot{\theta}$, the system can be written as

$$d/dt \begin{pmatrix} x_1 \\ x_2 \end{pmatrix} = \begin{pmatrix} 0 & 1 \\ 0 & 0 \end{pmatrix} \begin{pmatrix} x_1 \\ x_2 \end{pmatrix} + \begin{pmatrix} 0 \\ 1 \end{pmatrix} u + \begin{pmatrix} 0 \\ k\sin x_1 \end{pmatrix}.$$

Note that the linear part of the system is globally controllable and the nonlinear operator $G(x) \equiv \begin{pmatrix} 0 \\ k\sin x_1 \end{pmatrix}$ satisfies all the assumptions of Theorem 6.4.2. Hence the nonlinear system is globally controllable.

Time Varying Systems

The results given in this section for time invariant systems also apply to time varying systems. We present a result on nonlinear perturbation without proof. The system is governed by the following equations

$$\begin{aligned} \dot{x} &= A(t)x + B(t)u + G(t,x), t \geq s, \qquad (6.4.64) \\ x(s) &= x_0. \end{aligned}$$

Lemma 6.4.3. Suppose A is locally integrable, B is essentially bounded and G satisfies the following conditions

$$\begin{aligned} \| G(t,0) \| &\leq k(t), k \in L_1^+(I_{s,T}) \\ \| G(t,x) - G(t,y) \| &\leq \ell(t) \| x - y \|, \ell \in L_1^+(I_{s,T}). \end{aligned}$$

Then for each $x_0 \in R^n$ and $u \in L_2(I_{s,T}, R^d)$ the perturbed system (6.4.64) has a unique (absolutely continuous) solution $x \in C(I_{s,T}, R^n)$. Further, for fixed $x_0 \in R^n$, and any $t \in I_{s,T}$, the map $u \longrightarrow x(t,u) \equiv F_t(u)$, called the input-state or input-output map, denoted by $F_t(u)$, is continuous and Lipschitz from $L_2(I_{s,t}, R^d)$ to R^n.

Proof. The proof follows from standard results given in Chapter 3. □

Following result is an extension of Theorem 6.4.2 to time varying systems and the proof is similar to that of Theorem 6.4.2.

Theorem 6.4.4. Suppose A is locally integrable, B is essentially bounded and G satisfies the conditions given in Lemma 6.4.3 and the unperturbed system,

$$\dot{x} = A(t)x + B(t)u, x(s) = x_0,$$

is globally controllable after time s. Then the question of controllability of the perturbed system given by (6.4.64) is equivalent to the question of existence of a solution over some finite time interval $I_{s,T} \equiv [s,T]$, of the nonlinear functional equation,

$$u(t) = g_T(t) + \int_s^T L_T(t,\theta)G(\theta, F_\theta(u))d\theta, t \in I_{s,T}, \tag{6.4.65}$$

where

$$\begin{aligned} F_t(u) &\equiv x(t,u), t \in I_{s,T} \\ g_T(t) &\equiv B'(t)\Phi'(T,t)C_T(s)^{-1}(x_1 - \Phi(T,s)x_0) \\ &\equiv B'(t)\Phi'(T,t)C_T(s)^{-1}z_0, t \in I_{s,T} \\ L_T(t,\theta) &\equiv -B'(t)\Phi'(T,t)C_T(s)^{-1}\Phi(T,\theta), (\theta,t) \in I_{s,T} \times I_{s,T}. \end{aligned}$$

Further, if the constant $\hat{\ell} \equiv \int_s^T \ell(t)dt$ is sufficiently small, the integral equation has a unique solution for any given $T > 0$ and the system is globally controllable. Alternatively, if

$$\lim_{T \downarrow s} \left\{ \gamma_T \int_s^T \ell(t)dt \right\} = 0$$

where $\gamma_T \equiv ess - sup\{\| L_T(t,\tau) \|, s \leq \tau, t \leq T\}$, the functional equation has always a solution for T sufficiently small and the system is globally controllable.

6.5 Stabilizability and Dynamic Compensators

Here in this section we study the question of stabilizability of linear time invariant unstable systems. We show that if the system is controllable then it is stabilizable by a state feedback control. Also we show that if the system is stabilizable and observable but the state is not accessible, a state estimator can be constructed which stabilizes the system.

Theorem 6.5.1 (State Feedback)

If the system $\dot{x} = Ax + Bu$ is controllable, then it is stabilizable by linear state feedback. The feedback operator K and the feedback control are given by

$$K = -B'\tilde{C}_T^{-1}, \tag{6.5.66}$$

$$u = Kx, \tag{6.5.67}$$

where

$$\tilde{C}_T \equiv \int_0^T e^{-sA}BB'e^{-sA'}ds. \tag{6.5.68}$$

Proof. We show that $(A+BK)$ is a stability matrix, that is, all its eigenvalues have negative real parts. For any $T>0$, controllability of the system implies that $\tilde{C}_T$ is positive definite and hence the matrix K is well defined. Computing the expression

$$\int_0^T (d/ds)\Big(e^{-sA}BB'e^{-sA'}\Big)ds$$

once by differentiating under the integral sign and once by integrating by parts, we obtain

$$A\tilde{C}_T + \tilde{C}_T A' = BB' - e^{-TA}BB'e^{-TA'}. \tag{6.5.69}$$

Then using $K \equiv -B'\tilde{C}_T^{-1}$ and the above identity, we have

$$\begin{aligned}(A+BK)\tilde{C}_T + \tilde{C}_T(A+BK)' &= (A\tilde{C}_T + \tilde{C}_T A') + BK\tilde{C}_T + \tilde{C}_T K'B' \\ &= (A\tilde{C}_T + \tilde{C}_T A') - 2BB' \\ &= -(BB' + e^{-TA}BB'e^{-TA'}).\end{aligned} \tag{6.5.70}$$

Defining

$$\Gamma \equiv (BB' + e^{-TA}BB'e^{-TA'}), \tag{6.5.71}$$

and $\tilde{A} \equiv (A+BK)$, we can rewrite the above expression in the form of the familiar Lyapunov equation

$$\tilde{A}\tilde{C}_T + \tilde{C}_T(\tilde{A})' = -\Gamma.$$

It follows from global controllability of the system that the second term in the expression for Γ is positive definite and hence $\Gamma \in M_s^+$, the space of real symmetric positive definite square matrices. Thus it follows from Lyapunov theory (see Theorem 4.4.1) that the matrix $(\tilde{A})' = (A+BK)'$ is a stability matrix and hence its transpose is also a stability matrix. Thus we have shown that the matrix K, given by the expression (6.5.66), provides a stabilizing feedback control law. This completes the proof. □

Example 1. Consider the oscillatory system

$$\ddot{\xi} + \beta\xi = u, \beta > 0. \tag{6.5.72}$$

Objective is to find a state feedback control law that makes the system asymptotically stable. The answer is obvious. All that is required is to add a damping term. We show that the theoretical method presented above does just this. Writing the state equation and using the Laplace transform one can easily verify that

$$e^{-tA} = \begin{pmatrix} \cos\sqrt{\beta}t & -(1/\sqrt{\beta})\sin\sqrt{\beta}t \\ \sqrt{\beta}\sin\sqrt{\beta}t & \cos\sqrt{\beta}t \end{pmatrix} \tag{6.5.73}$$

and

$$e^{-tA}BB'e^{-tA'} = \begin{pmatrix} (1/\beta)\sin^2\sqrt{\beta}t & -(1/\sqrt{\beta})\sin\sqrt{\beta}t\cos\sqrt{\beta}t \\ -(1/\sqrt{\beta})\sin\sqrt{\beta}t\cos\sqrt{\beta}t & \cos^2\sqrt{\beta}t \end{pmatrix}. \tag{6.5.74}$$

Since the system is globally controllable, we can determine the controllability matrix by integrating the above expression over any interval of time of positive length. The most convenient one in this case is $[0, T]$ with $T = (\pi/\sqrt{\beta})$. Using this we find that

$$\int_0^T e^{-tA}BB'e^{-tA'}\,dt = \begin{pmatrix} (\pi/2\beta\sqrt{\beta}) & 0 \\ 0 & (\pi/2\sqrt{\beta}) \end{pmatrix}. \tag{6.5.75}$$

This is a diagonal matrix and so easily invertible. Then the reader can easily check that

$$K = -B'C_T^{-1} = [0 \quad -(2\sqrt{\beta}/\pi)] \tag{6.5.76}$$

Using this back in the state space model of equation (6.5.72) we arrive at the following system of equations

$$\begin{aligned} \dot{x}_1 &= x_2 \\ \dot{x}_2 &= -\beta x_1 - (2\sqrt{\beta}/\pi)x_2. \end{aligned}$$

This system is clearly asymptotically stable as expected.

Example 2. The reader should try the truly unstable system given by

$$\ddot{\xi} - \beta\xi = u, \beta > 0. \tag{6.5.77}$$

Since this system is globally controllable, the controllability matrix $C_T > 0$ for any $T > 0$. Thus one can choose $T = (1/2\sqrt{\beta})$. For this choice the controllability matrix is given by

$$C_T \equiv \begin{pmatrix} (1/4\beta\sqrt{\beta})(\sinh 1 - 1) & (-1/4\beta)(\cosh 1 - 1) \\ (-1/4\beta)(\cosh 1 - 1) & (1/4\sqrt{\beta})(\sinh 1 + 1) \end{pmatrix}. \tag{6.5.78}$$

This is a positive definite matrix. The determinant is given by

$$\alpha \equiv det\ C_T = (1/16\beta^2 e)(e^2 - 3e + 1).$$

Inverting the matrix C_T one can find

$$-BB'C_T^{-1} \equiv (-1/4\beta\alpha)\begin{pmatrix} 0 & 0 \\ (\cosh 1 - 1) & (1/\sqrt{\beta})(\sinh 1 - 1) \end{pmatrix}. \tag{6.5.79}$$

Finally the feedback system is given by

$$\begin{aligned} \dot{x}_1 &= x_2 \\ \dot{x}_2 &= -[(1/4\beta\alpha)(\cosh 1 - 1) - \beta]x_1 - (1/4\beta\alpha\sqrt{\beta})(\sinh 1 - 1)x_2. \end{aligned}$$

The reader can easily check that the coefficients have the proper sign. Notice that the method will automatically assign large negative feedback for large β.

Next, we wish to consider the problem of stabilization by output feedback, not state feedback. Since the state of a system is not always accessible, this is a more realistic problem. Before we can consider this problem, we must solve the following estimator problem. Consider the system

$$\dot{x} = Ax + Bu, \; y = Hx,$$

with the output y which is available to the observer, not the state x. Our goal is to produce an estimate $\tilde{x}$ of x from the observation y so that $\tilde{x} \longrightarrow x$ as $t \to \infty$. In other words we demand that the estimator has at least the asymptotic tracking property whatever the initial state may be. Towards this goal, we introduce an estimator of the form

$$\dot{\tilde{x}} = \tilde{A}\tilde{x} + Bu + Ly$$

where we must choose $\tilde{A}$ and L so that

$$\lim_{t\to\infty} \| \tilde{x}(t) - x(t) \| = 0.$$

This problem has a very easy solution.

Corollary 6.5.2. If the system is observable, that is $Rank(A|H) = n$, the estimator problem has a solution. The estimator is given by

$$\dot{\tilde{x}} = (A - LH)\tilde{x} + Bu + Ly, \text{ with } L = \mathcal{C}_T^{-1}H',$$

where

$$\mathcal{C}_T \equiv \int_0^T e^{-tA'}H'He^{-tA}dt$$

is the observability matrix.

Proof. Define the error vector $e \equiv x - \tilde{x}$. It is easy to see that this is governed by the differential equation

$$\dot{e} = (A - LH)e + (A - LH - \tilde{A})\tilde{x}.$$

Choosing $\tilde{A} = (A - LH)$, this reduces to the homogeneous equation $\dot{e} = (A - LH)e$ and the estimator takes the form

$$\dot{\tilde{x}} = (A - LH)\tilde{x} + Bu + Ly.$$

Since the pair (A, H) is observable, the pair (A', H') is controllable, that is the system

$$\dot{\xi} = A'\xi + H'v$$

is globally controllable. The controllability matrix for this system can be taken as

$$C_T \equiv \int_0^T e^{-tA'} H' H e^{-tA} dt.$$

Note that this is equivalent to the observability matrix as introduced in Chapter 5 and it is positive definite. It follows from the previous result (Theorem 6.5.1) that the control given by

$$v \equiv -HC_T^{-1}\xi$$

is a stabilizing control law stabilizing the system

$$\dot{\xi} = A'\xi + H'(-HC_T^{-1})\xi = (A' - H'HC_T^{-1})\xi.$$

In other words the matrix $(A' - H'HC_T^{-1})$ and hence its adjoint, $(A - C_T^{-1}H'H)$, is a stability matrix. Choosing

$$L \equiv C_T^{-1}H',$$

the dynamics of the error process is given by

$$\dot{e} = (A - C_T^{-1}H'H)e.$$

This system is asymptotically stable with respect to the origin. The estimator is given by

$$\dot{\tilde{x}} = (A - LH)\tilde{x} + Bu + Ly, L \equiv C_T^{-1}H'.$$

Asymptotic stability of the error dynamics implies asymptotic tracking property of the estimator. This ends the proof. □

Now we are prepared to consider the problem of stabilization by output feedback. Consider the system

$$\dot{x} = Ax + Bu, \qquad y = Hx, \qquad u = Ky, \tag{6.5.80}$$

where the matrices $\{A, B, H, K\}$ have compatible dimensions. This will give a output feedback system in the form

$$\dot{x} = (A + BKH)x. \tag{6.5.81}$$

The question is, given that the original system is unstable with the matrices $\{A, B, H\}$, is it possible to find a matrix K that stabilizes the feedback system ? Generally the answer to this question is negative even if the pair (A, B) is controllable and the pair (A, H) is observable. We shall see that under precisely these conditions one can construct a stabilizing compensator. The basic idea is to construct a parallel system (observer), similar to the original system, $\dot{x} = Ax + Bu$, but with accessible states which carry the same "information" as the true state x. In other words the compensated system should have the form

$$\dot{x} = Ax + Bu = Ax + BKz \tag{6.5.82}$$
$$\dot{z} = \tilde{A}z + Ly = \tilde{A}z + LHx, \tag{6.5.83}$$

where y is the true output of the original system driving the observer and u is the (linear feedback) control generated by the state z of the compensator not the inaccessible state x of the plant. Our problem is to choose the compensator matrices $\{\tilde{A}, L\}$ so that the $2n$ dimensional compensated system given by

$$(d/dt)\begin{pmatrix} x \\ z \end{pmatrix} = \begin{pmatrix} A & BK \\ LH & \tilde{A} \end{pmatrix}\begin{pmatrix} x \\ z \end{pmatrix} \tag{6.5.84}$$

is asymptotically stable. Consider the $(2n \times 2n)$ matrix

$$J \equiv \begin{pmatrix} I_n & 0 \\ I_n & -I_n \end{pmatrix}$$

where, for any positive integer k, I_k denotes the identity operator (matrix) in R^k. Note that $J^2 = I_{2n}$ and hence J is its own inverse. Now we introduce the coordinate transformation,

$$\begin{pmatrix} x \\ z \end{pmatrix} = J\begin{pmatrix} \xi \\ \eta \end{pmatrix}. \tag{6.5.85}$$

Substituting (6.5.85) into (6.5.84) we obtain

$$(d/dt)\begin{pmatrix}\xi\\ \eta\end{pmatrix} = \begin{pmatrix} A+BK & -BK \\ (A+BK-LH)-\tilde{A} & \tilde{A}-BK \end{pmatrix}\begin{pmatrix}\xi\\ \eta\end{pmatrix}. \tag{6.5.86}$$

Choosing $\tilde{A} = A + BK - LH$, this reduces to

$$(d/dt)\begin{pmatrix}\xi\\ \eta\end{pmatrix} = \begin{pmatrix} A+BK & -BK \\ 0 & A-LH \end{pmatrix}\begin{pmatrix}\xi\\ \eta\end{pmatrix}. \tag{6.5.87}$$

The eigenvalues of the matrix in equation (6.5.87) are determined by the matrices $A + BK$ and $A - LH$. Now if the pair (A, B) or equivalently, the linear system, $\dot{x} = Ax + Bu$, is controllable (suffices if stabilizable) then, as we have seen before, we can choose a matrix $K \in M(d \times n)$ so that $A+BK$ is a stability matrix. Similarly if the pair (A, H) is observable, that is, the system $\dot{x} = Ax, y = Hx$, is observable, then by virtue of the rank condition, (A', H') is controllable. In other words the system $\dot{\zeta} = A'\zeta + H'v$ is (globally) controllable. Hence there exists a matrix $\tilde{K} \in M(m \times n)$ such that $A' + H'\tilde{K}$ is a stability matrix. Clearly $(A' + H'\tilde{K})' = (A + \tilde{K}'H)$ is also a stability matrix. Thus one can choose the matrix $L = -\tilde{K}'$. This makes the compensated system asymptotically stable. Consequently the original system is also asymptotically stable. This is easily seen by writing explicitly the compensator equations

$$\begin{aligned}\dot{\xi} &= (A + BK)\xi - BK\eta \\ \dot{\eta} &= (A - LH)\eta.\end{aligned}$$

Since $(A - LH)$ is a stability matrix, the second equation is asymptotically stable with respect to the origin, that is, $\lim_{t\to\infty} \eta(t) = 0$. For the first variable ξ, note that the solution is given by

$$\xi(t) = e^{(A+BK)t}\xi(0) - \int_0^t e^{(A+BK)(t-s)} BK\eta(s)ds, t \geq 0.$$

Using the fact that $(A+BK)$ is a stability matrix, one can easily verify from the above expression that $\lim_{t\to\infty} \xi(t) = 0$ also. The reader is encouraged to verify this. From these we conclude that $\lim_{t\to\infty} x(t) = 0$ and $\lim_{t\to\infty} z(t) = 0$. Thus we have proved the following result.

Theorem 6.5.3 (Output Feedback) If the original system (6.5.80) is controllable and observable then there exists a stabilizing compensator (observer) given by

$$\dot{x} = Ax + BKz \tag{6.5.88}$$

$$\dot{z} = (A + BK - LH)z + LHx \tag{6.5.89}$$

where K can be taken as that given by the expression (6.5.66) and L is chosen as $L = \mathcal{C}_T^{-1}H'$ where $\mathcal{C}_T$ is the observability matrix as given in the preceding corollary.

6.6 Controllability of Linear Impulsive Systems

Consider the linear system

$$dx(t) = A(t)x(t)dt + B(t)u(dt), x(0) = x_0, \tag{6.6.90}$$

where A is locally integrable and B is bounded measurable $n \times d$ matrix valued function. The controls are elements of the space of countably additive bounded vector (R^d) valued measures on the sigma algebra $\mathcal{B}_0$ of subsets of the positive half of the real line $R_0 \equiv [0, \infty)$ having bounded total variation on bounded subsets of R_0. We denote this class by $\mathcal{M}^{\ell oc}(R_0, R^d)$. We wish to study the question of controllability of these measure driven systems. Note that our control measures may also contain purely impulsive forces located at arbitrary set of instants of time. We introduce the following admissible class. Let μ be a countably additive bounded positive measure having bounded total variation on bounded subsets. We call it the dominating measure. Denote the set

$$\mathcal{U}_\mu \equiv \{u \in \mathcal{M}^{\ell oc}(R_0, R^d) : \lim_{\mu(\sigma) \to 0} u(\sigma) = 0, \sigma \in \mathcal{B}_0\}.$$

Since the measure μ is not necessarily absolutely continuous with respect to the Lebesgue measure, the set $\mathcal{U}_\mu$ may contain sums of Dirac measures. Let $\Phi(t.s), 0 \leq s \leq t < \infty$, denote the transition operator corresponding to $A(t), t \geq 0$ and $x \in B^{\ell oc}(R_0, R^n)$ the solution given by

$$x(t) = \Phi(t, 0)x_0 + \int_0^t \Phi(t, s)B(s)u(ds), t \geq 0.$$

Theorem 6.6.1. Consider the impulsive system (6.6.90) with A locally integrable and B μ-essentially bounded, and define the controllability matrix

$$M_\mu(T) \equiv \int_0^T \Phi(T, s)B(s)B^*(s)\Phi^*(T, s)\mu(ds), T \in R_0, \tag{6.6.91}$$

corresponding to the dominating measure μ. The system is globally controllable if and only if there exists a $T \in R_0$ such that the controllability matrix $M_\mu(T)$ is positive definite.

Proof. Let $x_1 \in R^n$ and define $z = x_1 - \Phi(T,0)x_0 \in R^n$. Clearly, the system is globally controllable if, for any $x_1 \in R^n$ or equivalently any $z \in R^n$, we can find a control u^o from the set $\mathcal{U}_\mu$ such that

$$z = \int_0^T \Phi(T,s)B(s)u^o(ds).$$

Now suppose the controllability matrix $M_\mu(T)$ is positive definite. We prove that the system is globally controllable. Let $z \in R^n$ be arbitrary. Take

$$u^o(\sigma) \equiv \int_\sigma B^*(s)\Phi^*(T,s)M_\mu^{-1}(T)z\mu(ds), \sigma \in \mathcal{B}_0. \tag{6.6.92}$$

Since Φ is bounded and B is μ-essentially bounded, it is clear that $u^o \in \mathcal{U}_\mu$ and hence admissible. Substituting this in the expression

$$y(T,u) \equiv \int_0^T \Phi(T,s)B(s)u(ds)$$

we find that

$$\begin{aligned} y(T,u^o) &= \int_0^T \Phi(T,s)B(s)u^o(ds) \\ &= \int_0^T \Phi(T,s)B(s)B^*(s)\Phi^*(T,s)M_\mu^{-1}(T)z\mu(ds) \\ &= \left(\int_0^T \Phi(T,s)B(s)B^*(s)\Phi^*(T,s)\mu(ds)\right)M_\mu^{-1}(T)z \\ &= z. \end{aligned} \tag{6.6.93}$$

Clearly this shows that the control given by (6.6.92) takes the system from x_0 to any arbitrary state $x_1 \in R^n$. Thus positivity of the controllability matrix $M_\mu(T)$ implies global controllability. Now we prove the converse. Suppose the system is globally controllable with respect to the admissible controls $\mathcal{U}_\mu$. We show that $M_\mu(T)$ is positive definite. Since the system is globally controllable, it is clear that

$$\left(\int_0^T \Phi(T,s)B(s)u(ds), \xi\right) = 0 \;\forall\; u \in \mathcal{U}_\mu \Longrightarrow \xi = 0. \tag{6.6.94}$$

This is equivalent to

$$\int_0^T (B^*(s)\Phi^*(T,s)\xi, u(ds)) = 0 \;\forall\; u \in \mathcal{U}_\mu \Longrightarrow \xi = 0. \tag{6.6.95}$$

Let $L_1(\mu, R^d)$ denote the vector space of all μ-measurable functions $\{h\}$ on $I \equiv [0,T]$ with values in R^d such that $\int_0^T \parallel h(t) \parallel \mu(dt) < \infty$. This is a Banach space. Since $u \in \mathcal{U}_\mu$, u is μ continuous and hence there exists a $g \in L_1(\mu, R^d)$ such that the statement (6.6.95) is equivalent to

$$\int_0^T (B^*(s)\Phi^*(T,s)\xi, g(s))\mu(ds)) = 0 \; \forall \; g \in L_1(\mu, R^d) \Longrightarrow \xi = 0. \quad (6.6.96)$$

This is equivalent to the following statement,

$$B^*(s)\Phi^*(T,s)\xi = 0, \mu \text{ almost all } s \in [0,T] \Longrightarrow \xi = 0.$$

Since B is μ-essentially bounded this statement is equivalent to

$$\int_0^T \parallel B^*(s)\Phi^*(T,s)\xi \parallel^2 \mu(ds) = 0 \;\Longrightarrow \xi = 0,$$

which, in turn, is equivalent to,

$$(M_\mu(T)\xi, \xi) = 0 \;\Longrightarrow \xi = 0.$$

This shows that the controllability matrix $M_\mu(T)$ is positive definite. This concludes the proof. □

As a consequence of the above result we have the following Corollary.

Corollary 6.6.2. Consider the system (6.6.90) and let $\mathcal{U}^o \subset \mathcal{M}^{\ell oc}(R_0, R^d)$ be a nonempty set denoting the class of admissible controls. The system is globally controllable with respect to this admissible set, if there exists a dominating measure μ (positive, countably additive, and has bounded total variation on bounded intervals) such that

$$M_\mu(T) \equiv \int_0^T \Phi(T,s)B(s)B^*(s)\Phi^*(T,s)\mu(ds) \quad (6.6.97)$$

is positive definite.

Purely Impulsive Controls. Let $\{t_i\} \in R_0$ be an arbitrary sequence of time instants. Define the measure

$$\mu(\sigma) \equiv \sum_{i=1}^{\infty} \lambda_i \delta_{t_i}(\sigma), \sigma \in \mathcal{B}_0 \quad (6.6.98)$$

for any sequence $\{\lambda_i\}$ satisfying $0 \leq \lambda_i \leq 1$. Clearly this is a countably additive bounded positive measure having locally bounded total variation. We choose this as the dominating measure and introduce the set

$$\mathcal{U}_\mu \equiv \left\{ u : \mathcal{B}_0 \to R^d : u(\sigma) \equiv \sum_{i=1}^{\infty} \lambda_i u_i \delta_{t_i}(\sigma), u_i \in R^d \right\}. \quad (6.6.99)$$

Let $J \equiv I(a,b) \equiv [a,b]$ denote any bounded interval with $a, b \in R_0$. Suppose B is a continuous and bounded matrix valued function. Define the controllability matrix as

$$M_\mu(J) = \sum_{\{i:t_i \in J\}} \lambda_i \Big(\Phi(\tau, t_i) B(t_i) B^*(t_i) \Phi^*(\tau, t_i) \Big). \tag{6.6.100}$$

Clearly it follows from the above controllability theorem that if for any such interval, $M_\mu(J)$ is positive definite the system is controllable over the given time interval.

Remark. It is interesting to note that the controllability matrix defined above may turn out to be positive definite with only a few impulsive forces over a short period of time. Thus the desired target may be reachable in short time. On the other hand if the intensity of the impulses are allowed to take values only from a compact set $U(\ni 0) \subset R^d$, with admissible controls now given by

$$\mathcal{U}_\mu \equiv \left\{ u : \mathcal{B}_0 \to R^d : u(\sigma) \equiv \sum_{i=1}^{\infty} \lambda_i u_i \delta_{t_i}(\sigma), u_i \in U \subset R^d \right\}, \tag{6.6.101}$$

the time to reach the target may increase with decreasing size of U.

6.7 Exercises

(Q1): The smallest eigenvalue of the controllability matrix C_T is given by

$$\lambda_m(T) \equiv \inf_{\|\xi\|=1} (C_T \xi, \xi).$$

Find the growth rate of this eigenvalue as a function of T. How does $\lambda_m(T)$ approach zero as T approaches zero. This result should be very useful for improving theorem 6.4.1.

(Q2): Give the proof of Theorem 6.4.4 by filling in the details following the proof of Theorem 6.4.2.

(Q3): Prove that global controllability implies positive definiteness of the matrix Γ as given by the expression (6.5.71).

(Q4): Give a detailed proof of Corollary 6.3.2 using the fact that $X(t)$ is nonsingular.

(Q5): Show that for the observer design one can choose

$$K = -B'\tilde{C}_T^{-1}, \text{ where } \tilde{C}_T \equiv \int_0^T e^{tA}BB'e^{tA'}dt$$
$$L = \hat{C}_T^{-1}H', \text{ where } \hat{C}_T \equiv \int_0^T e^{tA'}H'He^{tA}dt.$$

(Q6): Consider the example 2 of section 5 and fill in all the missing details.

(Q7): Show that even if K in the observer-system (6.5.88) and (6.5.89) is chosen arbitrarily (not necessarily stabilizing matrix), the observer state z asymptotically converges to the state x provided (A, H) is observable and the observer gain L is chosen as specified in the proof of Theorem 6.5.2.

6.8 Bibliographical Notes

The concept of controllability is very natural and extremely useful for applications to engineering problems and socioeconomic and government planning process. For original contributions to this concept see Kalman and his coworkers [51, 176]. Controllability and non controllability tells what is do able and what is not. The basic results on controllability of linear systems can be found in many standard books on control theory and its applications such as Ogata [72], Brogan [23], Shinners [83], Ahmed [2], Isidori [49] and Kailath [50]. For infinite dimensional systems see Zabczyk [99]. Results on controllability of perturbed linear and nonlinear systems based on Banach fixed point theorems are presented here for the first time. Controllability under control constraints and controllability of linear systems driven by vector measures are new. For more on such results including controllability of systems governed by differential inclusions see Ahmed [134].

Chapter 7

Basic Calculus of Variations

7.1 Introduction

The precursor to modern control theory is the classical calculus of variations. The basic problem can be stated in words as follows. One is given a functional of certain process, satisfying certain boundary conditions, and also certain regularity properties which determine the function space on which the functional is well defined. The problem is to find a function that extremizes the given functional. Classical calculus of variations dealt with many interesting problems related to geometry where one wishes to minimize length, area, volume, shapes, etc. satisfying some constraints. In the case of dynamic systems, if the functional is a measure of performance or revenue, one is interested to maximize the functional and find the entity that does it. In case the functional is a measure of cost or losses, one is interested to minimize the functional and determine the corresponding element. In fact the extremum may or may not exist or there may be multiple extremals. Thus one is naturally interested in the questions of existence and uniqueness of solutions. Once this question is settled, one wants to develop necessary and sufficient conditions whereby one can compute the extremals. This is the subject matter of this chapter.

7.2 Finite Dimensional Problems

Let us first consider some elementary but instructive examples arising in classical optimization problems. Suppose $f : R^n \longrightarrow R$ is a function and Γ a subset of R^n. The question is: does f attain its minimum (or maximum) on Γ. Mathematically we are looking for a solution to the problem

$$\min\{f(x), x \in \Gamma\}. \tag{7.2.1}$$

The first question is, does it have a solution; and the second question is, if a solution does exist, how do we find it. Before we can answer these questions, we need the notions of lower and upper semi continuities which are more general than the notion of continuity.

Definition 7.2.1. Let E be any Banach space and φ a real valued function on E. The function φ is said to be lower semi continuous at $x \in E$ if, for every sequence $\{x_n\}$ converging to x,

$$\varphi(x) \leq \liminf_{n\to\infty} \varphi(x_n)$$

and it is said to be upper semi continuous at x if

$$\varphi(x) \geq \limsup_{n\to\infty} \varphi(x_n),$$

and it is said to be lower or upper semi continuous on a set $\Gamma \subset E$, if the corresponding statements hold for all $x \in \Gamma$.

Now we return to the question of existence. The simplest answer to the question of existence is, if f is continuous and Γ is compact, then f attains both its maximum and minimum on Γ. In fact continuity is not essential. For existence of a minimum, it suffices if f is merely lower semi continuous on Γ. This is stated in the following theorem.

Theorem 7.2.2. If $\Gamma \subset R^n$ is compact and $f : R^n \longrightarrow \bar{R}$ is lower (upper) semi continuous and bounded away from $-\infty(+\infty)$, then f attains its minimum (maximum) on Γ.

Proof. We prove for the minimum. For the maximum, the proof is identical. Let

$$\inf\{f(x), x \in \Gamma\} = m > -\infty,$$

and suppose $\{x_n\}$ is any minimizing sequence, that is,

$$\lim_{n\to\infty} f(x_n) = m.$$

Then, since Γ is compact, there is a subsequence $\{x_{n_k}\} \subset \{x_n\}$ and an element $x_o \in \Gamma$ such that $\lim_{k\to\infty} x_{n_k} = x_o$. By lower semi continuity of f and the preceding identity, we have

$$f(x_o) \leq \liminf_{k\to\infty} f(x_{n_k}) = \lim_{k\to\infty} f(x_{n_k}) = m.$$

Since $x_o \in \Gamma$, it is evident that $f(x_o) \geq m$. Hence we must have $f(x_o) = m$, that is, f attains its infimum (hence the minimum) on Γ. This completes the proof. □

If the set Γ is not compact, minimum may not exist. A simple example is

$$f(x) = x, \Gamma \equiv (0, 1].$$

Clearly $\inf\{f(x), x \in \Gamma\} = 0$, but $0 \notin \Gamma$.

However, the compactness of Γ is not a necessary condition. In fact, the lack of compactness can be compensated by some regularity properties of the function f. One such condition is given in the following result.

Theorem 7.2.3. Suppose $f : R^d \longrightarrow \bar{R}$ is lower semi continuous satisfying the following conditions:

$$-\infty < f(x) \not\equiv +\infty, \forall x \in R^d, \text{ and } \lim_{\|x\| \to \infty} f(x) = +\infty. \tag{7.2.2}$$

Then f attains its minimum.

Proof. The proof is as follows. Since $f(x) \not\equiv +\infty$, and $f(x) > -\infty$ for all $x \in R^d$, it is evident that $\inf\{f(x), x \in R^d\} = m$ exists with $|m| < \infty$. Let $\{x_n\} \in R^d$ be a minimizing sequence so that

$$\lim_{n \to \infty} f(x_n) = m.$$

If the sequence $\{x_n\}$ is unbounded, $\lim_{n \to \infty} f(x_n) = \infty$ by the property (7.2.2). This contradicts finiteness of m. Hence the sequence $\{x_n\}$ must be bounded. A bounded set in R^d is always relatively compact, that is, its closure is compact. Thus there exists a subsequence, $\{x_{n_k}\} \subset \{x_n\}$, and an element $x_o \in R^d$, such that $\lim_{k \to \infty} x_{n_k} = x_o$. Since f is lower semi continuous we have

$$f(x_o) \leq \lim_{k \to \infty} f(x_{n_k}).$$

Hence $f(x_o) = m$. In other words, f attains its minimum at x_o. This ends the proof. □

Note: Obviously, the condition (7.2.2) is satisfied if $f(x) \geq c$ for some $c \in R$ and coercive, that is,

$$\lim_{\|x\| \to \infty} \frac{f(x)}{\| x \|} \longrightarrow +\infty.$$

We present here few simple examples:

(E1):

$$f(x) = (b, x) + c, \quad \text{and} \quad \Gamma = B_r \equiv \{x \in R^n : \| x \| \leq r\} \tag{7.2.3}$$

where $b \in R^n, c \in R$ and $r \in (0, \infty)$. Clearly f is continuous and Γ is compact. The reader can easily verify that f attains its minimum at $x_o = -(r/ \parallel b \parallel)b$ and maximum at $z_o = (r/ \parallel b \parallel)b$. These are points on the boundary of the ball B_r.

(E2):

$$f(x) = (1/2)(Qx, x) + (b, x) + c, \quad \text{and} \quad \Gamma = R^n \tag{7.2.4}$$

where Q is a real symmetric positive definite $(n \times n)$ matrix and $b \in R^n$ and $c \in R$. Clearly the set Γ is not compact, but the function f satisfies the property

$$f(x) > -\infty, \lim_{\|x\| \to \infty} f(x) = +\infty,$$

and it is also bounded on bounded sets. Clearly, here the point at which f attains its minimum can be found by setting $\nabla f = 0$, that is, $Qx + b = 0$. Since Q is positive definite, it is invertible and the solution is given by $x_o = -Q^{-1}b$. In example (E2), if Q is negative definite, the property (7.2.2) is reversed and we have

$$\infty > f(x) \neq -\infty, \forall x \in R^n, \text{ and } \lim_{\|x\| \to \infty} f(x) = -\infty. \tag{7.2.5}$$

In this case f has a maximum and it is given by $x_o = (-Q)^{-1}b$. Note that for any arbitrary matrix Q, the function f of example (E2) is continuous and therefore, it attains both its maximum and minimum on any compact subset of R^n. However they can not be determined by setting $\nabla f = 0$. For example, considering (E1), it is evident that this will yield $\nabla f = b = 0$, which obviously does not make sense.

7.3 Existence of Solutions for Variational Problems

Most of the problems of calculus of variations involve infinite dimensional Banach spaces. One typical problem is the Lagrange problem. Let $I \equiv [a, b]$ be a closed bounded interval and consider the functional given by

$$\varphi(x) \equiv \int_I \ell(t, x(t), \dot{x}(t))dt, x(a) = \xi, x(b) = \eta, \xi, \eta \in R^n. \tag{7.3.6}$$

The basic problem is to find a suitable function x with values $x(t) \in R^n$ satisfying the boundary conditions specified so that it minimizes the functional φ. Clearly this involves infinite dimensional Banach spaces, like the space of continuous functions denoted by $C(I, R^n)$, the space of absolutely continuous functions $AC(I, R^n)$ or more generally, $BV(I, R^n)$, the space of functions of bounded variations and the Sobolev spaces $W^{m,p}$ to be defined shortly.

Definition 7.3.1. Let E be any Banach space and φ a real valued function on E. The function φ is said to be weakly lower semi continuous at $x \in E$ if, for every sequence $\{x_n\}$ converging weakly to x,

$$\varphi(x) \leq \liminf_{n \to \infty} \varphi(x_n)$$

and it is said to be weakly upper semi continuous at x if

$$\varphi(x) \geq \limsup_{n \to \infty} \varphi(x_n),$$

and it is said to be weakly lower or weakly upper semi continuous on a set $\Gamma \subset E$, if the corresponding statements hold for all $x \in \Gamma$. The function φ is said to be weakly continuous if it is both upper and lower semi continuous in the weak sense.

Again one can state similar results as in the finite dimensional case. We present this classical result.

Theorem 7.3.2. Let Γ be a weakly compact subset of a Banach space E, and φ a real valued weakly lower (upper) semi continuous function on E. Then φ attains it's minimum (maximum) on Γ.

Proof. The proof is exactly the same as in the finite dimensional case.

Theorem 7.3.3. Let E be a reflexive Banach space and Γ a nonempty weakly closed subset of E. Let φ be a real valued weakly lower semi continuous function on Γ satisfying (i): $-\infty < \varphi(x) \not\equiv +\infty$ and (ii): $\lim_{\|x\| \to \infty} \varphi(x) = +\infty$. Then φ attains it's minimum on Γ.

Proof. Since φ is bounded away from $-\infty$ and it is not identically $+\infty$, we have

$$\inf\{\varphi(x), x \in \Gamma\} = m, \text{ with } |m| < \infty.$$

Let $\{x_n\} \subset \Gamma$ be a minimizing sequence, that is,

$$\lim_{n \to \infty} \varphi(x_n) = m.$$

Clearly by virtue of the property (ii), $\{x_n\}$ is a bounded sequence. In a reflexive Banach space, every bounded set is conditionally or relatively weakly compact. Hence there exists a subsequence $\{x_{n_k}\} \subset \{x_n\}$ and an element x_o such that

$$x_{n_k} \xrightarrow{w} x_o \text{ in } E.$$

Since Γ weakly closed, and x_o is the weak limit of a sequence from Γ, we have $x_o \in \Gamma$. By our assumption, φ is weakly lower semi continuous and hence we have

$$\varphi(x_o) \leq \liminf_{k\to\infty} \varphi(x_{n_k}) = \lim_{k\to\infty} \varphi(x_{n_k}) = m.$$

Since $x_o \in \Gamma$ and m is the infimum of φ on Γ, $m \leq \varphi(x_o)$. Thus we have

$$m \leq \varphi(x_o) \leq \liminf_{k\to\infty} \varphi(x_{n_k}) = \lim_{k\to\infty} \varphi(x_{n_k}) = m.$$

This proves that φ attains its minimum at $x_o \in \Gamma$. This completes the proof. □

The basic problems of calculus of variations are (i) Lagrange problem, (ii) Bolza problem and (iii) Meyer's problem as described below:

(i) Lagrange Problem: $\varphi(x) \equiv \int_I \ell(t, x(t), \dot{x}(t))dt \longrightarrow \inf, x \in \Gamma \subset E$ where E is a suitable Banach space.
(ii) Bolza problem: $\varphi(x) = \int_I \ell(t, x(t), \dot{x}(t))dt + \Psi(x(b)) \longrightarrow \inf, x \in \Gamma \subset E$.
(iii) Meyer's problem: $\varphi(x) \equiv \Psi(a, x(a); b, x(b)) \longrightarrow \inf, x \in \Gamma \subset E$.

Note that in each of the above problems the constraint set may be different. We consider the Lagrange problem first. Our problem is to determine if the Lagrange problem has a solution. For this we need the notions of convexity, supporting hyperplanes etc.

Definition 7.3.4. (Convex Function) Let E be a Banach space and f a (possibly extended) real valued function defined on E. The function f is said to be convex if for every $x, y \in E$ and $\alpha \in [0, 1]$

$$f((1-\alpha)x + \alpha y) \leq (1-\alpha)f(x) + \alpha f(y). \tag{7.3.7}$$

Similarly, if E is replaced by a closed convex subset Γ of E and f satisfies the above inequality for all $x, y \in \Gamma$, we say that f is convex on Γ.

Definition 7.3.5. (Supporting hyperplane) Let D be a nonempty closed convex subset of the Banach space E and $x_0 \in \partial D$ a point on the boundary. A linear functional $e^* \in E^*$ having unit norm ($\| e^* \|_{E^*} = 1$) is said to be a supporting hyperplane for the set D at the point x_0, if

$$(e^*, x) \leq (e^*, x_0) = \ \| x_0 \| \ \forall x \in D.$$

In the following we wish to show how intimately the notions of convexity and weak lower semi continuity are related.

Definition 7.3.6. (Support functional) An element $e^* \in E^*$ is called a support functional of the function f at $x_0 \in E$ if

$$(e^*, x - x_0) \leq f(x) - f(x_0) \ \ \forall x \in E.$$

Definition 7.3.7. (Gateaux differentials and sub differentials) A real valued functional f defined on E is said to be Gateaux differentiable at the point $x_0 \in E$ if, for every $y \in E$, the limit

$$\lim_{s \to 0} (1/s)\{f(x_0 + sy) - f(x_0)\} \equiv df(x_0, y)$$

exists. Further, if $y \longrightarrow df(x_0, y)$ is continuous and linear, then there exists an $e^* \in E^*$, dependent on x_0, such that

$$df(x_0, y) = (e^*, y).$$

The element $e^* \in E^*$ satisfying the preceding identity, is called the Gateaux gradient of f at $x_0 \in E$. The function f is said to be Gateaux differentiable if it is so at every point $x_0 \in E$.

In case the functional is not Gateaux differentiable, it may be differentiable in some generalized sense. One such general notion is the concept of sub differentials. It is a subset of the dual E^* given

$$\partial f(x_0) = \left\{ e^* \in E^* : \lim_{s_n \to 0} (1/s_n)(f(x_0 + s_n y) - f(x_0)) = (e^*, y), \ \forall \ y \in E \right\}.$$

That is, the subdifferential is a family of continuous linear functionals on E which are given by the limits of the difference quotient along all subsequences $\{s_n\}$ for which they may exist.

Theorem 7.3.8. A real valued function f defined on a Banach space E is weakly lower semi continuous if it is convex and continuously (linearly) Gateaux differentiable.

Proof. Since f is convex, it follows from the definition that for any pair $x_0, x \in E$, we have

$$f((1-\alpha)x_0 + \alpha x) = f(x_0 + \alpha(x - x_0)) \leq (1-\alpha)f(x_0) + \alpha f(x), \ \forall \alpha \in [0,1].$$

Thus

$$(1/\alpha)(f(x_0 + \alpha(x - x_0)) - f(x_0)) \leq f(x) - f(x_0).$$

Now letting $\alpha \to 0$, we obtain

$$df(x_0, x - x_0) \leq f(x) - f(x_0).$$

Since by our hypothesis, the Gateaux differential $y \longrightarrow df(x_0, y)$ is continuous and linear we obtain

$$(e^*, x - x_0) = (Df(x_0), x - x_0) \leq f(x) - f(x_0), \tag{7.3.8}$$

where $e^* = Df(x_0) \in E^*$. Now let $x_n \xrightarrow{w} x_0$ in E. Then

$$0 = \liminf_{n\to\infty}(Df(x_0), x_n - x_0) \leq \liminf_{n\to\infty} f(x_n) - f(x_0)$$

implying that

$$f(x_0) \leq \liminf_{n\to\infty} f(x_n).$$

Thus we have proved that f is weakly lower semi continuous whenever it is convex and continuously Gateaux differentiable. □

It is interesting to note that the convexity of f and its Gateaux differentiability imply that the Gateaux derivative $Df(\xi) \equiv F(\xi), \xi \in E$, is a monotone operator mapping E to E^*, that is,

$$(F(x) - F(y), x - y) \geq 0.$$

Indeed, evaluating the differentials at two points x_1, x_2 it follows from (7.3.8) that

$$(Df(x_1), x_2 - x_1) \leq f(x_2) - f(x_1), \tag{7.3.9}$$

$$(Df(x_2), x_1 - x_2) \leq f(x_1) - f(x_2). \tag{7.3.10}$$

Adding these, one can easily check that

$$(Df(x_2) - Df(x_1), x_2 - x_1) \geq 0,$$

that is the operator $F(x) \equiv Df(x)$ from E to its dual E^* is monotone.

It is also interesting to note that if f is continuously Gateaux differentiable satisfying the inequality (7.3.8), then it is convex. Indeed, for $x_1, x_2 \in E$, it follows from (7.3.8) that

$$(Df(x_0), x_1 - x_0) \leq f(x_1) - f(x_0) \tag{7.3.11}$$

$$(Df(x_0), x_2 - x_0) \leq f(x_2) - f(x_0). \tag{7.3.12}$$

Multiplying the first inequality by $(1 - \alpha)$ and the second by α for $\alpha \in [0, 1]$ and then summing we obtain

$$(Df(x_0), (1 - \alpha)x_1 + \alpha x_2 - x_0) \leq (1 - \alpha)f(x_1) + \alpha f(x_2) - f(x_0).$$

This is true for all $x_0, x_1, x_2 \in E$. Choosing $x_0 = (1-\alpha)x_1 + \alpha x_2$ and substituting in the above expression, we have

$$0 \le (1-\alpha)f(x_1) + \alpha f(x_2) - f((1-\alpha)x_1 + \alpha x_2).$$

Hence

$$f((1-\alpha)x_1 + \alpha x_2) \le (1-\alpha)f(x_1) + \alpha f(x_2)$$

implying convexity of f. The implication of the result stated in Theorem 7.3.8 is far reaching. Here there is a deep interplay between geometry and topology. To see this we need the following definition.

Definition 7.3.9. The graph of any map F from a Banach space X to a Banach space Y is a subset of $X \times Y$ given by $\mathcal{G}(F) \equiv \{\{x,y\} \in X \times Y : y = F(x)\}$.

Corollary 7.3.10. Every convex function $f : E \longrightarrow R$, which is continuously Gateaux differentiable, is supported below at every point on its graph by an affine functional and consequently any such function f is given by the supremum of affine functions.

Proof. By definition, the graph of f is given by

$$\mathcal{G}(f) = \{\{e,r\} \in E \times \bar{R} : r = f(e)\}.$$

It follows from the expression (7.3.8) that

$$f(x) \ge f(x_0) + (Df(x_0), x - x_0) = f(x_0) + (e^*, x - x_0) \ \forall x, x_0 \in E. \quad (7.3.13)$$

Setting $f(x_0) = r_0$ and $f(x) = r$ it is clear that $\{x_0, r_0\}, \{x, r\} \in \mathcal{G}(f)$. Thus it follows from the above inequality that

$$r \ge r_0 + (e^*, x - x_0) = (e^*, x) + c_0,$$

and by Definition 7.3.6, $Df(x_0) = e^*$ is the support functional of f at x_0. Thus, for every point $x \in E$, the support functionals of f have the form

$$\{g_{c,e^*}(x) = c + (e^*, x), c \in R, e^* \in E^*\}$$

and that $f(x) \ge g_{c,e^*}(x)$ for all $x \in E$. Irrespective of whether E is separable or not we can take a countable family of support functionals of f like

$$\{g_i(x) = c_i + (e_i^*, x), c_i \in R \text{ and } e_i^* \in E^*, i \ge 1\}$$

in the neighborhood of x and define

$$f_m(x) = \sup\{g_i(x), 1 \leq i \leq m\}.$$

Clearly $\{f_m(x)\}$ is a sequence of nondecreasing functions satisfying $f(x) \geq f_m(x)$ for $x \in E$ and $f(x) = \lim_{m \to \infty} f_m(x)$. This shows that any convex function f which is continuously Gateaux differentiable on E is given, at every point $x \in E$, by the limit of a sequence of nondecreasing affine functionals evaluated at the point x. This completes the proof. □

Remark. The weak lower semi continuity of a function f satisfying the assumptions of Corollary 7.3.10 can also be proved from the approximation result as stated above. Indeed, let f_m denote the affine approximations of f and suppose $x_n \xrightarrow{w} x_0$. Clearly, for each integer m, f_m is weakly continuous and $f_m(x_n) \longrightarrow f_m(x_0)$ as $n \to \infty$. Thus

$$\begin{aligned}
\liminf_{n\to\infty} f(x_n) &= \liminf_{n\to\infty} = \left\{\lim_{m\to\infty} f_m(x_n)\right\} \\
&\geq \lim_{m\to\infty} \left\{\liminf_{n\to\infty} f_m(x_n)\right\} \\
&= \lim_{m\to\infty} f_m(x_0) = f(x_0).
\end{aligned}$$

Remark. It follows from the above remark that every convex functional which is continuously Gateaux differentiable is weakly lower semi continuous.

Now we are prepared to consider the problems of calculus of variations. Considering the Lagrange problem,

$$\varphi(x) = \int_I \ell(t, x(t), \dot{x}(t))dt,$$

we observe that ℓ is a function of x and its derivative. So a natural function space for this problem is the space of absolutely continuous functions $AC(I, R^n)$, or more generally, the space of functions of bounded variation, $BV(I, R^n)$. This is due to the fact that such functions are differentiable almost every where and the derivatives are locally integrable. But when we choose any such function space, we may encounter difficulties with proving that a minimizing sequence has a weakly convergent subsequence. We discuss this further in the sequel. A more convenient function space is the Sobolev space $W^{1,p} \equiv \{\phi \in L_p(I, R^n) : \dot{\phi} \in L_p(I, R^n)\}$. This space is given the norm topology,

$$\| \phi \|_{1,p} = \left(\| \phi \|_p^p + \| \dot{\phi} \|_p^p \right)^{1/p}.$$

With respect to this norm, it is a Banach space. For $1 < p < \infty$, it is a reflexive Banach space and consequently every (see Chapter 1, Theorem 1.5.3) closed bounded subset of it is weakly compact. This fact makes things easier and it is used to prove the existence of solutions of variational problems like the Lagrange problem and more. For more details on Sobolev spaces see Adams [1]. Our problem is to find a solution of the Lagrange problem satisfying the boundary conditions $x(a) = \xi, x(b) = \eta$ where $\xi, \eta \in R^n$. Towards this goal we define the set

$$\Gamma \equiv \{x \in W^{1,p} : x(a) = \xi, x(b) = \eta\}.$$

Theorem 7.3.11. Suppose the function $\ell = \ell(t, x, z)$ is measurable in t on I and continuous in the second argument x on R^n and convex in the third argument z on R^n and also once Gateaux differentiable in this variable. Further suppose that there exists an $h \in L_1(I)$, $\alpha, \beta > 0$ and $1 < p \leq \gamma, \delta < \infty$, such that

$$\ell(t, x, z) \geq h(t) + \alpha|x|^\gamma + \beta|z|^\delta \tag{7.3.14}$$

for all $x, z \in R^n$. Then the Lagrange problem has a solution $x_o \in \Gamma$.

Proof. If

$$\varphi(x) = \int_I \ell(t, x(t), \dot{x}(t))dt \equiv +\infty,$$

(for all $x \in \Gamma$) there is nothing to prove. So we may assume that there is at least one $x \in \Gamma$ such that $\varphi(x) < \infty$. Let $\{x_n\} \subset \Gamma$ be a minimizing sequence, that is

$$\lim_{n \to \infty} \varphi(x_n) = M \equiv \inf\{\varphi(x), x \in \Gamma\}.$$

The sequence $\{x_n\}$ is necessarily bounded. This is verified as follows. First note that by virtue of the inequality (7.3.14) we have

$$\varphi(x_n) \geq \int_I h(t)dt + \alpha \int_I |x_n(t)|^\gamma dt + \beta \int_I |\dot{x}_n(t)|^\delta dt. \tag{7.3.15}$$

Since for any set I of finite Lebesgue measure and $1 \leq p \leq r$, $L_r(I, R^n) \subset L_p(I, R^n)$, it follows from the above inequality that there exist constants $c_1, c_2 > 0$, possibly dependent on the parameters $\alpha, \beta, p, \gamma, \delta$ and the Lebesgue measure of the set I, such that

$$\varphi(x_n) \geq \int_I h(t)dt + c_1 \int_I |x_n(t)|^p dt + c_2 \int_I |\dot{x}_n(t)|^p dt. \tag{7.3.16}$$

Thus if the sequence $\{x_n\} \subset \Gamma \subset W^{1,p}(I, R^n)$ is not bounded it will follow from the above inequality that

$$\lim_{n\to\infty} \varphi(x_n) = +\infty.$$

This contradicts the very hypothesis that $\{x_n\}$ is a minimizing sequence. Therefore $\{x_n\}$ is a bounded sequence. Since $W^{1,p}$ is a reflexive Banach space, every bounded sequence has a weakly convergent subsequence. Relabeling this subsequence as the original sequence we may assume that $\{x_n\}$ itself converges weakly to some $x_o \in W^{1,p}$. This implies that

$$x_n \xrightarrow{w} x_o \text{ in } L_p(I, R^n)$$
$$\dot{x}_n \xrightarrow{w} \dot{x}_o \text{ in } L_p(I, R^n).$$

Since Γ is closed, the limit $x_o \in \Gamma$. Clearly it follows from the weak convergence of the sequence $\{\dot{x}_n\}$ to $\dot{x}_o$ and the identities

$$x_n(t) = \xi + \int_a^t \dot{x}_n(s)ds, \quad x_o(t) = \xi + \int_a^t \dot{x}_o(s)ds, t \in I,$$

that $\{x_n(t)\}$ converges point wise to $x_o(t)$, for each $t \in I$. Since $z \longrightarrow \ell(t, x, z)$ is convex and once differentiable, it follows from Corollary 7.3.10, that there exists a nondecreasing sequence of affine functions $\{\ell_m\}$ such that $\ell(t, x, z) \geq \ell_m(t, x, z)$ for almost all $t \in I$ and all $x, z \in R^n$ and

$$\ell(t, x, z) = \lim_{m\to\infty} \ell_m(t, x, z), \ \forall\, x, z \in R^n, \text{ and } a.e\ t \in I.$$

Thus for every finite integer m, we have

$$\varphi(x_n) = \int_I \ell(t, x_n(t), \dot{x}_n(t))dt \geq \int_I \ell_m(t, x_n(t), \dot{x}_n(t))dt \equiv \varphi_m(x_n). \quad (7.3.17)$$

Since ℓ_m is continuous in the second variable and affine in the third argument and $x_n(t) \to x_o(t)$ point wise and $\dot{x}_n \to \dot{x}_o$ weakly in $L_p(I, R^n)$ we have

$$\liminf_{n\to\infty} \varphi(x_n) \geq \liminf_{n\to\infty} \varphi_m(x_n) = \varphi_m(x_o) = \int_I \ell_m(t, x_o(t), \dot{x}_o(t))dt, \quad (7.3.18)$$

for every finite integer m. Note that $\ell_m(\cdot, x_o(\cdot), \dot{x}_o(\cdot))$ is a monotone increasing sequence of extended real valued functions on I. Hence, it follows from the generalized monotone convergence theorem (see Appendix, Theorem A.6.2) that

$$\lim_{m\to\infty} \int_I \ell_m(t, x_o(t), \dot{x}_o(t))$$
$$= \int_I \lim_{m\to\infty} \ell_m(t, x_o(t), \dot{x}_o(t))dt = \int_I \ell(t, x_o(t), \dot{x}_o(t))dt.$$

In view of this, it follows from (7.3.18) that

$$\liminf_{n\to\infty} \varphi(x_n) \geq \varphi(x_o).$$

Since $x_o \in \Gamma$ and $\{x_n\}$ is a minimizing sequence, we have

$$M \leq \varphi(x_o) \leq \liminf_{n\to\infty} \varphi(x_n) = \lim_{n\to\infty} \varphi(x_n) = M.$$

This shows that φ attains its minimum at $x_o \in \Gamma$ and hence the Lagrange problem has a solution. This completes the proof. □

Now we consider the Bolza problem:

$$\varphi(x) = \int_I \ell(t, x(t), \dot{x}(t))dt + \Psi(x(b)) \longrightarrow \inf, x \in \Gamma$$

where $\Gamma \subset \{x \in W^{1,p}(I, R^n) : x(a) = \xi\}$.

Corollary 7.3.12. Suppose the function ℓ satisfies the assumptions of Theorem 7.3.11 and Ψ is real valued lower semi continuous on R^n satisfying

$$\Psi(z) \geq c_1 + c_2|z|^r, \forall z \in R^n$$

for certain constants $c_1 \in R, c_2 \geq 0, r \in (0, \infty)$. Then the Bolza problem has a solution $x_o \in \Gamma$.

Proof. In view of the proof of the previous theorem, it suffices to give a brief outline. Again if $\varphi(x) \equiv +\infty$ there is nothing to prove. So assume the contrary. Suppose $\{x_n\} \subset \Gamma$ is a minimizing sequence. By virtue of the assumptions on ℓ and Ψ, it is clear that this is a bounded sequence in $W^{1,p}(I, R^n)$ and hence there exists a weakly convergent subsequence. Naming this subsequence as $\{x_n\}$, it follows from the weak convergence of the derivatives $\{\dot{x}_n\}$ in $L_p(I, R^n)$ to $\dot{x}_o$ that

$$x_n(b) = \xi + \int_a^b \dot{x}_n(s)ds \longrightarrow \xi + \int_a^b \dot{x}_o(s)ds = x_o(b)$$

as $n \to \infty$. By lower semi continuity of Ψ, we have

$$\liminf_{n\to\infty} \Psi(x_n(b)) \geq \Psi(x_o(b)).$$

Clearly

$$\begin{aligned}\liminf_{n\to\infty} \varphi(x_n) &\geq \liminf_{n\to\infty} \int_I \ell(t, x_n(t), \dot{x}_n(t))dt + \liminf_{n\to\infty} \Psi(x_n(b)) \\ &\geq \int_I \ell(t, x_o(t), \dot{x}_o(t))dt + \Psi(x_o(b)) = \varphi(x_o).\end{aligned}$$

The rest of the argument being identical as in the proof of Theorem 7.3.11, we conclude that the Bolza problem has a solution. □

Now we consider the Meyer's problem,

$$\varphi(x) \equiv \Psi(x(a), x(b)) \longrightarrow \inf, x \in \Gamma \subset W^{1,p}. \tag{7.3.19}$$

Corollary 7.3.13. Consider the Meyer's problem. Let $\Gamma \subset W^{1,p}(I, R^n)$ be a closed subset. Suppose $\Psi : R^n \times R^n \longrightarrow \bar{R}$ is an extended real valued lower semi continuous function and there exist constants $c_1 \in R, c_2, c_3 \geq 0, c_2 + c_3 > 0, r_2, r_3 \in (0, \infty)$ so that

$$\Psi(\xi, \zeta) \geq c_1 + c_2|\xi|^{r_2} + c_3|\zeta|^{r_3} \;\forall\; \xi, \zeta \in R^n.$$

Then the Meyer's problem has a solution.

Proof. If $\varphi(x) \equiv +\infty$ there is nothing to prove. So, as before, we assume that there is at least one element in Γ for which it is finite. Let $\{x_n\} \subset \Gamma$ be a minimizing sequence. Then the sequence is bounded, since otherwise

$$\Psi(x_n(a), x_n(b)) \geq c_1 + c_2|x_n(a)|^{r_2} + c_3|x_n(b)|^{r_3} \longrightarrow +\infty$$

contradicting that $\{x_n\}$ is a minimizing sequence from Γ. Hence there exists a subsequence, relabeled as the original sequence, and an element $x_o \in W^{1,p}$ such that $x_n \xrightarrow{w} x_o$ in $W^{1,p}$. Since Γ is closed, $x_o \in \Gamma$. Then it follows from similar arguments as presented in the proof of Theorem 7.3.11, that $x_n(t) \longrightarrow x_o(t)$ point wise in t on I and hence $x_n(a) \longrightarrow x_o(a)$ and $x_n(b) \longrightarrow x_o(b)$. By lower semi continuity of Ψ, we have

$$\Psi(x_o(a), x_o(b)) \leq \liminf_{n\to\infty} \Psi(x_n(a), x_n(b)).$$

Thus we have $\varphi(x_o) \geq \liminf_{n\to\infty} \varphi(x_n)$. The rest of the proof is identical to that of Theorem 7.3.11. Hence the Meyer's problem has a solution. □

Next we consider a Meyer's problem with variable end points. Let I be any closed bounded interval with $s, \tau \in I, s \leq \tau$. The problem is to find a triple $\{x_o, s_o, \tau_o\}$ with $x_o \in W^{1,p}([s_o, \tau_o], R^n)$ $s_o \leq \tau_o, s_o, \tau_o \in I$ such that the functional

$$\varphi(x, s, \tau) \equiv \Psi(s, x(s), \tau, x(\tau))$$

attains its minimum at the point $\{x_o, s_o, \tau_o\}$. Before we consider this problem we note that by setting the values of elements of $W^{1,p}(J, R^n)$ outside the interval $J \subset I$ constants equal to their boundary values, we can embed this space in $W^{1,p}(I, R^n)$. Hence the above problem can be reformulated as an extremal problem on $W^{1,p}(I, R^n) \times I \times I$.

Corollary 7.3.14. Consider the Meyer's problem with variable end points as described above. Suppose $\Gamma \subset W^{1,p}(I, R^n)$ is closed and $\Psi : I \times R^n \times I \times R^n \longrightarrow \bar{R}$ is an extended real valued lower semi continuous function and there exist constants $c_1 \in R, c_2, c_3 \geq 0, c_2 + c_3 > 0$ and $r_2, r_3 \in (0, \infty)$ such that

$$\Psi(s, \xi, \tau, \zeta) \geq c_1 + c_2|\xi|^{r_2} + c_3|\zeta|^{r_3} \ \forall\ \xi, \zeta \in R^n.$$

Then the Meyer's problem with variable end points has a solution.

Proof. Let $\{x_n\} \in \Gamma$ with end points $\{s_n, \tau_n\} \subset I$, be a minimizing sequence. That is,

$$\begin{aligned} M &= \inf\{\Psi(s, x(s), \tau, x(\tau)), s, \tau \in I, s \leq \tau, x \in \Gamma\} \\ &= \lim_{n \to \infty} \Psi(s_n, x_n(s_n), \tau_n, x(\tau_n)). \end{aligned}$$

Without loss of generality we may assume that s_n is a nondecreasing sequence while τ_n is a non increasing sequence converging to s_o and τ_o respectively. Again by virtue of the coercivity assumption on Ψ, the sequence $\{x_n\}$ is bounded in $W^{1,p}$ and hence there exists a subsequence and an element $x_o \in W^{1,p}$ to which x_n converges weakly in $W^{1,p}$. By similar arguments as presented many times before, we have point wise convergence of $x_n(t) \longrightarrow x_o(t)$ for each $t \in I$. Using Holder inequality, the reader can easily verify that

$$|x_n(s_n) - x_o(s_o)| \leq (s_o - s_n)^{(p-1)/p} \left(\int_{s_n}^{s_o} |\dot{x}_n|^p \right)^{1/p} + |x_n(s_o) - x_o(s_o)|.$$

Since the sequence $\{\dot{x}_n\}$ is bounded in $L_p(I, R^n)$ norm, it follows from this inequality and the point wise convergence stated above that $x_n(s_n) \longrightarrow x_o(s_o)$ as $n \to \infty$. Using similar arguments one can verify that $x_n(\tau_n) \to x_o(\tau_o)$ as $n \to \infty$. Then it follows from lower semi continuity of Ψ that

$$\begin{aligned} \varphi(x_o, s_o, \tau_o) &= \Psi(s_o, x_o(s_o), \tau_o, x_o(\tau_o)) \\ &\leq \liminf_{n \to \infty} \Psi(s_n, x_n(s_n), \tau_n, x_n(\tau_n)) = \liminf_{n \to \infty} \varphi(x_n, s_n, \tau_n). \end{aligned}$$

Again using similar arguments as presented before, it follows from the above inequality and the coercivity condition that the generalized Meyer's problem has a solution. This ends the brief outline of our proof. □

Note that for all the variational problems (Lagrange, Bolza, Meyer's) considered here, we assumed coercivity condition. This condition can be dispensed with if the set Γ is weakly compact. This is stated in the following theorem.

Theorem 7.3.15. Consider any of the Lagrange, Bolza, and the Meyer's problems as in Theorem 7.3.11 to Corollary 7.3.14. Suppose all the functionals considered there are lower semi continuous but not necessarily coercive and that Γ is a weakly compact subset of $W^{1,p}$. Then the Lagrange, Bolza, and the Meyer's problems have solutions.

Proof. This is a special case of the abstract result given by Theorem 7.3.3.

Now we consider a problem that involves constraints described by differential inclusions. Find $x \in W^{1,p}$ that imparts a minimum to the functional subject to the differential constraints indicated

$$\begin{aligned} \varphi(x) &\equiv \int_a^b \ell(t, x(t), \dot{x}(t))dt + \Psi(x(a), x(b)) \longrightarrow \inf, \\ \dot{x}(t) &\in Q(t, x(t)) \ a.e\ t \in I. \end{aligned} \tag{7.3.20}$$

Theorem 7.3.16. Suppose the following assumptions hold:

(A1): $Q(t, \xi)$ is a set valued map, measurable in $t \in I$ and upper semi continuous (usc) in $\xi \in R^n$, taking values from the class of nonempty, closed, convex subsets of R^n, denoted by $cc(R^n)$.
(A2): the functions ℓ and Ψ satisfy the assumptions of Corollary 7.3.12.

Then there exists an element $x_o \in W^{1,p}$ that solves the problem (7.3.20).

Proof. We give a brief outline of the proof. Let d denote the standard Euclidean metric on R^n. Introduce the functions d_Q and ℓ_Q as defined below:

$$d_Q(t, x, z) = \begin{cases} 0, & \text{if } z \in Q(t, x); \\ \infty, & \text{otherwise,} \end{cases} \quad \text{and} \quad \ell_Q(t, x, z) \equiv d(z, Q(t, x)). \tag{7.3.21}$$

It is easy to show that both these functions are measurable in t on I, continuous in x on R^n, which follow from the measurability in t and usc in x of the set valued function Q on $I \times R^n$. Both these functions are convex in the third argument and the first function is sub differentiable and the second one is Gateaux differentiable with respect to this variable. Now modify the functional to be minimized by the new functional

$$\tilde{\varphi}(x) \equiv \int_a^b \{\ell(t, x(t), \dot{x}(t)) + d_Q(t, x(t), \dot{x}(t))\}dt + \Psi(x(a), x(b)).$$

Thus the problem (7.3.20) has been reduced to a Bolza problem as treated in Corollary 7.3.12. But d_Q is not Gateaux differentiable in the third argument and so we can not apply Corollary 7.3.12 directly. We introduce another family of functionals approximating $\tilde{\varphi}$ as follows

$$\varphi_\varepsilon(x) \equiv \int_a^b \{\ell(t,x(t),\dot{x}(t)) + (1/\varepsilon)\ell_Q(t,x(t),\dot{x}(t))\}dt + \Psi(x(a),x(b)). \tag{7.3.22}$$

It is obvious that for each $x \in W^{1,p}$

$$\tilde{\varphi}(x) \geq \varphi_\varepsilon(x), \forall \varepsilon > 0, \text{ and } \lim_{\varepsilon \downarrow 0} \varphi_\varepsilon(x) = \tilde{\varphi}(x).$$

If $\tilde{\varphi} \equiv +\infty$, there is nothing to prove. So we assume that $\inf\{\tilde{\varphi}(x), x \in W^{1,p}\} = M < +\infty$. It follows from Corollary 7.3.12 that, for every $\varepsilon > 0$, the functional φ_ε attains its minimum at some point $x_\varepsilon \in W^{1,p}$. Since $\ell_Q \geq 0$, coercivity of ℓ and Ψ alone implies that the set $\{x_\varepsilon, \varepsilon > 0\}$ is a bounded subset of $W^{1,p}(I, R^n)$. Let $\varepsilon_n \longrightarrow 0$ as $n \to \infty$. Since $W^{1,p}$ is a reflexive Banach space, there exists a subsequence of the sequence $\{x_{\varepsilon_n}\}$, relabeled as $\{x_n\}$, and an element $x_o \in W^{1,p}$ to which $x_n \xrightarrow{w} x_o$. Hence one can deduce that $\tilde{\varphi}(x_o) = M$. This concludes the outline of our proof. □

Remark. Note that, for the existence of a solution of the problem (7.3.20), it is necessary that the differential inclusion has at least one solution. We have tacitly made this assumption for the proof of the above result.

Another class of problems of significant interest is as follows. For a closed bounded interval $I \equiv [a,b]$, one may consider any of the Lagrange, Bolza or Meyer's problem with general isoperimetric constraints like:

$$\Gamma \equiv \Big\{x \in W^{1,p} : \int_I f_i(t,x(t),\dot{x}(t))dt \leq c_i, 1 \leq i \leq m_1,$$
$$\int_I g_i(t,x(t),\dot{x}(t))dt = d_i, 1 \leq i \leq m_2, \quad (x(a),x(b)) \in G \subset R^{2n}\Big\}$$

where G is a closed bounded set and $\{c_i, d_i\}$ are arbitrary constants so that Γ is a nonempty subset of $W^{1,p}$. As in Theorem 7.3.11, under the assumptions of Gateaux differentiability and coercivity of the functions $\{f_i, g_j; 1 \leq i \leq m_1, 1 \leq j \leq m_2\}$ one may show that Γ is a weakly compact subset of $W^{1,p}$. Using this one can prove similar existence results for the extremal problems based on weak lower(upper) semi continuity and weak compactness arguments. The reader is encouraged to carry out the details.

7.4 Necessary Conditions

In this section we wish to study necessary conditions for extremal problems as discussed in the preceding section. First we consider the Lagrange problem subject to terminal constraints,

$$\varphi(x) = \int_0^T \ell(t, x(t), \dot{x}(t))dt \longrightarrow \inf,$$
$$x(0) = \xi \text{ and } x(T) = \eta. \tag{7.4.23}$$

The solution of this problem is given by the celebrated Euler-Lagrange equation as stated below.

Theorem 7.4.1. Suppose $\ell(t, u, v)$ is measurable in t on I and once continuously differentiable in both u and v on R^{2n} denoted by ℓ_u, ℓ_v respectively. Then, in order that x_o minimizes the functional φ, it is necessary that it satisfies the following 2-point boundary value problem given by:

$$(d/dt)\ell_v(t, x, \dot{x}) = \ell_u(t, x, \dot{x}), \tag{7.4.24}$$
$$x(0) = \xi, x(T) = \eta. \tag{7.4.25}$$

Proof. Suppose φ attains its minimum at $x_o \in AC(I, R^n)$ satisfying the given boundary conditions. Denote by Γ the linear subspace

$$\Gamma \equiv \{x \in AC(I, R^n) : x(0) = x(T) = 0\}.$$

Define the function

$$g(\theta) \equiv \varphi(x_o + \theta y), y \in \Gamma, \theta \in R.$$

Clearly $x_o + \theta y$ satisfies the boundary conditions for all choices of $\theta \in R$ and $y \in \Gamma$. Since φ attains its minimum at x_o, g attains its minimum at $\theta = 0$ and consequently

$$(d/d\theta)g(\theta)|_{\theta=0} \equiv g^{'}(\theta)|_{\theta=0} = 0.$$

Differentiating g with respect to θ and letting $\theta \rightarrow 0$, we obtain

$$g^{'}(0) = \int_I \{< \ell_u(t, x_o(t), \dot{x}_o(t)), y(t) > + < \ell_v(t, x_o(t), \dot{x}_o(t)), \dot{y}(t) >\}dt$$
$$= 0 \; \forall y \in \Gamma, \tag{7.4.26}$$

where, for clarity, we have used $<,>$ to denote the scalar product in R^n. If ℓ_v is differentiable almost every where along the path x_o, it follows from integration by parts applied to the second term that

$$\int_I \{< (d/dt)\ell_v(t, x_o(t), \dot{x}_o(t)) - \ell_u(t, x_o(t), \dot{x}_o(t)), y(t) >\}dt = 0 \; \forall y \in \Gamma.$$

Let $AC_0(I,R)$ denote the space of all absolutely continuous real valued functions on I vanishing at the boundaries. Choose any $z \in R^n$ and $h \in AC_0(I,R)$, and note that $zh(\cdot) \in \Gamma$ for all such choices. Choosing $y = zh$, it follows from the above expression that,

$$\int_I \Xi_z(t)h(t)dt = 0 \ \forall \ h \in AC_0(I,R), \tag{7.4.27}$$

where

$$\Xi_z(t) \equiv < (d/dt)\ell_v(t, x_o(t), \dot{x}_o(t)) - \ell_u(t, x_o(t), \dot{x}_o(t)), z > . \tag{7.4.28}$$

The identity (7.4.27) implies that $\Xi_z(t) = 0$ $a.e.$ This is easily verified as follows. Suppose, on the contrary, the set $J \equiv \{t \in I : \Xi_z(t) \neq 0\}$ has positive Lebesgue measure. Let $\sigma \subset J$ be such that $\Xi_z(t) > 0$, for all $t \in \sigma$. For any $\varepsilon > 0$, choose a nonempty subset $\sigma_\varepsilon \subset \sigma$ so that the Lebesgue measure $\lambda(\sigma \setminus \sigma_\varepsilon) \leq \varepsilon$. Now choose a function $h \in AC_0(I,R)$ such that

$$h(t) = 1, t \in \sigma_\varepsilon, 0 \leq h(t) \leq 1 \text{ for } t \in \sigma, \text{ and } h(t) = 0 \text{ for } t \notin \sigma.$$

Using this choice of h, we have

$$\int_I \Xi_z(t)h(t)dt > 0$$

which contradicts the statement (7.4.27). Thus we conclude that $\Xi_z(t) = 0, a.e.$ Since $z \in R^n$ is also arbitrary, it follows from this that

$$(d/dt)\ell_v(t, x_o(t), \dot{x}_o(t)) - \ell_u(t, x_o(t), \dot{x}_o(t)) = 0 \ a.e \ t \in I. \tag{7.4.29}$$

This is the Euler-lagrange equation. Now rejecting the a priori hypothesis of differentiability of ℓ_v, we may assume that ℓ_u is locally integrable. In that case, we integrate the first term of (7.4.26) by parts and arrive at the following identity,

$$\begin{aligned} g'(0) &= \int_I \left\{ < -\int_0^t ds \ \ell_u(s, x_o(s), \dot{x}_o(s)) + \ell_v(t, x_o(t), \dot{x}_o(t)), \dot{y}(t) > \right\} dt \\ &= 0 \ \forall y \in \Gamma. \end{aligned}$$

Considering any open subset J of I, we can show that

$$L_0 \equiv \{f \in L_1(J, R^n) : f(t) = \dot{y}(t)|_J, y \in \Gamma\}$$

is dense in $L_1(J, R^n)$ for every such J. Hence from the above expression we can conclude that

$$\ell_v(t, x_o(t), \dot{x}_o(t)) = \int_0^t ds\ \ell_u(s, x_o(s), \dot{x}_o(s)) \text{ for almost all } t \in \text{Int } I. \tag{7.4.30}$$

But this identity implies that ℓ_v is differentiable almost every where and so it brings us back to the Euler-Lagrange equation. This completes the proof. □

In the preceding problem we did not impose any constraint other than the terminal constraint. Now we consider a similar problem with additional constraints. Let Γ be a closed convex subset of $AC(I, R^n)$ absorbing also the boundary conditions. Thus the problem can be restated as

$$\begin{aligned} \varphi(x) &= \int_0^T \ell(t, x(t), \dot{x}(t))dt \longrightarrow \inf, \\ x &\in \Gamma. \end{aligned} \tag{7.4.31}$$

Theorem 7.4.2. Suppose ℓ satisfies the assumptions of Theorem 7.4.1 and let Γ be a closed convex subset of $AC(I, R^n)$ and suppose the minimum is attained at $x_o \in \Gamma$. Then x_o must satisfy the following Euler-Lagrange variational inequality:

$$\int_0^T < \ell_u(t, x_o(t), \dot{x}_o(t)) - (d/dt)\ell_v(t, x_o(t), \dot{x}_o(t)), x(t) - x_o(t) > dt \geq 0, \quad \forall x \in \Gamma. \tag{7.4.32}$$

In case $x_o \in$ Int Γ, this reduces to the Euler-Lagrange equation.

Proof. Suppose the minimum is attained at $x_o \in \Gamma$. Since Γ is convex, for any $x \in \Gamma$ and $\varepsilon \in [0, 1]$, $x_o + \varepsilon(x - x_o) \in \Gamma$. Thus by virtue of minimality of x_o, $\varphi(x_o + \varepsilon(x - x_o)) \geq \varphi(x_o)$. This implies that

$$(D\varphi(x_o), x - x_o) \geq 0, \forall x \in \Gamma,$$

where $D\varphi(x_o)$ denotes the Gateaux or the directional derivative of φ at the point x_o. Using the differentiability assumptions and integration by parts, one can easily arrive at the following inequality,

$$\int_0^T < \ell_u(t, x_o, \dot{x}_0) - (d/dt)\ell_v(t, x_o, \dot{x}_o), x - x_o > dt \geq 0, \forall x \in \Gamma. \tag{7.4.33}$$

For the last statement, note that if x_o is in the interior of the set Γ, one can choose a $\delta > 0$ such that the δ-neighborhood of x_o, denoted by $N_\delta(x_o)$, is contained in the interior of Γ. Since the above expression must hold for all $x \in N_\delta(x_o)$, it is evident that the inequality holds in both the directions and hence reduces to an equality. Thus the result follows from similar arguments as given in the proof of Theorem 7.4.1. This completes the proof. □

Remark. In view of the above results, it is clear that Theorem 7.4.1 is a special case of Theorem 7.4.2.

Suppose D is a closed convex, possibly unbounded, subset of R^n, and suppose we require that the solution lies in this set for all $t \in I$. Thus the problem can be stated as follows:

$$\begin{aligned} \varphi(x) &= \int_0^T \ell(t, x(t), \dot{x}(t))dt \longrightarrow \inf, \\ x(t) &\in D \; \forall \; t \in I. \end{aligned} \tag{7.4.34}$$

This is solved in the following theorem.

Theorem 7.4.3. Suppose the assumptions of Theorem 7.4.2 hold. Then, in order that x_o be a solution of the problem stated above, it is necessary that it satisfies the following differential inclusion

$$(d/dt)\ell_v(t, x, \dot{x}) - \ell_u(t, x, \dot{x}) \in \partial \mathcal{X}_D(x(t)) \; a.e. t \in I, \tag{7.4.35}$$

where $\partial \mathcal{X}_D$ is the sub differential of the indicator function $\mathcal{X}_D$.

Proof. This is a special case of Theorem 7.4.2. For non triviality of the problem, it is necessary that $\xi, \eta \in D$. Define

$$\Gamma = \Gamma_D \equiv \{x \in AC(I, R^n) : x(0) = \xi, x(T) = \eta, \xi, \eta \in D, x(t) \in D, t \in I\}.$$

Clearly this is a nonempty set. Then it follows from the previous theorem that

$$\int_0^T < \ell_u(t, x_o, \dot{x}_o) - (d/dt)\ell_v(t, x_o, \dot{x}_o), x - x_o > dt \geq 0, \forall x \in \Gamma_D.$$

Choose any $t \in (0, T)$ and $\varepsilon > 0$, sufficiently small, so that the interval given by $J_\varepsilon \equiv [t - (\varepsilon/2), t + (\varepsilon/2)]$ is contained $[0, T]$. Since the above inequality must hold for all $x \in \Gamma_D$, we may choose x such that $x(s) = x_o(s), s \in I \setminus J_\varepsilon$. Substituting this in the preceding expression, we obtain

$$\int_{J_\varepsilon} < \ell_u(t, x_o, \dot{x}_o)) - (d/dt)\ell_v(t, x_o, \dot{x}_o), x - x_o > dt \geq 0, \forall x \in \Gamma_D.$$

Dividing this by the Lebesgue measure of the set J_ε and letting $\varepsilon \downarrow 0$, we obtain

$$< \ell_u(t, x_o(t), \dot{x}_0(t)) - (d/dt)\ell_v(t, x_o(t), \dot{x}_o(t)), x(t) - x_o(t) > \ \geq 0 \ \forall\, x \in \Gamma_D. \tag{7.4.36}$$

Clearly this implies that both $x_o(t), x(t) \in D$ and that the above inequality must hold for almost all $t \in I$. Now by definition, the sub differential of the indicator function $\mathcal{X}_D$ at any point $w \in D$ is given by

$$\partial \mathcal{X}_D(w) = \{z \in R^n : (z, v - w) \leq 0, \forall\, v \in D\}. \tag{7.4.37}$$

Thus it follows from the above two expressions that

$$(d/dt)\ell_v(t, x_o(t), \dot{x}_o(t)) - \ell_u(t, x_o(t), \dot{x}_o(t)) \in \mathcal{X}_D(x_o(t)) \text{ for almost all } t \in I. \tag{7.4.38}$$

This completes the proof. □

Remark. An alternative proof is much straight forward. Remove the constraint by including it in the integrand yielding a new functional $\tilde{\varphi}$ given by

$$\tilde{\varphi}(x) \equiv \int_0^T \{\ell(t, x, \dot{x}) + \mathcal{X}_D(x(t))\}dt,$$

and use Theorem 7.4.1 to obtain

$$(d/dt)\ell_v - \ell_u - \partial \mathcal{X}_D \ni 0.$$

This is precisely the differential inclusion (7.4.35) we obtained by direct approach.

Similarly one can consider problems with terminal constraints such as

$$\varphi(x) = \int_0^T \ell(t, x, \dot{x})dt \longrightarrow \inf$$
$$x(0) \in D_1, x(T) \in D_2, \tag{7.4.39}$$

where D_1 and D_2 are two nonempty subsets of R^n.

Theorem 7.4.4. Suppose ℓ satisfy the assumptions of Theorem 7.4.1, and D_1, D_2 are nonempty closed convex subsets of R^n. Then, in order that x be a solution of the problem (7.4.39), it is necessary that it satisfies the Euler-Lagrange equation subject to the terminal conditions (transversality conditions)

$$(d/dt)\ell_v(t, x, \dot{x}) - \ell_u(t, x, \dot{x}) = 0, a.e.\ t \in I, \tag{7.4.40}$$
$$0 \in \partial \mathcal{X}_{D_1}(x(0)), 0 \in \partial \mathcal{X}_{D_2}(x(T)), \tag{7.4.41}$$

where $\partial \mathcal{X}_D$ is the sub differential of the indicator function $\mathcal{X}_D$.

Proof. Again the problem can be reformulated as one without constraints by introducing the functional

$$\tilde{\varphi}(x) \equiv \int_0^T \ell(t, x, \dot{x})dt + \mathcal{X}_{D_1}(x(0)) + \mathcal{X}_{D_2}(x(T)) \longrightarrow \inf, \tag{7.4.42}$$

and use the procedure of Theorem 7.4.1 in conjunction with the sub differentials of the indicator functions. □

Let us now consider the Bolza problem

$$\varphi(x) = \int_0^T \ell(t, x, \dot{x})dt + \Psi_1(x(0)) + \Psi_2(x(T)) \longrightarrow \inf. \tag{7.4.43}$$

Corollary 7.4.5. Consider the Bolza problem as stated above. Suppose ℓ satisfy the assumptions of Theorem 7.4.1, and the functions Ψ_1, Ψ_2 are continuously Gateaux differentiable on R^n. Then, in order that x be a solution of the problem (7.4.43), it is necessary that it satisfies the Euler-Lagrange equation subject to the terminal conditions

$$(d/dt)\ell_v(t, x, \dot{x}) - \ell_u(t, x, \dot{x}) = 0, a.e.\ t \in I, \tag{7.4.44}$$

$$D\Psi_1(x(0)) = 0, D\Psi_2(x(T)) = 0, \tag{7.4.45}$$

where $D\Psi$ denotes the Gateaux derivative of Ψ.

Proof. The proof follows immediately from the previous theorem. Here the sub differential reduces to Gateaux differential. □

At the end of the previous section we considered the question of existence of solutions for variational problems with integral equality and inequality constraints. Here we consider a similar problem for necessary conditions:

$$\varphi(x) \equiv \int_a^b \ell(t, x, \dot{x})dt \longrightarrow \inf,$$

subject to the constraints

$$\begin{aligned} & x(a) = \xi, \ x(b) = \eta; \\ & \int_a^b f_i(t, x, \dot{x})dt = c_i, i = 1, 2, \cdots, r: \\ & \int_a^b f_i(t, x, \dot{x})dt \leq d_i, i = r+1, r+2, \cdots, m. \end{aligned} \tag{7.4.46}$$

For the solution of this problem we introduce a set of new variables $\{z_i, 1 \leq i \leq m\}$ given by

$$z_i(t) \equiv \int_a^t f_i(s, x(s), \dot{x}(s))ds, i = 1, 2, \cdots, m.$$

Thus the original integral constraints are equivalent to the differential and terminal constraints of $\{z_i\}$ given by the following conditions

$$\begin{aligned} &\dot{z}_i(t) = f_i(t, x(t), \dot{x}(t)), z_i(a) = 0, 1 \leq i \leq m; \\ &z_i(b) = c_i, 1 \leq i \leq r;\ z_i(b) \leq d_i, r+1 \leq i \leq m. \end{aligned} \qquad (7.4.47)$$

Defining $y \equiv col(x, z) \in R^{n+m}$ and using the method of Lagrange multipliers we introduce the extended Lagrangian

$$\tilde{\ell}(t, x, z, \lambda; \dot{x}, \dot{z}, \dot{\lambda}) \equiv \ell(t, x, \dot{x}) + \sum_{i=1}^{m} \lambda_i(t)(f_i - \dot{z}_i) \equiv \ell(t, y, \lambda, \dot{y}, \dot{\lambda}).$$

Since there are inequality constraints on the boundary, it is also convenient to remove them by adding appropriate penalties on the terminal condition. We do this by using the indicator function $\mathcal{X}_D$ of the set D given by

$$D \equiv \{w = col(u, v) \in R^{n+m} : u = \eta, v_i = c_i, 1 \leq i \leq r; v_i \leq d_i, r+1 \leq i \leq m\}.$$

Thus instead of minimizing the original functional φ subject to the (integral equality and inequality) constraints, we minimize the extended functional

$$\tilde{\varphi}(y, \lambda) \equiv \int_a^b \tilde{\ell}(t, y, \lambda, \dot{y}, \dot{\lambda})dt + \mathcal{X}_D(y(b)),$$

subject only to the conditions $y(a) = col(\xi, 0), \lambda(a) \not\equiv 0$. Again for necessary conditions, one can write the Lagrange equations similar to those of Theorem 7.4.4.

Multivariable Case:

So far we have considered problems where the functions solving the variational problems are functions of only one variable $t \in I$. However this is easily extended to multivariable case. Let Σ be a bounded, open, connected subset of R^n with smooth boundary $\partial\Sigma$, and $I \equiv (0, T)$ is a bounded interval. We wish to consider the Lagrange problem:

$$\varphi(u) \equiv \int_{I\times\Sigma} \ell(t, \xi, u(t, \xi), u_t(t, \xi), D_\xi u, D_\xi^2 u)d\xi dt \longrightarrow \inf \qquad (7.4.48)$$

$$u(0, \xi) = u_0(\xi),\ u(T, \xi) = u_2(\xi),\ \mathcal{L}_b u(t, \xi) = 0, \qquad (7.4.49)$$

where the initial and final conditions are specified by some suitable functions $\{u_0, u_1\}$ defined on Σ and further u must also satisfy certain boundary conditions where we have used $\mathcal{L}_b$ to denote the boundary operator. Here we have used $D_\xi u$ and $D_\xi^2 u$ to denote the vector and the matrix of first and second partials respectively of u with respect to the spatial variables in Σ. For rigorous formulation of the problem one must invoke appropriate Sobolev spaces, for example,

$$W_p^{1,2} \equiv \{u : u, u_t, D_{\xi_i} u, D^2_{\xi_i,\xi_j} u \in L_p(I \times \Sigma), 1 \leq i, j \leq n\},$$

which is furnished with the norm topology,

$$\| u \|_{W_p^{1,2}} \equiv \left(|u|^p_{L_p} + |u_t|^p_{L_p} + \| D_\xi u \|^p_{L_p} + \| D^2_\xi u \|^p_{L_p} \right)^{(1/p)}.$$

Define

$$\Gamma \equiv \{u \in W_p^{1,2} : u(0,\xi) = u_0(\xi), u(T,\xi) = u_1(\xi), \mathcal{L}_b u(t,\xi) = 0, \\ \text{for } (t,\xi) \in I \times \partial\Sigma\}.$$

We assume that $\ell = \ell(t, \xi, d, v, r, s) : I \times \Sigma \times R \times R \times R^n \times R^{n^2} \longrightarrow R$ is measurable in the first two arguments on $I \times \Sigma$, and continuously differentiable with respect to all the other arguments on $R \times R \times R^n \times R^{n^2}$. Again for $p > 1$, we can prove the following result.

Theorem 7.4.6. Suppose $p > 1$, and Γ is a non empty closed convex subset of $W_p^{1,2}$ and the Lagrangian ℓ is measurable in $\{t, \xi\}$ on $I \times \Sigma$ for each $\{d, v, r, s\} \in R \times R \times R^n \times R^{n^2}$ and, for almost all $\{t, \xi\} \in I \times \Sigma$, it is continuously differentiable on $R \times R \times R^n \times R^{n^2}$ and further ℓ is convex in the fourth and the sixth argument $\{v, s\} \in R \times R^{n^2}$. Then the Lagrange problem has a solution.

Proof. The proof is similar to the single variable case. □

Following similar procedure as in the single variable case we can prove the following necessary conditions.

Theorem 7.4.7. Suppose the assumptions of Theorem 7.4.6 hold. Then, in order for $u^o \in \Gamma$ to be optimal, it is necessary that u^o satisfy the following Euler-Lagrange equation

$$(\partial/\partial t)\ell_v + \sum_{i=1}^{n} D_{\xi_i} \ell_{r_i} - \sum_{i,j=1}^{n} D^2_{\xi_i,\xi_j} \ell_{s_{i,j}} = \ell_d, \ (t,\xi) \in I \times \Sigma. \qquad (7.4.50)$$

7.5 Basic Algorithm for Numerical Computation

Here we present a simple algorithm for numerical solution of the variational problems. For reference, we consider the problem stated in Theorem 7.4.2,

$$\varphi(x) = \int_0^T \ell(t, x(t), \dot{x}(t))dt \longrightarrow \inf, \qquad x \in \Gamma. \tag{7.5.51}$$

Suppose at the n-th stage of computation, $x_n \in \Gamma$ has been already computed.
Step 1: Compute the gradient of φ at x_n as

$$D\varphi(x_n)(t) \equiv \ell_u(t, x_n(t), \dot{x}_n(t)) - (d/dt)\ell_v(t, x_n(t), \dot{x}_n(t)), t \in I.$$

Step 2: Define $x_{n+1} \equiv x_n - \varepsilon D\varphi(x_n)$ for $\varepsilon > 0$, sufficiently small so that $x_{n+1} \in \Gamma$.
Step 3: Compute

$$\begin{aligned}\varphi(x_{n+1}) &= \varphi(x_n) + < D\varphi(x_n), x_{n+1} - x_n > + o(\varepsilon)\\ &= \varphi(x_n) - \varepsilon \parallel D\varphi(x_n) \parallel^2 + o(\varepsilon).\end{aligned}$$

This step shows that for ε sufficiently small x_{n+1} should provide an improvement.
Step 4: Check if $\varphi(x_{n+1}) < \varphi(x_n)$, if not, reduce the step size ε and repeat the process till this is satisfied. Fix the corresponding ε and use in the following steps.
Step 5: If $\varphi(x_{n+1}) < \varphi(x_n)$ compute the difference and check if $|\varphi(x_{n+1}) - \varphi(x_n)| \leq \delta$ for some pre-specified tolerance δ. Print the results if so, if not, go to step 1 with x_n replaced by x_{n+1}, thereby closing the loop.

If in step 4, it is impossible to find an $\varepsilon > 0$ for which the strict inequality is satisfied, one must conclude that x_n is a local minimum. In fact if x_n is a local minimum, it follows from the necessary conditions of optimality that

$$(D\varphi(x_n), x - x_n) \geq 0 \;\forall x \in \Gamma \cap N(x_n),$$

where $N(x_n)$ is a neighborhood of x_n.

Remark on the Algorithm. We may caution the reader that, unlike in linear quadratic problems, $\{\varphi(x_n)\}$ may not be a monotone decreasing sequence. Whenever it increases, one must reduce the step size ε and continue. This will guarantee improvement in the cost and convergence to at least a local minimum.

Remark. For the isoperimetric problems or problems with constraints, one must also include the multiplier λ in computing the gradient for search direction. □

For illustration of the algorithm, consider the simple problem: Find $x \in \Gamma \equiv [1,2]$ that minimizes the function $f(x) = ax^2$ where $a > 0$. Obviously the result is $x_o = 1$. One can easily check that $(Df(x^*), x - x^*) = Df(x^*)(x - x^*) \geq 0$, $\forall x \in \Gamma$, if and only if $x^* = x_0$ and for all other points $x^*(\neq x_0) \in \Gamma$ this is violated. Clearly this result does not follow from setting derivative of f to zero and solving the resulting equation.

Similarly one may consider the classical quadratic problem

$$f(x) = (1/2)(Qx, x) + (y, x) + c \longrightarrow \inf, \quad x \in \Gamma = B_r$$

where B_r is the ball in R^d of radius r, or any other closed convex set, Q is a symmetric positive definite matrix, $y \in R^d$ and c is a constant. If $z \equiv -Q^{-1}y \in \Gamma$, this is the solution and obtained by setting the derivative to zero. If $z \notin \Gamma$ this makes no sense. The algorithm presented above will work. Note that $df(x_n) = Qx_n + y$. Given $x_n \in \Gamma$, choose $\varepsilon > 0$ sufficiently small so that $x_{n+1} \equiv x_n - \varepsilon(Qx_n + y) \in \Gamma$. Then we have

$$f(x_{n+1}) - f(x_n) = -\varepsilon \parallel Qx_n + y \parallel^2 + (\varepsilon^2/2) < Q(Qx_n + y), (Qx_n + y) > .$$

Clearly if ε is sufficiently small x_{n+1} is an improvement over x_n. On the other hand if no positive ε can be found for which the inclusion $x_{n+1} \equiv x_n - \varepsilon(Qx_n + y) \in \Gamma$ is satisfied, x_n must be on the boundary of the set Γ and the vector $Qx_n + y$ is directed outward of the set Γ. In this case we have

$$(Qx_n + y, x) \geq (Qx_n + y, x_n) \ \forall \ x \in \Gamma$$

and hence x_n is the solution and the algorithm stops.

In case $\Gamma = R^d$, the optimal step size can be used to speed up the rate of convergence and it is given by

$$\varepsilon_n \equiv \frac{\parallel Qx_n + y \parallel^2}{< Q(Qx_n + y), (Qx_n + y) >}.$$

Then

$$f(x_{n+1}) - f(x_n) = -(1/2)\frac{\parallel Qx_n + y \parallel^4}{< Q(Qx_n + y), (Qx_n + y) >}.$$

This gives the maximum drop of the cost at the $(n+1)$-st step and this is the reason why this algorithm is known as the steepest decent algorithm. Since

Q is positive definite, it follows from this expression that the algorithm can stop at this step if and only if $Qx_n + y = 0$. This result is also valid for infinite dimensional Hilbert space H provided Q is self adjoint positive and there exists a $\gamma > 0$ such that $(Q\xi, \xi) \geq \gamma \parallel \xi \parallel^2$ for all $\xi \in H$. For example, consider the case

$$f(x) \equiv (1/2) \int_{I \times I} < K(t,s)x(t), x(s) > dtds + \int_I < b(t), x(t) > dt + c,$$

where $K \in L_2(I \times I, M_s^+)$ is a symmetric positive definite kernel, $b \in L_2(I, R^n)$ and c is a constant. Unfortunately, for the general problems of Calculus of variations, it is not easy to choose the step size optimally as in the quadratic case, since it involves another level of optimization on the top of the original one (see Ahmed, Dabbous and Wong [152]).

7.6 Some Examples

For illustration of the results of Section 4, we present here some elementary examples.

(E1): $\varphi(x) = \int_0^1 x^2(t)dt \longrightarrow \inf$, $x(0) = 0, x(1) = 1$. Here $\ell(t, x, \dot{x}) = x^2$. The Euler-Lagrange equation is given by $(d/dt)\ell_v = \ell_u$ leads to the equation $x(t) \equiv 0$. Clearly a continuous function x can not satisfy this identity as well the boundary condition simultaneously. This problem has no solution. It fails because the Lagrangian ℓ is not convex in the third variable. A minimizing sequence is $\{x_n \equiv t^n\}$ which satisfies the boundary conditions, but the limit function is discontinuous at $t = 1$.

(E2): $\varphi(x) = \int_0^1 \dot{x}^2 dt, x(0) = 0, x(1) = 1$. The Euler-Lagrange equation is given by $\ddot{x} = 0$. In this case ℓ is convex in the third argument and a unique solution exists which is given by $x(t) = t$.

(E3): $\varphi(x) = \int_0^1 (\dot{x}^2 - x^2)dt, x(0) = 0, x(1) = 1$. In this case the Euler-Lagrange equation is given by $\ddot{x} + x = 0$. Solving this equation with the given boundary conditions, one finds that

$$x(t) = (1/\sin 1) \sin t.$$

(E4): $\varphi(x) = (1/2) \int_0^1 (\dot{x}^2 + x^2)dt, x(0) = 0, x(1) = 1$. In this case the Euler-Lagrange equation is given by $\ddot{x} - x = 0$. solving this equation subject to the boundary conditions, the reader can verify that the solution is given by

$$x(t) = (2e/(e^2 - 1)) \sinh t.$$

(E5): It is evident that the curve of minimum length joining any two points on the plane is the straight line. For proof, draw any curve between any two points $A = (x_1, y_1)$ and $B = (x_2, y_2)$ on the plane and check that the length of an infinitesimal part of it is given by $ds = dx\sqrt{(1 + \dot{y}(x)^2)}$ and hence the length of the entire curve is given by

$$L(A, B) = \int_{x_1}^{x_2} \sqrt{(1 + \dot{y}^2)}ds.$$

Using the Euler-Lagrange equation $(d/dx)\ell_{\dot{y}} = \ell_y$, we arrive at the following expression

$$\dot{y} = c\sqrt{(1 + \dot{y}^2)}$$

where $c < 1$ is a constant. This leads to the following expression

$$\dot{y} = \sqrt{c^2/(1 - c^2)} = \text{ constant}$$

and hence the curve of minimal length is the straight line joining the two points. The constants are determined by using the positions of the points $\{A, B\}$.

(E6): Now we consider an example of an isoperimetric problem. Find a curve of length L, joining points P_a and P_b on the plane that maximizes the area under the curve. Let this curve be denoted by $y = u(\xi)$ as a function of ξ along the x-axis. We already know that the length is given by

$$L = \int_a^b \sqrt{(1 + \dot{u}^2)}d\xi.$$

The area under the curve is given

$$A = \int_a^b u(\xi)d\xi.$$

We must maximize A for the pre-specified length L. As treated in Section 4, (7.4.47), one takes the Lagrangian as

$$\tilde{\ell} = u + \lambda(\sqrt{(1 + \dot{u}^2)} - \dot{z}),$$

where λ is the Lagrange multiplier. Writing the Lagrange equation, one finds that λ is constant and that $\dot{z} = \sqrt{1 + \dot{u}^2}$. Thus one may just consider

$$\ell \equiv u + \lambda(\sqrt{1 + \dot{u}^2}).$$

When ℓ is independent of the ξ variable, the reader can verify that the Lagrange equation is equivalent to

$$(d/d\xi)(\ell - \dot{u}\ell_{\dot{u}}) = 0.$$

This is the so called integral curve. Hence $\ell - \dot{u}\ell_{\dot{u}} = c_1$ for some constant c_1. Substituting the expression for ℓ in this equation we find that

$$(c_1 - u)\sqrt{1 + \dot{u}^2} = \lambda.$$

Introducing a parameter θ and setting $\dot{u} = tan\theta$ it follows from this that

$$u = c_1 - \lambda \cos\theta.$$

Note also that from $(d/d\xi)u = tan\theta$ we have $d\xi = du/(\tan\theta) = +\lambda\cos\theta$. Hence $\xi = c_2 + \lambda\sin\theta$. Thus we have arrived at the following pair of parametric curves

$$(u - c_1) = -\lambda\cos\theta \quad (\xi - c_2) = \lambda\sin\theta. \tag{7.6.52}$$

Eliminating the variable θ we obtain

$$(u - c_1)^2 + (\xi - c_2)^2 = \lambda^2.$$

This is a family of circles centered at the point (c_1, c_2) on the $u-\xi$ plane with radius λ. These constants must be determined from the boundary conditions

$$u(a) = u_0, u(b) = u_1, \text{ and } \int_a^b \sqrt{1 + \dot{u}^2}\, d\xi = L,$$

where $\{u_0, u_1, L\}$ are known constants which determine the constants $\{c_1, c_2, \lambda\}$. The conclusion is that the solution u is the part of a circle with the given arc length L whose ends are attached to the points $(a, u_0), (b, u_1)$.

(E7): Consider a ball of mass m suspended from the ceiling by an elastic string of stiffness k. Focussing only on vertical motion, the kinetic energy is given by $K.E = (1/2)m\dot{x}^2$ and the potential energy is $P.E = (1/2)kx^2$ where x denotes the displacement from the rest state and $\dot{x}$ the displacement rate. According to the basic principles of classical mechanics (Hamilotn's principle, D'Alembert's principle of least action etc.) the Lagrangian is given by the sum of kinetic energies minus the sum of potential energies of the system. This is valid for conservative systems. For nonconservative (or dissipative) systems one must add to the Lagrangian, the work done by external forces, giving

$$(d/dt)\ell_v - \ell_u = Q$$

where Q denotes the dissipative forces. Thus here the Lagrangian is given by

$$\ell \equiv (1/2)\{m\dot{x}^2 - kx^2\}.$$

Using the Euler-Lagrange equation, one arrives at the equation for harmonic oscillator given by,

$$m\ddot{x} + kx = 0.$$

In the presence of dissipative forces like friction, $Q = -b\dot{x}$ and the Euler equation now takes the form

$$m\ddot{x} + b\dot{x} + kx = 0.$$

(E8): (Elasto Dynamics) Consider a metallic string or electric cable of mass density $\rho(t,x)$ with stiffness coefficient $k(t,x)$ stretched and attached to the top ends of two vertical poles separated by a horizontal distance L. We are interested in the dynamics of vibration of the cable in the vertical plane. The kinetic and the elastic potential energies are given by

$$K.E = \int_0^L \rho(t,\xi)u_t^2 d\xi \quad P.E. = \int_0^L k(t,\xi)(u_\xi)^2 d\xi.$$

We may note that the expression for the potential energy is valid for small displacements only. The external force (wind or otherwise) acting on the cable is $f(t,\xi)$ per unit mass at position $\xi \in (0,L)$. The Lagrangian is given by

$$\ell \equiv \int_0^L \{\rho u_t^2 - k(u_\xi)^2 + \rho f u(t,\xi)\} d\xi.$$

Using the Euler-Lagrange equation (7.4.50), we obtain the dynamics of vibration of the cable

$$(\partial/\partial t)(\rho u_t) + (\partial/\partial\xi)(k u_\xi) = \rho f.$$

If both ρ and k are independent of the space and time variables, this equation reduces to a second order wave equation:

$$\rho(\partial^2/\partial t^2)u - k(\partial^2/\partial\xi^2)u = \rho f.$$

The natural boundary condition for this problem is $u(t,0) = u(t,L) = 0$. Clearly, to solve this equation one must provide the initial conditions.

Remark. We must mention the fact that using the Lagrangian technique, one can derive the dynamics of almost all complex systems of the classical mechanics. For example, using this technique, the dynamics of space craft with flexible appendages such as antenna, and space stations were derived by Biswas and Ahmed, Lim and Ahmed [160, 181] which involve both rigid and flexible bodies. Study of Lyapunov type stability conditions for such distributed parameter systems were derived and stabilizing controls were designed.

7.7 Exercises

(Q1): Consider the Lagrange problem,

$$\varphi(x) = \int_0^T \ell(t, x(t), \dot{x}(t))dt \longrightarrow \inf$$

subject to the constraint that

$$x \in \Gamma \equiv \{x \in AC(I, R^n) : x(0) = \xi, x(T) = \eta, \text{ and } |\dot{x}(t)| \leq \beta(t) \ a.e\}$$

where $\xi, \eta \in R^n$ and $\beta \in L_1^+(I)$ are given. Give sufficient conditions for the existence of a solution.

(Q2): Consider the same Lagrange problem with $\Gamma \equiv \{x \in AC(I, R^n) : \dot{x}(t) \in K\}$ where K is a compact subset of R^n. Give sufficient conditions for the existence of a solution.

(Q3): Consider the example (E6) and set $a = 0, b = 1, u(0) = u(1) = 0$ and $L = \pi$. Use the variational calculus to construct the circle.

(Q4) Consider the functional

$$\varphi(x) \equiv \int_{-1}^{+1} t^2(\dot{x})^2 dt.$$

Show that the function x that minimizes this is given by $x(t) = -1$ for $t \in [-1, 1)$ and $x(1) = 1$. Clearly $x \notin AC(I)$ but it is in $BV(I)$.

(Q5): (Elasto Dynamics) Assuming small displacement show that the equations of vibration of elastic rods and beams is given by

$$(\partial/\partial t)(\rho u_t) + (\partial^2/\partial \xi^2)(k u_{\xi,\xi}) = \rho f.$$

This is the well known Euler Beam equation. In order to solve such equations, one must add the appropriate boundary conditions.

(Hint) The Lagrangian is given by

$$\ell(t, \xi, u_t, u_\xi, u_{\xi,\xi}) = (1/2)\rho u_t^2 - (1/2)k u_{\xi,\xi}^2 + \rho f u.$$

Remark. Example 4 shows that solutions for variational problems may lie in a much larger space than those used here.

7.8 Bibligraphical Notes

Calculus of variation is one of the oldest branch of mathematics. For historical reasons it should be exciting to read from the books of the pioneers who created the subject in the first place. There are many outstanding names and my limited knowledge prevents me from listing all of them. In the text we have mentioned Lagrange, Meyer, Bolza, Wierstrass. For an excellent historical account the reader is referred to the extensive references given in Cesari's book [24]. The reader may enjoy reading Bolza [22] and Elsgolc [34]. For recent developments see Cesari [24], Vainberg [92], Clark [26] and Zeidler [100]. For applications see Petrov [76], Ahmed [2], Biswas [160], Lim [181]. Here we have presented several existence theorems and necessary conditions for extremals. I learnt from my personal contact with Cesari, that Tonelli was his teacher and that he was the first who gave rigorous proof of existence of solutions of variational problems in the 20th century (around 1930). Cesari himself made outstanding contributions in this area since that time. The book of Cesari [24] and the references therein provide a monument of information.

Chapter 8

Optimal Control: Necessary Conditions and Existence

8.1 Introduction

In many social, economic, management and physical problems it is very important to use limited resources in a judicious way while seeking the goal. For example, in production industry certain goods are produced using available man power and material resources including some form of energy. The objective is to use the resources in a way so as to minimize the cost of production while maintaining the production level and the quality of products. Similarly in public sector, the government wishes to attain certain social goals in a given period of time under given resource constraints. The problem is to find a policy that achieves the goal within the given period of time with minimum cost of resources used. It is evident that controllability problem arises here. The set goal may not be attainable within the given time period under the given resource constraints. In that case one must redefine the goal, reduce the expectations, or use the limited resources in a way that minimizes the discrepancy between the set goal and the attainable state.

In this chapter we develop methods whereby optimal policies can be computed for these or more general problems of this nature. These methods are given in terms of necessary conditions of optimality. Necessary conditions of optimality without the support of existence of optimal policies have no meaning. Thus the question of existence of optimal controls is very important and it is treated in the final section.

8.2 General Problem

Let us consider the control system along with the specified initial state x_i and target state x_T and the class of admissible controls $\mathcal{U}_{ad}$:

$$\dot{x} = f(t, x, u), t \in I \equiv [0, T], \tag{8.2.1}$$

$$x_i = \xi \ \in R^n, x_T = \eta \in R^n \tag{8.2.2}$$

$$u \in \mathcal{U}_{ad} \equiv \{u(t), t \in I, u \ \text{ measurable and } u(t) \in U \ a.e\}, \tag{8.2.3}$$

where $U \subset R^m$ is generally a closed bounded convex set. This specifies the resource constraints.
The problem is to find a control that takes the system from the given initial state $x(0) = \xi$ to the desired final state $x(T) = \eta$ while minimizing the cost of doing so (cost of operation) measured by the following functional, called the cost functional, given by

$$J(u) \equiv \int_0^T \ell(t, x(t), u(t))dt, \tag{8.2.4}$$

where ℓ is a suitable, possibly nonnegative, function of its arguments defined on $I \times R^n \times R^m$. The very first question that one must settle is: does the problem have a solution. Clearly it involves the question of controllability. Given that the system is controllable, that is, there is a control policy $\tilde{u} \in \mathcal{U}_{ad}$ that transfers the system from the initial state ξ to the target state η at time T, there arises the next question if there is one that does it with minimum cost, that is one that minimizes the functional (8.2.4). This problem is known as Lagrange problem.

In certain problems the terminal state is free, not specified as in the above problem. Here one wishes to find a control policy that only minimizes a functional of the form

$$J(u) \equiv \varphi(x(T)). \tag{8.2.5}$$

In other words there is no hard and fast specification of the goal state; one only wishes to reach as close as possible to the target. For example, φ may be given by

$$\varphi(x(T)) \equiv d(x(T), K) \equiv \inf\{d(x(T), \zeta), \zeta \in K\}, \tag{8.2.6}$$

where K is a closed, possibly bounded, set in R^n signifying the desirable set of targets. Following the terminology of the classical Calculus of variation, as in Chapter 7, this problem is known as the Bolza problem, popularly known as the terminal control problem. Combining the Lagrange and the Bolza problem one obtains the Meyer's problem

$$J(u) \equiv \int_I \ell(t, x(t), u(t))dt + \varphi(x(T)), \tag{8.2.7}$$

where the first term is some times called the running cost and the last term is known as the terminal cost.

Remark. Note that both the Lagrange problem (8.2.4) and the Meyer's problem (8.2.7) can be reduced to the Bolza problem by augmenting the state vector as follows

$$\dot{x}_{n+1} = \ell(t, x, u), t \in I. \tag{8.2.8}$$

With this extended state space $R^n \times R$ or very often, $R^n \times R_0$ in case ℓ is nonnegative, one obtains the Bolza problem with

$$\varphi(\tilde{x}(T)) = x_{n+1}(T)$$

corresponding to the Lagrange problem (8.2.4). Similarly the Meyer's problem (8.2.7) reduces to the equivalent Bolza problem,

$$\varphi(\tilde{x}(T)) = x_{n+1}(T) + \tilde{\varphi}(x(T))$$

where $\tilde{\varphi}$ plays the same role as that of φ of the Meyer's problem.

8.3 Necessary Conditions of Optimality

In this section we develop three principal necessary conditions of optimality with increasing generality. Theorem 8.3.1 is based on the assumption of differentiability with respect to both the state and the control variables. In Theorem 8.3.3, differentiability assumption with respect to the control variables is dropped. Theorem 8.3.5 deals with relaxed (measure valued) controls giving generalized necessary conditions of optimality. In this theorem the assumption of convexity of the control constraint set is dropped unlike the first two results. This is very important in many practical applications. In addition to this we have several corollaries representing special cases.

8.3.1 Ordinary Controls

First we consider the Meyer's problem for smooth functions $\{f, \ell, \varphi\}$. Precisely we use the following assumption.

Assumption A1: Both the vector field $f(t, x, u)$ and the scalar function $\ell(t, x, u)$ along with their hessians $\{f_x, \ell_x, f_u, \ell_u\}$ are continuous on $R^n \times U$ for almost all $t \in I$ and Lebesgue integrable in t on I for all $(x, u) \in R^n \times U$.

First we consider the free terminal problem, in the sense that the terminal state is not specified. Instead, a penalty is imposed through the terminal cost φ.

Define the Hamiltonian as follows

$$H(t,x,y,v) =< f(t,x,v), y > +\ell(t,x,v), (t,x,y,v) \in I \times R^n \times R^n \times R^m. \tag{8.3.9}$$

We denote by φ_x and H_u the gradients of φ with respect to x and that of H with respect to u respectively while f_x^* denotes the adjoint of the Hessian f_x.

Theorem 8.3.1. Consider the Meyer's problem (8.2.7) and suppose the assumption (A1) holds, $U \subset R^m$ is a closed bounded convex set with $\mathcal{U}_{ad}$ denoting the admissible controls and that the optimal control problem has a solution. Let $u^o \in \mathcal{U}_{ad}$ be a control with the associated solution trajectory $x^o \in C(I, R^n)$. Then, in order that the pair $\{u^o, x^o\}$ be optimal it is necessary that there exists a function $\psi \in C(I, R^n)$ satisfying the following (necessary) conditions:

$$\int_I < H_u(t, x^o(t), \psi(t), u^o(t)), u(t) - u^o(t) > dt \geq 0 \ \forall u \in \mathcal{U}_{ad}, \tag{8.3.10}$$

$$\dot{x}^o(t) = H_\psi = f(t, x^o(t), u^o(t)), \ x^o(0) = \xi, \tag{8.3.11}$$

$$\dot{\psi}(t) = -H_x = -f_x^*(t, x^o(t), u^o(t))\psi(t) - \ell_x(t, x^o(t).u^o(t)),$$
$$\psi(T) = \varphi_x(x^o(T)). \tag{8.3.12}$$

Proof. Let $x \in AC(I, R^n)$ denote any solution of the system (8.2.1) corresponding to the control policy $u \in \mathcal{U}_{ad}$. Since $u^o \in \mathcal{U}_{ad}$ is optimal with x^o being the associated trajectory, it is clear that

$$J(u^o) \equiv \int_0^T \ell(t, x^o(t), u^o(t))dt + \varphi(x^o(t))$$
$$\leq \int_0^T \ell(t, x(t), u(t))dt + \varphi(x(t)) \equiv J(u) \tag{8.3.13}$$

for all $u \in \mathcal{U}_{ad}$. For any $\varepsilon \in [0,1]$, define $u^\varepsilon = u^o + \varepsilon(u - u^o)$. Since U is a closed convex set, $\mathcal{U}_{ad}$ is also a closed convex subset of $L_\infty(I, R^m)$ and therefore $u^\varepsilon \in \mathcal{U}_{ad}$. Thus

$$J(u^o) \leq J(u^\varepsilon) \ \forall \ \varepsilon \in [0,1] \text{ and } \forall \ u \in \mathcal{U}_{ad}. \tag{8.3.14}$$

Letting $dJ(u^o, u - u^o)$ denote the Gateaux (directional) differential of J at u^o in the direction $u - u^o$, it follows from (8.3.14) that

$$dJ(u^o, u - u^o) \geq 0 \ \forall \ u \in \mathcal{U}_{ad}. \tag{8.3.15}$$

Let x^ε be the solution of equation (8.2.1) corresponding to the control policy u^ε and the same initial state $x^\varepsilon(0) = \xi$. Under the assumptions of regularity of f in x and u it is easy to verify that

$$\lim_{\varepsilon \to 0} u^\varepsilon(t) \longrightarrow u^o(t) \text{ } a.e \text{ on } I \equiv [0, T]$$
$$\lim_{\varepsilon \to 0} x^\varepsilon(t) \longrightarrow x^o(t) \text{ uniformly on } I.$$

Note that corresponding to the same initial state $x^\varepsilon(0) = x^o(0) = \xi$ the state trajectories x^ε and x^o satisfy the following equations

$$\dot{x}^\varepsilon = f(t, x^\varepsilon(t), u^\varepsilon(t)), t \in I$$
$$\dot{x}^o = f(t, x^o(t), u^o(t)), t \in I$$

almost every where. Subtracting one from the other and recalling the C^1 regularity of f both in x and u we arrive at the following equation

$$(d/dt)(x^\varepsilon(t) - x^o(t)) = f_x(t, x^o, u^o)(x^\varepsilon - x^o) + f_u(t, x^o, u^o)(u^\varepsilon - u^o) + o(\varepsilon)$$

where $o(\varepsilon)$ denotes the remainder terms in the approximation. In other words

$$\lim_{\varepsilon \to 0}\{(1/\varepsilon)o(\varepsilon)\} = 0.$$

Dividing through by ε and letting $\varepsilon \to 0$ and denoting by y the limit

$$y(t) \equiv \lim_{\varepsilon \to 0}(1/\varepsilon)(x^\varepsilon(t) - x^o(t))$$

it follows from the above expression that y must satisfy the following initial value problem

$$\dot{y}(t) = f_x(t, x^o(t), u^o(t))y(t) + f_u(t, x^o(t), u^o(t))(u(t) - u^o(t)), \quad (8.3.16)$$
$$y(0) = 0.$$

We call (8.3.16) the variational equation. Using C^1 regularity of ℓ in both x and u and C^1 regularity of φ in x, it follows from simple computations using (8.3.14) that

$$dJ(u^o, u - u^o) = \int_I \{< \ell_x(t, x^o, u^o), y(t) > + < \ell_u(t, x^o, u^o), u - u^o >\}dt$$
$$+ < \varphi_x(x^o(T)), y(T) > \geq 0 \; \forall \; u \in \mathcal{U}_{ad}. \quad (8.3.17)$$

For convenience of analysis we reorganize this inequality in the form

$$dJ(u^o, u - u^o) = \int_I < \ell_x(t, x^o, u^o), y(t) > dt + < \varphi_x(x^o(T)), y(T) >$$
$$+ \int_I < \ell_u(t, x^o, u^o), u - u^o > dt \geq 0 \; \forall \; u \in \mathcal{U}_{ad}, \quad (8.3.18)$$

where the first two terms are dependent on y linearly and the third term is a linear functional of control. We wish to write this expression as a functional of the control $u - u^o$. Note that the variational equation (8.3.16) is linear in $u - u^o$. Since $f_u(t, x^o(t), u^o(t))$ is fixed, clearly the map

$$f_u(\cdot, x^o(\cdot), u^o(\cdot)))(u(\cdot) - u^o(\cdot)) \longrightarrow y(\cdot)$$

is linear. In fact this is a continuous linear map from $L_1(I, R^n)$ to $C(I, R^n)$. This is justified as follows. Since $x^o \in AC(I, R^n)$ and I is a compact interval we have $\sup\{\| x^o(t) \|, t \in I\} < \infty$ and since U is a closed bounded set, it follows from our assumption (A1), in particular, C^1 regularity of f in $u \in U$, and integrability, that

$$f_u(\cdot, x^o(\cdot), u^o(\cdot)))(u(\cdot) - u^o(\cdot)) \in L_1(I, R^n).$$

We know from the theory of differential equations that in this case the variational equation has a unique solution $y \in AC(I, R^n) \subset C(I, R^n)$. Hence the map

$$f_u(\cdot, x^o(\cdot), u^o(\cdot)))(u(\cdot) - u^o(\cdot)) \longrightarrow y(\cdot)$$

is a continuous linear map or equivalently a bounded linear map from L_1 (I, R^n) to $C(I, R^n)$. Further, because of C^1 regularity of ℓ and φ in x and integrability of ℓ_x, we have

$$L(y) \equiv \int_I < \ell_x(t, x^o, u^o), y(t) > dt + < \varphi_x(x^o(T)), y(T) > \quad \in R,$$

where, for convenience, we have used the notation $L(y)$ to denote the functional as displayed. Thus the map

$$\begin{aligned} &f_u(\cdot, x^o(\cdot), u^o(\cdot)))(u(\cdot) - u^o(\cdot)) \\ &\quad \longrightarrow \int_I < \ell_x(t, x^o, u^o), y(t) > dt + < \varphi_x(x^o(T)), y(T) > \end{aligned}$$

is a bounded linear map. Hence by the Riesz representation theorem or by the duality between $L_1(I, R^n)$ and $L_\infty(I, R^n)$ we may conclude that there exists a $\psi \in L_\infty(I, R^n)$ so that

$$\begin{aligned} L(y) &= \int_I < \ell_x(t, x^o, u^o), y(t) > dt + < \varphi_x(x^o(T)), y(T) > \\ &= \int_I < f_u(\cdot, x^o(\cdot), u^o(\cdot)))(u(t) - u^o(t)), \psi(t) > dt \\ &= \int_I < f_u^*(t, x^o(t), u^o(t))\psi(t), u(t) - u^o(t) > dt, \quad (8.3.19) \end{aligned}$$

where f_u^* denotes the transpose of the Hessian f_u. Using this representation, it follows from (8.3.18) that

$$\begin{aligned} dJ(u^o, u - u^o) \\ &= \int_I < \ell_u(t, x^o, u^o) + f_u^*(t, x^o(t), u^o(t))\psi(t), u - u^o > dt \\ &= \int_I < H_u(t, x^o(t), \psi(t), u^o(t)), u(t) - u^o(t) > dt \ \geq 0 \ \forall \ u \in \mathcal{U}_{ad}. \end{aligned} \tag{8.3.20}$$

Thus we have derived the first term of the necessary conditions of optimality provided we can justify that ψ can be chosen from the class of absolutely continuous functions, that is, $\psi \in AC(I, R^n)$. Using the variational equation for y it follows from (8.3.19) that

$$\begin{aligned} L(y) &= \int < f_u(t, x^o, u^o)(u(t) - u^o(t)), \psi(t) > dt \\ &= \int_0^T < \dot{y} - f_x(t, x^o, u^o)y(t), \psi(t) > dt. \end{aligned} \tag{8.3.21}$$

Integrating by parts, assuming the regularity required, and recalling that $y(0) = 0$, it follows from the above expression that

$$L(y) =< y(T), \psi(T) > + \int_0^T < y(t), -\dot{\psi}(t) - f_x^*(t, x^o, u^o)\psi(t) > dt. \tag{8.3.22}$$

On the other hand the functional L is also given by

$$L(y) =< y(T), \varphi_x(x^o(T)) > + \int_0^T < y(t), \ell_x(t, x^o, u^o) > dt. \tag{8.3.23}$$

From these two equivalent expressions, it follows that we may choose ψ as the solution of the following Cauchy problem

$$\begin{aligned} \dot{\psi} &= -f_x^*(t, x^o, u^o)\psi - \ell_x(t, x^o, u^o) \\ \psi(T) &= \varphi_x(x^o(T)). \end{aligned} \tag{8.3.24}$$

Since by our assumption (A1), both $\ell_x(\cdot, x^o(\cdot), u^o(\cdot))$ and $f_x^*(\cdot, x^o(\cdot), u^o(\cdot))$ are integrable, and this equation is linear along the optimal trajectories, it has a unique absolutely continuous solution and hence $\psi \in AC(I, R^n)$. Thus we have derived the last part of the necessary conditions (8.3.12). The middle part is just the state equation along the optimal path. This completes the proof. □

Corollary 8.3.2. Under the assumptions of Theorem 8.3.1, the necessary condition (8.3.10) is equivalent to the point wise necessary condition given by

$$< H_u(t, x^o(t), \psi(t), u^o(t)), v - u^o(t) > \;\geq 0 \, \forall \, v \in U \text{ and for almost all } t \in I. \tag{8.3.25}$$

Proof. By assumption (A1), the derivative of the Hamiltonian H is measurable in t on I and continuous in the rest of the variables. Thus the function $H_u(t, x^o(t), \psi(t), u^o(t))$ is measurable in t on I and that it is also in $L_1(I, R^m)$. By our choice of the admissible class, the controls $\{u^o, u\} \in \mathcal{U}_{ad}$ and so bounded measurable functions belonging to the class $L_\infty(I, R^m)$. Thus one may conclude that almost all points t of the interval I are regular or Lebesgue density points for the function $t \longrightarrow < H_u(t, x^o(t), \psi(t), u^o(t)), u(t) - u^o(t) >$. Let t be any one such regular point different from the terminal points $\{0, T\}$ and consider the interval $J_\varepsilon \equiv [t - (\varepsilon/2), t + (\varepsilon/2)]$ for $\varepsilon > 0$ sufficiently small so that $J_\varepsilon \subset I$. For any $v \in U$, choose the control

$$u(r) = \begin{cases} u^o(r), & \text{for } r \in I \setminus J_\varepsilon; \\ v, & \text{for } r \in J_\varepsilon. \end{cases}$$

Since (8.3.10) holds for all admissible controls and u, as defined above, is admissible, substituting this control there we obtain

$$\int_{J_\varepsilon} < H_u(s, x^o(s), \psi(s), u^o(s)), v - u^o(s) > dt \geq 0 \, \forall \, v \in U.$$

Dividing this by ε and letting $\varepsilon \downarrow 0$, we arrive at the inequality (8.3.26). This completes the proof. □

The result of Theorem 8.3.1 is based on the assumption that both the functions $\{f, \ell\}$ are continuously differentiable with respect to the control variable. In the following result we relax this condition.

Assumption A2: Both the vector field $f(t, x, u)$ and the scalar function $\ell(t, x, u)$ are measurable in t and Lebesgue integrable on I for all $(x, u) \in R^n \times U$, and continuous in $\{x, u\} \in R^n \times U$ for almost all $t \in I$. With respect to x, the functions f, ℓ and φ are twice differentiable with the first partials being measurable in t and continuous with respect to the rest of the arguments and second partials uniformly bounded.

Here we shall use spike variation. In other words we shall perturb the optimal control arbitrarily only on a small set to obtain another control and compare this against the optimal control. This is known as Ekland's variational principle.

Theorem 8.3.3. (Pontryagin Minimum Principle) Consider the system

$$\dot{x} = f(t, x, u), t \in I, x(0) = \xi \tag{8.3.26}$$

with the cost functional (8.2.7) and the admissible controls $\mathcal{U}_{ad}$ as defined by (8.2.3) and suppose the assumption (A2) holds. Then, if the pair $\{u^o, x^o\} \in \mathcal{U}_{ad} \times AC(I, R^n)$ is optimal, there must exist a multiplier $\psi \in AC(I, R^n)$ satisfying the following necessary conditions:

$$H(t, x^o(t), \psi(t), u^o(t)) \leq H(t, x^o(t), \psi(t), v) \ \forall v \in U, \ a.e\ t \in I, \tag{8.3.27}$$

$$\dot{x}^o(t) = H_\psi = f(t, x^o(t), u^0(t)), \ x^o(0) = \xi, \tag{8.3.28}$$

$$\begin{aligned} \dot{\psi}(t) &= -H_x = -f_x^*(t, x^o(t), u^o(t))\psi(t) \\ &\quad - \ell_x(t, x^o(t), u^o(t)), \ \psi(T) = \varphi_x(x^o(T)). \end{aligned} \tag{8.3.29}$$

Proof. Let $\{u^o, x^o\}$ be the optimal pair and $\sigma \in \mathcal{B}_I$ be arbitrary where $\mathcal{B}_I$ denotes the sigma algebra of Borel subsets of the interval I and $v \in U$. We shall use λ to denote the Lebesgue measure. Define

$$u^\sigma(t) = \begin{cases} u^o(t), & \text{for } t \in I \setminus \sigma; \\ v, & \text{for } t \in \sigma. \end{cases}$$

Let x^σ denote the solution of equation (8.3.26) corresponding to the control policy u^σ. Define

$$y^\sigma \equiv x^\sigma - x^o.$$

Then we have

$$\dot{y}^\sigma(t) = f_x(t, x^0, u^\sigma)y^\sigma(t) + (f(t, x^o, u^\sigma) - f(t, x^o, u^o)) + R_1(\lambda(\sigma)) \tag{8.3.30}$$

where R_1 denotes the remainder. We shall prove later that

$$\lim_{\lambda(\sigma)\to 0} \{(1/\lambda(\sigma)) \parallel R_1 \parallel\} = 0. \tag{8.3.31}$$

Again by virtue of optimality of the pair $\{u^o, x^o\}$

$$J(u^o) \leq J(u^\sigma), \ \forall\ \sigma \in \mathcal{B}_I \text{ and } \forall v \in U. \tag{8.3.32}$$

Using ℓ and φ one can easily verify that the inequality (8.3.32) is equivalent to

$$\int_I (\ell(t, x^o, u^\sigma) - \ell(t, x^o, u^o))dt + L(y^\sigma) + R_2(\lambda(\sigma)) \geq 0, \ \forall \sigma \in \mathcal{B}_I, v \in U, \tag{8.3.33}$$

where we have used $L(y^\sigma)$ to denote the following expression

$$L(y^\sigma) \equiv \int_I < \ell_x(t, x^o, u^\sigma), y^\sigma > dt + < \varphi_x(x^o(T)), y^\sigma(T) >, \tag{8.3.34}$$

and R_2 for the remainder which is also of small order $o(\lambda(\sigma))$ like R_1 as shown later. For $\lambda(\sigma)$ sufficiently small one may ignore the remainder term and conclude from the variational equation (8.3.30) that

$$f(\cdot, x^o(\cdot), u^\sigma(\cdot)) - f(\cdot, x^o(\cdot), u^o(\cdot)) \longrightarrow y^\sigma(\cdot)$$

is continuous linear. Further it follows from (8.3.34) that

$$y^\sigma(\cdot) \longrightarrow L(y^\sigma)$$

is continuous linear. Hence the composition map is a continuous linear functional on $L_1(I, R^n)$ and by similar arguments as in Theorem 8.3.1, we may conclude the existence of a multiplier or adjoint vector $\psi^\sigma \in L_\infty(I, R^n)$ such that $L(y^\sigma)$ has the alternate representation in terms of the adjoint vector given by

$$L(y^\sigma) \equiv \int_I < (f(t, x^o(t), u^\sigma(t)) - f(t, x^o(t), u^o(t))), \psi^\sigma(t) > dt. \tag{8.3.35}$$

Using the expression (8.3.35) in (8.3.33) we obtain the following inequality

$$\begin{aligned} &\int_I (\ell(t, x^0, u^\sigma) - \ell(t, x^o, u^o))dt \\ &\quad + \int_I < (f(t, x^o(t), u^\sigma(t)) - f(t, x^o(t), u^o(t))), \psi^\sigma(t) > dt \\ &\quad + R_2(\lambda(\sigma)) \geq 0, \ \forall \sigma \in \mathcal{B}_I, v \in U. \end{aligned} \tag{8.3.36}$$

We show later that, as $\lambda(\sigma) \to 0$, $\psi^\sigma \longrightarrow \psi$ uniformly on I which is a solution of the adjoint equation (8.3.29). Assuming this to be true for the moment and letting $t \in I$ be any regular point and $\sigma \in \mathcal{B}_I$ with σ containing t and $\sigma \longrightarrow \{t\}$ as $\lambda(\sigma) \longrightarrow 0$, and dividing the expression (8.3.36) by $\lambda(\sigma)$ and letting $\lambda(\sigma) \to 0$ and using the fact that R_2 is of small order $o(\lambda(\sigma))$, we arrive at the following expression

$$\ell(t, x^o(t), v) - \ell(t, x^o(t), u^o(t)) + < f(t, x^o(t), v) - f(t, x^o(t), u^o(t)), \psi(t) > \geq 0,$$

which holds for all $v \in U$ and almost all $t \in I$. In terms of the Hamiltonian H, as defined by the expression (8.3.9), this is the same as

$$H(t, x^o(t), \psi(t), u^o(t)) \leq H(t, x^o(t), \psi(t), v) \ \forall v \in U, \ a.e\ t \in I.$$

This is (8.3.27) and so we have proved the minimum principle. Now we show that ψ, as defined above, can be chosen as the absolutely continuous solution of the adjoint equation (8.3.29). Using the variational equation (8.3.30), in (8.3.36) and integrating by parts, we arrive at the following equivalent expression

$$\int_I (\ell(t,x^o,u^\sigma) - \ell(t,x^o,u^o))dt + \int_I < (-\dot{\psi}^\sigma - f_x^*(t,x^o,u^\sigma)\psi^\sigma), y^\sigma(t) > dt$$
$$+ < \psi^\sigma(T), y^\sigma(T) > + R_3(\lambda(\sigma)) \geq 0, \ \forall \sigma \in \mathcal{B}_I, v \in U. \tag{8.3.37}$$

Thus if we choose ψ^σ as the solution of the following Cauchy problem

$$\begin{aligned} \dot{\psi}^\sigma &= -f_x^*(t,x^o,u^\sigma)\psi^\sigma - \ell_x(t,x^o,u^\sigma) \\ \psi^\sigma(T) &= \varphi_x(x^o(T)), \end{aligned} \tag{8.3.38}$$

we arrive at the same expression as given by (8.3.33) modulo the terms of small order $o(\lambda(\sigma))$. Now we show that the solution ψ^σ of the above Cauchy problem converges to the solution of our adjoint equation (8.3.29). First note that since by assumption (A2), the functions $t \longrightarrow f_x(t,x^o(t),u^\sigma(t))$ and $t \longrightarrow \ell_x(t,x^o(t),u^\sigma(t))$ are integrable, and the system is linear, the reader can easily verify that there is a finite positive number b such that

$$\sup\{\| \psi^\sigma(t) \|, t \in I\} \leq b$$

independently of $\sigma \in \mathcal{B}_I$ and $v \in U$. Using equation (8.3.38) and the adjoint equation (8.3.29) and subtracting one from the other and rearranging terms suitably we find that

$$\begin{aligned} \| \psi^\sigma(t) - \psi(t) \| \leq & \int_t^T \| f_x^*(s,x^o,u^\sigma) - f_x^*(s,x^o,u^o) \| \| \psi^\sigma(s) \| ds \\ & + \int_t^T \| \ell_x(s,x^o,u^\sigma) - \ell_x(s,x^o,u^o) \| ds \\ & + \int_t^T \| f_x^*(s,x^o,u^o) \| \| \psi^\sigma(s) - \psi^o(s) \| ds. \end{aligned} \tag{8.3.39}$$

Defining $g(s) \equiv \| f_x^*(s,x^o,u^o) \|$ and using Gronwall inequality it follows from this that

$$\begin{aligned} \| \psi^\sigma(t) - \psi(t) \| \leq C \int_\sigma \Big\{ & b \| f_x^*(s,x^o,u^\sigma) - f_x^*(s,x^o,u^o) \| \\ & + \| \ell_x(s,x^o,u^\sigma) - \ell_x(s,x^o,u^o) \| \Big\} ds \end{aligned} \tag{8.3.40}$$

for all $t \in I$ where $C = exp\{\int_I g(s)ds\}$. By virtue of assumption (A2), $g \in L_1$ and the rest of the integrands are also in L_1. Thus $C < \infty$ and all the integrals are also finite. Hence letting $\lambda(\sigma) \to 0$, we arrive at the conclusion that ψ^σ converges to ψ uniformly on I. Thus we have proved that the multiplier ψ is given by the solution of the adjoint equation (8.3.29). Clearly equation (8.3.28) is just the state equation corresponding to the optimal control. This proves all the necessary conditions of optimality as stated in the theorem. It remains now to justify that the remainders like $\{R_1, R_2, R_3\}$ are all of small order $o(\lambda(\sigma))$. All of them are proved in the same way. We prove that R_1 has the property as claimed. First we prove that $y^\sigma \to 0$ uniformly on I as $\lambda(\sigma) \to 0$. Clearly the exact expression for y^σ is given by

$$\dot{y}^\sigma = f(t, x^o, u^\sigma) - f(t, x^o, u^o) + \int_0^1 < f_x(t, x^o + \theta y^\sigma, u^\sigma), y^\sigma(t) > d\theta. \quad (8.3.41)$$

Hence

$$\begin{aligned} \| y^\sigma(t) \| \; &\leq \int_0^t \| f(s, x^o(s), u^\sigma(s)) - f(s, x^o(s), u^o(s)) \| \, ds \\ &+ \int_0^t \sup\{\| f_x(s, x^o + \theta y^\sigma, u^\sigma) \|; \theta \in [0,1]\} \, \| y^\sigma(s) \| \, ds. \end{aligned} \quad (8.3.42)$$

Since both x^o and x^σ are continuous and bounded so also is y^σ. Therefore it follows from assumption (A2) that there exists a $K \in L_1^+(I)$ such that

$$\sup\{\| f_x(s, x^o + \theta y^\sigma, u^\sigma) \|\} \leq K(s), s \in I.$$

Using Gronwall inequality once again, it follows from the above expression that

$$\begin{aligned} \| y^\sigma(t) \| \; &\leq exp\left\{\int_0^t K(s)ds\right\}\left\{\int_I \| f(s, x^o, u^\sigma) - f(s, x^o, u^o) \| \, ds\right\} \\ &= exp\left\{\int_0^t K(s)ds\right\}\left\{\int_\sigma \| f(s, x^o, u^\sigma) - f(s, x^o, u^o) \| \, ds\right\}. \end{aligned} \quad (8.3.43)$$

Note that as $\lambda(\sigma) \to 0$, $f(t, x^o(t), u^\sigma(t)) \to f(t, x^o(t), u^o(t))$ *a.e.* Hence it follows from Lebesgue dominated convergence theorem that $y^\sigma \longrightarrow 0$ uniformly on I as $\lambda(\sigma) \to 0$. It follows from the expressions (8.3.30) and (8.3.41) that the remainder is given by

$$R_1(\lambda(\sigma))(t) \equiv \int_0^1 \Big(f_x(t, x^o + \theta y^\sigma, u^\sigma) - f_x(t, x^o, u^\sigma)\Big) y^\sigma(t) d\theta. \quad (8.3.44)$$

Taking the norm on either side and using the assumption that the second partial of f with respect to x is uniformly bounded by some number, say, $\gamma > 0$, we obtain

$$\| R_1(\lambda(\sigma))(t) \| \leq \gamma \| y^\sigma(t) \|^2, \ \forall\, t \in I. \tag{8.3.45}$$

Since x^o is bounded and the controls take values from the bounded set U, the integrands appearing in the last factor of (8.3.43) are in L_1 and so finite almost every where. Hence it follows from the above estimate (8.3.45) that as $\lambda(\sigma) \to 0$,

$$(1/\lambda(\sigma))R_1 \to 0.$$

This completes the proof. □

8.3.2 Transversality Conditions:

There are interesting control problems where the initial and the final states may not be specified as certain fixed points in the state space. Instead they may be required to belong to certain specified manifolds. An example is the moon landing problem where the final state is the moon surface rather than any specified location on it. We consider two situations.

(TC1) Let K be a closed convex subset of the state space R^n and suppose it is required to find a control that drives the system from a fixed initial state x_0 to the target set K while minimizing the cost functional

$$J_0(u) \equiv \int_0^T \ell(t, x, u)dt.$$

This is a Lagrange problem subject to the terminal constraint, $x(T) \in K$. Assuming controllability from x_0 to the set K at time T, this can be converted to a Meyer's problem given by

$$J(u) \equiv \int_0^T \ell(t, x, u)dt + I_K(x(T)) \longrightarrow \inf,$$

where $I_K(\xi)$ denotes the indicator function of the set K, that is, $I_K(\xi) = 0$ if $\xi \in K$ and $+\infty$ if $\xi \notin K$. Since K is closed the indicator function is lower semi continuous and since it is convex the function I_K is convex and so it is sub differentiable with the sub differential denoted by $\partial I_K(\cdot)$. Given that this problem has a solution, the optimal control must necessarily drive the system from the given initial state x_0 to some final state $x^* \in K$ at time T. Thus given the end points, the optimal control that realizes this transfer must satisfy the same necessary conditions of optimality as given by Theorem 8.3.1, Corollary 8.3.2 and Theorem 8.3.3. The only difference is that now the adjoint systems (8.3.12) and (8.3.29) must satisfy the boundary condition

$$\psi(T) \in \partial I_K(x^o(T)).$$

This is equivalent to the condition

$$(\psi(T), y) \leq (\psi(T), x^o(T)) \;\forall\; y \in K.$$

This coincides with the definition of normal cone (to the set K at the point $x \in K$) given by:

$$N_K(x) \equiv \{z \in R^n : (z, y) \leq (z, x) \;\forall\; y \in K\}.$$

Accordingly the transversality condition can be also stated as $\psi(T) \in N_K(x^o(T))$. This is equivalent to

$$\psi(T) \perp T_K(x^o(T))$$

where $T_K(\xi)$ denotes the tangent plane to the set K passing through the point $\xi \in K$. Strict convexity of the set K guarantees the existence of a unique tangent plane to each point on K.

Remark. This result can also be derived by choosing a sequence of smooth functions φ_n approximating the indicator function I_K and converging to it in the limit.

(TC2) More generally, let M_0 and M_1 be two smooth manifolds in R^n with dimensions $m_0 < n$ and $m_1 < n$ respectively. Suppose the system

$$\dot{x} = f(t, x, u), t \in I, x(0) \in M_0, x(T) \in M_1, u \in \mathcal{U}_{ad},$$

controllable from M_0 to M_1 in time T. The problem is to find a control that drives the system from M_0 to M_1 while minimizing the cost functional

$$J(u) \equiv \int_0^T \ell(t, x, u)dt.$$

Again the necessary conditions of optimality given by Theorems 8.3.1-8.3.3 remain valid with the boundary conditions for the adjoint equations given by

$$-\psi(0) \in N_{M_0}(x^o(0)), \psi(T) \in N_{M_1}(x^o(T)),$$

where $N_M(\xi)$ denotes the normal cone to the manifold M with vertex at the point ξ. Equivalently, in terms of tangent manifolds, the transversality conditions can be described as

$$\psi(0) \perp T_{M_0}(x(0)), \quad \psi(T) \perp T_{M_1}(x(T))$$

where $T_M(\xi)$ denotes the tangent plane passing through the point $\xi \in M$. In general the normal and the tangent cones denoted by N_M and T_M are defined in the sense of Clark ([26] p 50). Thus the complete set of necessary conditions (including the terminal conditions) for the Lagrange problem is given by (NC1),(NC2) and (NC3) as stated below:

$$\begin{aligned}
&(NC1) : H(t, x^o(t), \psi(t), u^o(t)) \leq H(t, x^o(t), \psi(t), v) \ \forall v \in U, \ a.e\ t \in I, \\
&(NC2) : \dot{x}^o(t) = H_\psi = f(t, x^o(t), u^0(t)), \ x^o(0) \in M_0, x^o(T) \in M_1, \\
&(NC3) : \dot{\psi}(t) = -H_x = -f_x^*(t, x^o(t), u^o(t))\psi(t) - \ell_x(t, x^o(t), u^o(t)), \\
&\qquad \psi(0) \in N_{M_0}(x^o(0)), \psi(T) \in N_{M_1}(x^o(T)).
\end{aligned}$$

For illustration, let us consider the manifolds given by

$$M_0 \equiv \bigcap_{k=1}^{n-m_0} \{x \in R^n : g_k(x) = 0\}$$

$$M_1 \equiv \bigcap_{k=1}^{n-m_1} \{x \in R^n : h_k(x) = 0\}$$

where the functions $\{g_k, h_r, 1 \leq k \leq n - m_0, 1 \leq r \leq n - m_1\}$ are assumed to be once continuously differentiable. Further, both the sets of gradient vectors $\{\nabla g_k(x), 1 \leq k \leq n - m_0\}$ and $\{\nabla h_k(x), 1 \leq k \leq n - m_1\}$ are linearly independent and do not vanish simultaneously any where in R^n. Under these assumptions, the manifolds M_0 and M_1 are smooth and at every point $\xi \in M_0$ and $\eta \in M_1$, the normal and the tangent cones $N_{M_0}(\xi), T_{M_0}(\xi)$ and $N_{M_1}(\eta), T_{M_1}(\eta)$ are well defined. The transversality conditions are then given by

$$\psi(0) \perp T_{M_0}(x^o(0)), \quad \psi(T) \perp T_{M_1}(x^o(T)).$$

The adjoint state ψ is said to satisfy the transversality conditions if the above expressions hold. These expressions provide $m_0 + m_1$ initial-boundary conditions whereby the exact positions of the initial and final states, $x^o(0) \in M_0$ and $x^o(T) \in M_1$, can be determined. Thus a complete set of $2n$ boundary conditions required for solving the optimality system (the minimum principle, the state equation and the adjoint equation) are now fully specified.

Remarks. In the presence of a terminal cost $\varphi(x(T))$, including the boundary condition $x(T) \in M_1$, the transversality condition takes the form

$$\psi(T) \in \varphi_x(x^o(T)) + N_{M_1}(x^o(T)).$$

This generalizes the transversality condition of Theorem 8.3.3 given by (8.3.29). Clearly if the terminal state is free, $M_1 = R^n$ and hence the normal cone $N_{M_1} = \{0\}$. As a consequence $\psi(T) = \varphi_x(x^o(T))$.

For illustration we present some elementary examples.

E1: Consider a system of population consisting of prey and predators governed by the celebrated Lotka-Volterra system of equations

$$\dot{x}_i = \alpha_i x_i + \sum_{j=i}^{n} \beta_{i.j} x_i x_j + \sum_{j=1}^{d} \gamma_{i,j} x_i u_j, i = 1, 2, \cdots, n.$$

We rewrite this in the canonical form $\dot{x} = f(x, u), t \geq 0$. The controls are constrained to take values from a closed convex bounded set $U(\ni 0) \subset R^d_+$ where R^d_+ denotes the positive orthant of the real Euclidian space R^d. The problem is to minimize the co-lateral damage to the prey population including the cost of administering controls given by the functional

$$J(u) \equiv \int_0^T \ell(x, u)dt$$

while realizing the goal of reducing the predator population to a specified level. Let the first m components of the population vector x denote the prey population and the last $n - m$ components the predator population. Let $x(0) = x_0$ be the initial population. The desired terminal manifold is given by

$$M_1 \equiv \{x \in R^n : x_{m+1} = z_1, x_{m+2} = z_2, \cdots, x_n = z_{n-m}\}$$

where $z_i, i = 1, 2, \cdots, n - m$ are the desired levels of the predator population. Clearly for this problem $M_0 = \{x_0\}$. Here the tangent plane T_{M_1} which coincides with the terminal manifold M_1 is of dimension m. Thus the transversality condition $\psi(T) \perp T_{M_1}$ reduces to $\psi_1(T) = \psi_2(T) = \cdots = \psi_m(T) = 0$. Hence for the optimality system we have $2n$ necessary boundary conditions given by $x(0) = x_0, \psi_1(T) = 0, \cdots, \psi_m(T) = 0, x_{m+1}(T) = z_1, x_{m+2}(T) = z_2, \cdots, x_n(T) = z_{n-m}$. This makes the $2n$ boundary value problem $\dot{x} = H_\psi, -\dot{\psi} = H_x$ mathematically consistent and complete.

E2: The dynamics of a geosynchronous satellite is given by

$$\begin{aligned}
&I_1\dot{p}_1 + (I_3 - I_2)(p_2 - \omega_0)p_3 = K_1u_1\\
&I_2\dot{p}_2 + (I_1 - I_3)p_1p_3 = K_2u_2\\
&I_3\dot{p}_3 + (I_2 - I_1)p_1(p_2 - \omega_0) = K_3u_3\\
&\dot{\theta}_1 = p_1 + (p_2 \sin\theta_1 + p_3\cos\theta_1)\tan\theta_2\\
&\dot{\theta}_2 = p_2\cos\theta_1 - p_3\sin\theta_1\\
&\dot{\theta}_3 = (p_2\sin\theta_1 + p_3\cos\theta_1)sec\theta_2,
\end{aligned}$$

where $\{p_i, i = 1,2,3\}$ and $\{\theta_i, i = 1,2,3\}$ denote the angular momenta and the Euler angles respectively with ω_0 denoting the spin rate of the earth. The parameters $\{I_i, i = 1,2,3\}$ denote the moment of inertias around the $\{x,y,z\}$ axes of the satellite, $\{u_i, i = 1,2,3\}$ denote the control torques. The problem is to transfer the system from a given initial state to the terminal manifold

$$M_1 \equiv \{(p,\theta) \in R^6 : p = 0 \text{ and } \theta_1 = \theta_3 = 0\}$$

at time T while minimizing the cost functional $J(u) \equiv \int_0^T \ell(p,\theta,u)dt$. In particular, the fuel cost is the predominant component of the cost functional since the satellite becomes useless once the fuel is exhausted. In this case the tangent plane to M_1 contains only the unit vector e_5 and the transversality condition reduces to $(\psi(T), e_5) = 0$ giving $\psi_5(T) = 0$.

8.3.3 Relaxed Controls

All the necessary conditions presented so far require strong regularity conditions in terms of the state and control variables. As observed already, Theorem 8.3.1 requires differentiability of f and ℓ with respect to the control variable. This condition is relaxed in Theorem 8.3.3 where only continuity is required. However in that theorem stronger regularity conditions are imposed with respect to the state variable. Also crucial is the assumption that the control constraint set U is convex. In the following result many of these assumptions are relaxed. This is achieved by relaxing the class of admissible controls to a broader class. This class consists of probability measure valued functions.

Let U be a compact subset of R^m and $C(U)$ the vector space of real valued continuous functions on U with the sup norm topology,

$$\| \phi \| = \sup\{|\phi(v)|, v \in U\}.$$

This makes $C(U)$ a Banach space. Denote by $M(U)$ the space of countably additive bounded signed measures on the Borel sigma field $\mathcal{B}_U$ of U having bounded total variation,

$$|\mu|_v \equiv \sup_{\pi} \sum_{\sigma \in \pi} |\mu(\sigma)|$$

where π is any partition of U into a finite number of disjoint members of $\mathcal{B}_U$ and the sup is taken over all such partitions. With respect to this norm topology, $M(U)$ is also a Banach space. It is known that the dual of $C(U)$ is $M(U)$, that is, for any continuous linear functional on $C(U)$, denoted by $L \in C^*(U)$, there exists a $\mu \in M(U)$ such that

$$L(\phi) = \int_U \phi(\zeta)\mu(d\zeta) \equiv \int_U \phi(\zeta) d\mu(\zeta).$$

We are interested in the subset $M_1(U) \subset M(U)$, the space of probability measures on U, that is, for $\mu \in M_1(U)$, $\mu(\Gamma) \geq 0$ for all $\Gamma \in \mathcal{B}_U$ and that $\mu(U) = 1$. Proceeding further, we have the vector space $L_1(I, C(U))$ with the standard norm topology given by,

$$\| w \| \equiv \int_I \| w(t) \|_{C(U)} \, dt$$

where we have used $w(t)$ to denote the function, $w(t)(\cdot)$, that is, $w(t)(\xi) = w(t, \xi)$. The dual of this space is given by

$$L_\infty^w(I, M(U))$$

which consists of weak star measurable functions on I with values in $M(U)$. By this we mean that for any $\varphi \in C(U)$ and $u \in L_\infty^w(I, M(U))$

$$t \longrightarrow \int_U \varphi(v) u_t(dv)$$

is an essentially bounded measurable function. We are interested in the subset

$$\mathcal{U}_{ad}^r \equiv L_\infty^w(I, M_1(U)) \subset L_\infty^w(I, M(U))$$

which we choose as the admissible class of generalized or relaxed controls. It is easily seen that $\mathcal{U}_{ad}^r$ is the unit sphere of $L_\infty^w(I, M(U))$.

Now we consider the system

$$\dot{x} = f(t, x(t), u_t) \equiv \int_U f(t, x(t), v) u_t(dv), x(0) = x_0, \tag{8.3.46}$$

with control $u \in \mathcal{U}_{ad}^r = L_\infty^w(I, M_1(U))$. Under standard assumptions, one can prove existence and uniqueness of solutions. We present here one such result.

Theorem 8.3.4. Consider the system (8.3.46) and suppose f is Borel measurable in all the variables and there exists a $K \in L_1^+(I)$ such that

$$\| f(t,\xi,v) \| \ \leq \ K(t)\{1+ \| \xi \|\} \ \forall\ v \in U, \tag{8.3.47}$$

and for each finite $R > 0$ there exists a $K_R \in L_1^+(I)$ such that

$$\| f(t,\xi,v) - f(t,\eta,v) \| \ \leq \ K_R(t) \| \xi - \eta \| \ \ \forall \xi,\eta \in B_R \ \forall\ v \in U, \tag{8.3.48}$$

where B_R is the closed ball of radius R with the center at the origin. Then for every $x_0 \in R^n$ and $u \in \mathcal{U}_{ad}^r$, the system has a unique solution $x \in AC(I, R^n)$.

Proof. Let $u \in \mathcal{U}_{ad}^r$ be given. First we prove an a priori bound. Let $x \in AC(I, R^n)$ be a solution. Then it follows from the growth condition that

$$\begin{aligned} \| x(t) \| &\leq \| x_0 \| + \int_0^t \| \int_U f(s, x(s), v) u_s(dv) \| \, ds \\ &\leq \| x_0 \| + \int_0^t K(s)[1+ \| x(s) \|] ds. \end{aligned} \tag{8.3.49}$$

The last inequality follows from the fact that u is a probability measure valued function and hence $u_s(U) = 1$ for all $s \in I$. Using Gronwall inequality it follows from this that there exists a finite positive number R such that

$$\sup\{\| x(t) \|, t \in I\} \leq \left(\| x_0 \| + \int_I K(s)ds\right) exp\left\{\int_I K(s)ds\right\} \leq R < \infty.$$

This shows that if the Cauchy problem (8.3.46) has a solution it is contained in a bounded subset of $C(I, R^n)$. Thus it suffices to prove that it has a unique solution in the metric space $C(I, R^n)$. Define the operator G as follows

$$Gx(t) \equiv x_0 + \int_0^t f(s, x(s), u_s) ds = x_0 + \int_0^t \int_U f(s, x(s), z) u_s(dz) ds.$$

It is clear that G maps $C(I, R^n)$ into itself. We show that G has a fixed point in $C(I, R^n)$, that is an element $x^* \in C(I, R^n)$ such that $x^* = Gx^*$. We use Banach fixed point theorem to prove this. Let $x, y \in C(I, R^n)$ with $x(0) = y(0) = x_0$. Since $x, y \in C(I, R^n)$ and I is a compact interval there exists a finite positive number r such that $x(t), y(t) \in B_r$ for all $t \in I$. Define

$$d_t(x,y) \equiv \sup\{\| x(s) - y(s) \|, 0 \leq s \leq t\} \tag{8.3.50}$$

$$W(t) \equiv \int_0^t K_r(s) ds, \tag{8.3.51}$$

where $K_r \in L_1^+(I)$ is the local Lipschitz variable as stated in the theorem. Using this notation, and the definition of the operator G, we find that

$$d_t(Gx, Gy) \leq \int_0^t K_r(s) d_s(x, y) ds = \int_0^t d_s(x, y) dW(s). \tag{8.3.52}$$

By repeated substitution into itself using the expression (8.3.52), one can verify that

$$d_t(G^n x, G^n y) \leq \Big(W^n(t)/\Gamma(n)\Big) d_t(x, y) \ \forall t \in I, \tag{8.3.53}$$

where Γ denotes the gamma function. Hence we have

$$d_T(G^n x, G^n y) \leq \Big(W^n(T)/\Gamma(n)\Big) d_T(x, y). \tag{8.3.54}$$

Since $W(T)$ is finite, it is clear that, for n sufficiently large, $\alpha_n \equiv (W^n(T)/\Gamma(n)) < 1$. In other words, for n sufficiently large the operator G^n is a contraction on the complete metric space $C(I, R^n)$. Thus by Banach fixed point theorem [see Chapter 1] it has a unique fixed point $x^* \in C(I, R^n)$. But then G has the same fixed point since

$$d_T(Gx^*, x^*) = d_T(GG^n x^*, G^n x^*) = d_T(G^n(Gx^*), G^n x^*) \leq \alpha_n d_T(Gx^*, x^*)$$

where $\alpha_n \in (0, 1)$. This inequality is satisfied only if $Gx^* = x^*$, that is x^* is also a fixed point of G. The fact that $x^* \in AC(I, R^n)$ follows from its integral expression, $x^* = Gx^*$, where G is the integral operator with integrands from $L_1(I, R^n)$. This completes the proof. □

Remark. According to the above theorem $x^* \in AC(I, R^n)$ is the unique solution of equation (8.3.46). Since every solution must satisfy the a priori bound as shown in the proof, we have $\| x \| \leq R$. If one constructs a solution by Piccard approximation starting from any point in the space $C(I, R^n)$, the sequence ultimately converges to one and the same fixed point.

Now we are prepared to prove necessary conditions of optimality using relaxed controls. Let the cost functional be given by

$$J(u) \equiv \int_0^T \ell(t, x(t), u_t) dt + \varphi(x(T)) = \int_0^T \int_U \ell(t, x(t), v) u_t(dv) dt + \varphi(x(T)) \tag{8.3.55}$$

where x is the solution of equation (8.3.46) corresponding to the control $u \in \mathcal{U}_{ad}^r$. For the relaxed problem we introduce the Hamiltonian H as follows

$$H(t, x, y, \mu) \equiv \int_U \{\ell(t, x, v) + < f(t, x, v), y >\} \mu(dv) \tag{8.3.56}$$

defined on

$$I \times R^n \times R^n \times M_1(U).$$

We introduce the following assumptions on the data $\{f, \ell, \varphi\}$.

Assumption A3: Both f and ℓ are Borel measurable in all the variables, and continuous in the last two arguments. Further, they are once differentiable in the state variable on R^n.

In the case of non convex control problems, for example non convex U or more generally non convex contingent set, Pontryagin minimum principle does not hold. Here we present a general minimum principle (necessary conditions of optimality) which holds for convex as well as non convex problems. This is certainly a far reaching generalization covering wider class of problems where Pontryagin minimum principle fails.

Theorem 8.3.5. Consider the system

$$\dot{x} = \int_U f(t, x, v) u_t(dv) \equiv f(t, x, u_t), t \in I, x(0) = \xi \tag{8.3.57}$$

with the cost functional (8.3.55) and the admissible controls $\mathcal{U}_{ad}^r$ where U is a closed bounded set (convexity not required) and suppose the assumption (A3) holds. Then, if the pair $\{u^o, x^o\} \in \mathcal{U}_{ad}^r \times AC(I, R^n)$ is optimal, there exists a multiplier $\psi \in AC(I, R^n)$ satisfying the following necessary conditions:

$$\int_I H(t, x^o(t), \psi(t), u_t^o)\, dt \le \int_I H(t, x^o(t), \psi(t), u_t)\, dt \ \forall\ u \in \mathcal{U}_{ad}^r, \tag{8.3.58}$$

$$\dot{x}^o(t) = H_\psi = f(t, x^o(t), u_t^0), \ x^o(0) = \xi, \tag{8.3.59}$$

$$\dot{\psi}(t) = -H_x = -f_x^*(t, x^o(t), u_t^o)\psi(t) - \ell_x(t, x^o(t).u_t^o),$$
$$\psi(T) = \varphi_x(x^o(T)). \tag{8.3.60}$$

Proof. Let $\{u^o, x^o\}$ be the optimal pair. Take any other admissible control $u \in \mathcal{U}_{ad}^r$ and define

$$u^\varepsilon \equiv u^o + \varepsilon(u - u^o), \text{ for } \varepsilon \in [0, 1].$$

Note that the set $\mathcal{U}_{ad}^r$ is convex even though U is not a convex set and the system is linear in control. Thus $u^\varepsilon \in \mathcal{U}_{ad}^r$. Let x^ε denote the solution of equation (8.3.57) corresponding to the control u^ε. One can easily verify that the limit

$$y \equiv \lim_{\varepsilon \to 0}((x^\varepsilon - x^o)/\varepsilon)$$

exists and is given by the solution of the variational equation

$$\begin{aligned}\dot{y} &= f_x(t, x^o(t), u_t^o)y + \int_U f(t, x^o(t), z)(u_t(dz) - u_t^o(dz)) \\ &= f_x(t, x^o(t), u_t^o)y + f(t, x^o(t), u_t - u_t^o), y(0) = 0, t \in I. \quad (8.3.61)\end{aligned}$$

Since u^o is optimal it is evident that

$$J(u^o) \leq J(u^o + \varepsilon(u - u^o)), \forall\ u \in \mathcal{U}_{ad}^r \text{ and } \varepsilon \in [0, 1].$$

Since both ℓ and φ are differentiable in x, J is Gateaux differentiable in u and the Gateaux derivative, in the direction $u - u^o$, is given by

$$dJ(u^o, u - u^o) = \int_0^T \ell(t, x^o(t), u_t - u_t^o)\ dt + L(y) \quad \geq 0,\ \forall\ u \in \mathcal{U}_{ad}^r \quad (8.3.62)$$

where $L(y)$ is given by

$$L(y) \equiv \int_0^T < \ell_x(t, x^o(t), u_t^o), y(t) > \ dt + < \varphi_x(x^o(T)), y(T) > . \quad (8.3.63)$$

Note that $y \longrightarrow L(y)$ is a continuous linear functional on $C(I, R^n)$ and that the map $f(\cdot, x^o(\cdot), u_\cdot - u_\cdot^o) \longrightarrow y$ is a continuous linear map from $L_1(I, R^n)$ to $C(I, R^n)$. Hence the composition map is a continuous linear functional on $L_1(I, R^n)$. This fact allows us to conclude that there exists a multiplier $\psi \in L_\infty(I, R^n)$ so that $L(y)$ has an alternate representation given by

$$L(y) = \int_0^T < f(t, x^o(t), u_t - u_t^o), \psi(t) > dt. \quad (8.3.64)$$

Using this in (8.3.62) we obtain

$$\int_0^T \{\ell(t, x^o(t), u_t - u_t^o) + < f(t, x^o(t), u_t - u_t^o), \psi(t) >\}dt \quad \geq 0,\ \forall\ u \in \mathcal{U}_{ad}^r. \quad (8.3.65)$$

In terms of Hamiltonian this is equivalent to

$$\int_I H(t, x^o(t), \psi(t), u_t^o)\ dt \leq \int_I H(t, x^o(t), \psi(t), u_t)\ dt\ \forall u \in \mathcal{U}_{ad}^r,$$

which is the necessary condition (8.3.58), once we have shown that the multiplier ψ is necessarily given by the solution of the adjoint equation ($\psi \in AC(I, R^n)$). Using the variational equation (8.3.61) in (8.3.64) and integrating by parts, we obtain

$$L(y) = < y(T), \psi(T) > - \int_0^T < \dot{\psi}(t) + f_x^*(t, x^o, u_t^o)\psi(t), y(t) > dt. \quad (8.3.66)$$

Comparing the equations (8.3.63) and (8.3.66) and arguing that these must hold for all solutions of the variational equation corresponding to all admissible controls, we find that ψ must satisfy the following Cauchy problem

$$-\dot{\psi}(t) = f_x^*(t, x^o(t), u_t^o)\psi(t) + \ell_x(t, x^o(t), u_t^o), \ \psi(T) = \varphi_x(x^o(T)), t \in I. \tag{8.3.67}$$

This is precisely the adjoint equation (8.3.60). Thus we have proved all the necessary conditions. This completes the proof. □

From the above theorem we can derive point wise necessary conditions of optimality similar to that of Corollary 8.3.2.

Corollary 8.3.6. Under the assumptions of Theorem 8.3.5, the necessary condition (8.3.58) is equivalent to the point wise necessary condition given by

$$< H(t, x^o(t), \psi(t), u_t^o) \leq H(t, x^o(t), \psi(t), \nu) \ \forall \ \nu \in M_1(U) \ \text{ for a.a } \ t \in I. \tag{8.3.68}$$

while the other necessary conditions (8.3.59), (8.3.60) remain unchanged.

Proof. The proof is again based on Lebesgue density argument and it is similar to that of Corollary 8.3.2. □

From this point wise necessary condition we observe that the right hand expression is given by

$$H(t, x^o(t), \psi(t), \nu) = \int_U H(t, x^o(t), \psi(t), \zeta)\nu(d\zeta).$$

Since U is assumed to be compact and H is continuous on U, H attains its minimum at some point, say, $w \in U$ which is dependent on t. Thus the measure ν that minimizes the right hand expression of the inequality (8.3.68) is given by the Dirac measure δ_{w_t}. This may give the impression that the control can be chosen as an ordinary function w_t. But this is entirely false since point wise selection may lead to a nonmeasurable function and all the integrals loose their meaning. However, if a measurable selection can be found then the necessary condition reduces to the classical necessary condition given by Theorem 8.3.3 as stated in Corollary 8.3.7 below.

Remark. Another interesting observation is, if optimal control exists from $L_\infty(I, U) \subset L_\infty^w(I, M(U))$ then the necessary condition given by (8.3.58) can be specialized to the class of controls which are simply Dirac measures concentrated along ordinary controls giving

$$\int_I H(t, x^o(t), \psi(t), \delta_{u^o(t)})dt$$
$$\leq \int_I H(t, x^o(t), \psi(t), \delta_{u(t)})dt \ \forall u \in \mathcal{U}_{ad} = L_\infty(I, U). \quad (8.3.69)$$

Clearly this simply leads to the classical inequality

$$\int_I H(t, x^o(t), \psi(t), u^o(t))dt \leq \int_I H(t, x^o(t), \psi(t), u(t)) \ \forall \ u \in \mathcal{U}_{ad}.$$

From this one obtains the point wise necessary condition given by

$$H(t, x^o(t), \psi(t), u^o(t)) \leq H(t, x^o(t), \psi(t), v) \ \forall \ v \in U.$$

Thus we have the following corollary giving the Pontryagin minimum principle.

Corollary 8.3.7. If the control problem (8.3.55)-(8.3.57) has a solution in the class of ordinary controls $\mathcal{U}_{ad} = L_\infty(I, U)$, then the necessary conditions of optimality of Theorem 8.3.5 reduce to the classical (standard) minimum principle of Pontryagin [77, 21, 38] as given in Theorem 8.3.3.

It is interesting to note that the proof of the Pontryagin minimum principle is substantially simplified by use of relaxed controls and then specializing to regular (measurable) controls.

Remark. It is important to emphasize once again that the convexity assumption of the set U is crucial. Without this, the problem may have no solution from the class of ordinary controls, $L_\infty(I, U)$. This assumption is required by Theorems 8.3.1, 8.3.3 while it is not required for Theorem 8.3.5. Even the boundedness assumption is not necessary for the relaxed problem. These are extremely important in many applications.

Some time the control constraint set may be given by a set of discrete points in R^m like

$$U \equiv \{e_i, i = 1, 2, 3, \cdots, s\}$$

where s is any finite integer. This is clearly a nonconvex set. Define

$$\Lambda \equiv \left\{\alpha \in R^s : \alpha_i \geq 0, \sum_{i=1}^{s} \alpha_i = 1\right\}.$$

In this case $M_1(U)$ is of the form

$$M_0(U) \equiv \left\{\mu \in M(U) : \mu(d\xi) = \sum_{i=1}^{s} p_i \ \delta_{e_i}(d\xi), \ p \in \Lambda\right\} \subset M_1(U),$$

where δ_e denotes the Dirac measure concentrated at the single point $\{e\}$ in the sense that for every Borel set B, $\delta_e(B) = 1$ if $e \in B$ and it is 0 otherwise. In this case the set of admissible controls $\mathcal{U}_s$ is given by

$$\mathcal{U}_s \equiv \left\{ u : u_t(d\xi) = \sum_{i=1}^{s} p_i(t)\delta_{e_i}(d\xi), p \in L_\infty(I, \Lambda) \right\}$$

and so the necessary conditions given by Corollary 8.3.6 reduce to the following one.

Corollary 8.3.8. Let $\mathcal{U}_s$ denote the class of admissible controls. Then, under the assumptions of Theorem 8.3.5, the necessary condition (8.3.58) is equivalent to the following one given by

$$\sum_{i=1}^{s} p_i^o(t) H(t, x^o(t), \psi(t), e_i)$$
$$\leq \sum_{i=1}^{s} p_i H(t, x^o(t), \psi(t), e_i) \ \forall\ p \in \Lambda \ \text{ for a.a } \ t \in I. \qquad (8.3.70)$$

while the other necessary conditions (8.3.59), (8.3.60) remain unchanged.

Proof. The proof is again based on Lebesgue density argument and it is similar to that of Corollary 8.3.2.

An Example. Here is a simple example illustrating the importance of relaxed controls. Consider the system

$$\begin{aligned} \dot{x}_1 &= (1 - x_2^2), \ x_1(0) = 0 \\ \dot{x}_2 &= u, x_2(0) = 0, \ t \in [0,1], \end{aligned}$$

with the control constraint set

$$U \equiv \{-1, +1\}$$

which consists of only two points as indicated. Clearly this is a nonconvex set. The problem is to find a control that drives the systems from the initial state $(0,0)$ to the required final state $(1,0)$. This problem does not have a solution if the admissible class of controls is taken as $L_\infty([0,1], U)$ while it has a solution if the admissible class is given by $L_\infty^w([0,1], M_1(U))$. Indeed to reach the target, we must have a nontrivial control u ($u \not\equiv 0, 0 \notin U$) such that

$$x_1(1) = \int_0^1 (1 - x_2^2(t))dt = 1 \text{ and } x_2(1) = \int_0^1 u(t)dt = 0. \qquad (8.3.71)$$

But this is impossible, because if $u \not\equiv 0$, $x_2(t) \not\equiv 0$ and this means that $x_1(1) < 1$. On the other hand a relaxed control exists. This is given by

$$u_t(d\xi) = p_1(t)\delta_1(d\xi) + p_2(t)\delta_{-1}(d\xi), 0 \le p_i(t) \le 1, p_1(t) + p_2(t) = 1,$$

where $\delta_a(d\xi)$ denotes the Dirac measure concentrated at a. If we take $p_1(t) = p_2(t) = (1/2)$, the reader can easily verify that

$$x_2(t) = \int_0^t \{p_1(s) - p_2(s)\} ds \equiv 0 \ \forall \ t \in [0,1].$$

Thus $x_1(1) = 1$ while $x_2(1) = 0$. This shows the power of relaxed controls. The reader can construct many such examples where optimal controls from the class of measurable functions do not exist if the set is non convex. See [58, 20, 24] for many more interesting examples.

A question that is often asked is: is it possible to physically construct or realize relaxed controls. It appears that it is not possible with the current technology. However the Corollary 8.3.8 suggests that it may be possible to approximate a relaxed control by a convex combination of Dirac measures. If U is compact, $M_1(U)$ is compact with respect to the weak topology. Then by Krein-Milman theorem (see Appendix A.7.8 and [33]) the closed convex hull of its extremals coincides with $M_1(U)$. The extremals are Dirac measures

$$D \equiv \{\delta_{e_i}, \{e_i\} \text{ dense in } U\}$$

and hence

$$\ell co \ D = M_1(U).$$

This immediately tells us that

$$L_\infty^w(I, M_1(U)) = c\ell \left\{ \sum_{i\ge 1} p_i(t)\delta_{e_i}, p_i(t) \ge 0, \sum_{i \ge 1} p_i(t) = 1 \right\}.$$

For further discussion on realizability of such controls see section 8.6.2.

8.4 Some Special Cases involving constraints

In this section we present some problems which involve state constraints. Some of these problems can be transformed into the standard problems treated in the preceding section. Consider the problem

$$\dot{x} = f(t, x, u), x(0) = x_0 \tag{8.4.72}$$

$$J(u) = \int_0^T \ell(t, x, u)dt + \varphi(x(T)) \longrightarrow \min \tag{8.4.73}$$

subject to the equality constraints

$$\int_0^T \ell_i(t, x, u)dt = c_i, 1 \le i \le k. \tag{8.4.74}$$

Define

$$f_{n+i} \equiv \ell_i, 1 \le i \le k \le n$$

and

$$x_{n+i}(t) \equiv \int_0^t f_{n+i}(s, x(s), u(s))ds, x_{n+i}(0) = 0, x_{n+i}(T) = c_i, 1 \le i \le k.$$

We may now introduce the augmented state variable as

$$z = col\{x_j, 1 \le j \le n + k\}$$

and the velocity or the vector field as

$$F \equiv col\{f_j, 1 \le j \le n + k\}.$$

This reduces the original problem into a standard problem treated in the previous sections. Indeed now we have the system

$$\begin{aligned} &\dot{z} = F(t, z, u), z(0) = z_0 = col\{x_0, 0\}, \\ &z(T) \in \{\xi \in R^{n+k} : z_{n+i} = c_i, i = 1, 2, \cdots k\}. \end{aligned} \tag{8.4.75}$$

The cost functional remains unchanged,

$$J(u) \equiv \int_0^T \ell(t, z(t), u(t))dt + \varphi(z(T))$$

except that we write this now using the state variable z though ℓ and φ are independent of the last k components of z. Define the augmented Hamiltonian as

$$H(t, z, u, \psi) \equiv (F(t, z, u), \psi) + \ell(t, z, u).$$

Necessary conditions may be stated as follows.

Theorem 8.4.1. In order that the pair $\{u^o, z^o\}$ be optimal it is necessary that there exists an absolutely continuous $n + k$ dimensional vector function ψ^o such that the triple $\{u^o, z^o, \psi^o\}$ satisfy the following inequality and equations:

$$H(t, z^o(t), u^o(t), \psi^o(t)) \le H(t, z^o(t), v, \psi^o(t)) \text{ for almost all } t \in I,$$

$$\text{and all } v \in U - \dot{\psi}_i^o = \begin{cases} H_{z_i}(t, z^o(t), u^o(t), \psi^o), & 1 \le i \le n, \\ 0, & n + 1 \le i \le n + k \end{cases}$$

$$\psi_i^o(T) = \varphi_{z_i}(z^o(T)), \quad 1 \le i \le n,$$

$$\dot{z}^o = H_\psi = F(t, z^o, u^o), z^o(0) = z_0, \quad z_{n+i}^o(T) = c_i, 1 \le i \le k.$$

Remark. Note that for the adjoint equations, there are n terminal (final) conditions specified while for the state equation there are $n+k$ initial conditions and k terminal conditions specified. Thus in all, there are $2(n+k)$ boundary conditions specified for $2(n+k)$ dimensional system which makes the system consistent.

Continuous equality or inequality constraints: Some control problems may have additional state and control constraints. Here we discuss two such problems.

(A): Equality constraints of the form

$$\ell_i(t, x(t), u(t)) = h_i(t), t \in I$$

can be handled by introducing additional Lagrange multipliers and redefining the cost functional as

$$\tilde{J}(u) \equiv \int_0^T \left\{ \ell(t, x(t), u(t)) + \sum_{i=1}^{k} \lambda_i(t)[\ell_i(t, x(t), u(t)) - h_i(t)] \right\} dt + \varphi(x(T)).$$

In this case the necessary conditions are given by

$$\begin{aligned} &H(t, x^o(t), u^o(t), \psi^o(t), \lambda^o(t)) \leq H(t, x^o(t), v, \psi^o(t), \lambda^o(t)) \text{ for a.a } t \in I, \\ &\quad \forall\, v \in U - \dot{\psi}_i^o = f_x^* \psi + \ell_x + \sum \lambda_i \ell_{i,x}, \quad \psi^o(T) = \varphi_x(x^o(T)), \\ &\dot{x}^o = H_\psi = f(t, x^o, u^o), x^o(0) = x_0 \\ &\ell_i(t, x^o(t), u^o(t)) = h_i(t), i = 1, 2, \cdots k. \end{aligned}$$

(B): Inequalities of the form

$$\ell_i(t, x(t), u(t)) \leq h_i(t), t \in I$$

can be handled by introducing penalty functions and augmenting the cost functional by

$$J(u) \equiv \int_I \left\{ \ell(t, x, u) + \sum \alpha_i I(\ell_i(t, x, u) > h_i(t)) \right\} dt + \varphi(x(T))$$

where

$$I(S) = \begin{cases} 1, S \text{ true} \\ 0, \text{otherwise} \end{cases}$$

and $\{\alpha_i\}$ are positive weights given to any violation of constraints. This, however, introduces discontinuities in the cost integrand. For approximation of such constraints see Teo et al [89].

8.5 Basic Algorithm for Numerical Computation

Here we present a simple algorithm for numerical solution of the control problem based on Corollary 8.3.2. It is clear from the expression (8.3.20) and (8.3.26) that the gradient of the cost functional J at any point w in the space of admissible controls evaluated at time t is given by

$$DJ(w)(t) = H_u(t, z(t), \phi(t), w(t)), t \in I$$

where z is the solution of the system equation corresponding to the control w and ϕ is the solution of the adjoint equation

$$\dot{\phi} = -f_x^*(t, z(t), w(t))\phi - \ell_x(t, z(t), w(t)), \phi(T) = \varphi_x(z(T)), t \in I.$$

Using this notation the necessary condition (8.3.10) can be rewritten as

$$\int_I < DJ(u^o)(t), u(t) - u^o(t) > dt \geq 0, \ \forall \ u \in \mathcal{U}_{ad}.$$

On the basis of this we may now develop an algorithm for computing the optimal controls.

Suppose at the n-th stage of computation, $u_n \in \mathcal{U}_{ad}$ has been already found.

Step 1: Solve the state equation (8.3.11) corresponding to the control u_n giving x_n.

Step 2: Replacing the pair $\{u^o, x^o\}$ by the pair $\{u_n, x_n\}$ solve the adjoint equation (8.3.12) to obtain ψ_n.

Step 3: Using the triple $\{x_n, \psi_n, u_n\}$ compute $DJ(u_n)(t) = H_u(t, x_n(t), \psi_n(t), u_n(t)), t \in I$.

Step 4: Define $u_{n+1}(t) \equiv u_n(t) - \varepsilon DJ(u_n)(t)$ for $\varepsilon > 0$, sufficiently small so that $u_{n+1}(t) \in U$.

Step 5: Compute

$$\begin{aligned} J(u_{n+1}) &= J(u_n) + < DJ(u_n), u_{n+1} - u_n > + o(\varepsilon) \\ &= J(u_n) - \varepsilon \parallel DJ(u_n) \parallel^2_{L_2(I,R^m)} + o(\varepsilon). \end{aligned}$$

This step shows that for $\varepsilon > 0$, sufficiently small, u_{n+1} should provide an improvement.

Step 6: Check if $J(u_{n+1}) < J(u_n)$, if not, reduce the step size ε and repeat the process till this is satisfied. Fix the corresponding ε and use in the following steps.

Step 7: If $J(u_{n+1}) < J(u_n)$ compute the difference and check if $|J(u_{n+1}) - J(u_n)| \leq \delta$ for some pre specified tolerance δ. Print the results if so, if not, go to step 1 with u_n replaced by u_{n+1}, thereby closing the loop.

If in step 5, it is impossible to find an $\varepsilon > 0$ for which the strict inequality is satisfied, one must conclude that u_n is a local minimum. In fact if u_n is a local minimum, it follows from the necessary conditions of optimality that

$$(DJ(u_n), u - u_n) \geq 0 \ \forall u \in \mathcal{U}_{ad} \cap N(u_n),$$

where $N(u_n)$ is a neighborhood of u_n.

For more elaborate and power full algorithms the reader is referred to [89, 2]. In particular, we found the conjugate gradient technique given in [89, 2] is very efficient.
For application of necessary conditions of optimality, the algorithm presented here is sufficient for many practical problems [175, 199, 193, 185, 186]. Recently He and Ahmed [140] used this algorithm to solve an optimal control problem involving a demographic model. They determine optimal immigration and job creation policies that can be used as guidelines for Government policy formulation.

The same algorithm was used to solve a related identification (inverse) problem. In general an identification problem can formulated as follows:

$$\dot{x} = f(t, x, \alpha), t \in I, y(t) \equiv h(t, x(t)), t \in I$$

$$J(\alpha) = \int_I \| h(t, x(t)) - y_o(t) \|^2 \, dt \longrightarrow \inf_{\alpha \in \Lambda},$$

where y_o is the measured (observed) output and y the model output dependent on the choice of $\alpha \in \Lambda$. If the parameters are time varying one can use the minimum principle given by Theorem 8.3.3 and the algorithm given above to determine the parameters. If the parameters are time invariant one must use the integral minimum principle which follows after the expression (8.3.69) in the remark following Corollary 8.3.6. This is given by

$$\int_I H(t, x^o(t), \psi(t), \alpha^o) dt \leq \int_I H(t, x^o(t), \psi(t), \alpha) \ \forall \ \alpha \in \Lambda,$$

where ψ is the adjoint variable given by the solution of the adjoint equation $\dot{\psi} = -H_x$. This method was successfully used to determine all the unknown infinitesimal parameters (such as birth and death rate, transition rates of one age group to the next, and migration and emigration rates) that determine the dynamics of demography. This was used to develop a demographic model for Canada. The raw data was obtained from STAT-CANADA. For detailed numerical results interested reader is referred to the original paper [140].

8.6 Existence of Optimal Controls

The question of existence of optimal controls is crucial. In the absence of existence, necessary conditions of optimality do not make much of a sense. In fact, it is equivalent to characterizing an entity that does not exist in the first place. In the previous sections we tacitly assumed existence. Here in this section we prove some basic existence results for convex and non convex problems. We do so for ordinary as well as relaxed controls.

8.6.1 Ordinary Controls

Consider the control system

$$\begin{aligned} &\dot{x} = f(t,x,u), t \in I \equiv [0,T], \\ &x(0) = x_0, u \in \mathcal{U}_{ad} \end{aligned} \tag{8.6.76}$$

with the cost functional

$$J(u) \equiv C(u,x) \equiv \int_I \ell(t,x,u)dt + \varphi(x(T)), \tag{8.6.77}$$

where x is the solution corresponding to the control u. The admissible controls are as defined in (8.2.3). Our main concern here is to prove existence of optimal controls, that is, controls at which the functional J attains its minimum. Towards this goal we introduce the following set valued function.

For $(t,x) \in I \times R^n$, define the set

$$Q(t,x) \equiv \{(\zeta,\eta) \in R \times R^n : \zeta \geq \ell(t,x,v), \eta = f(t,x,v) \text{ for some } v \in U\}.$$

This set valued function $(t,x) \longrightarrow Q(t,x)$ is known as the contingent set, because this describes the set of admissible directions the system can take given the current state (t,x).

Definition 8.6.1 (1): A multi function $G : R^n \longrightarrow 2^{R^n} \setminus \emptyset$ is said to be upper semi continuous at the point $x \in R^n$ if for every $\varepsilon > 0$, there exists a $\delta > 0$ such that $G(\xi) \subset G_\varepsilon(x)$ for all $\xi \in N_\delta(x)$, where $G_\varepsilon(x)$ denotes the ε neighborhood of the set $G(x)$ and $N_\delta(x)$ denotes the δ-neighborhood of x. The multi function G is said to be upper semi continuous if it is so at every point in its domain of definition.

(ii): A multi function $Q : I \times R^n \longrightarrow 2^{R \times R^n} \setminus \emptyset$ is said to satisfy the weak Cesari property [24] if for each $t \in I$

$$\bigcap_{\varepsilon>0} clcoQ(t, N_\varepsilon(x)) \subset Q(t,x). \tag{8.6.78}$$

Remark. It is clear from the above inclusion that $Q(t,x)$ must be closed convex valued for the Cesari property to hold. A sufficient condition for Q to satisfy this property is that $x \longrightarrow Q(t,x)$ is upper semi continuous with closed convex values.

Another important fact is that if the vector function $F(t,x,u) \equiv col\{\ell(t,x,u), f(t,x,u)\}$ is such that $Q(t,x) \equiv F(t,x,U)$ is closed convex valued and upper semi continuous in x, then Q has the weak Cesari property. This is off course a stronger condition (requirement). Our first existence result is as follows.

Theorem 8.6.2. Consider the control problem as stated above with f satisfying the assumptions of Theorem 8.3.4. Suppose ℓ is measurable in the first variable and continuous in the second and third arguments and φ is continuous and bounded on bounded subsets of R^n. Further, suppose that for every finite positive number b, there exists an $h_b \in L_1(I,R)$ such that

$$\ell(t,x,u) \geq h_b(t) \ \ a.e \ \ t \in I, x \in B_b, u \in U,$$

and that the multifunction Q satisfies the weak Cesari property (8.6.78). Then there exists an optimal control.

Proof. For notational convenience, let $u \rightarrow x(u)(\cdot)$ denote the unique solution of the system equation (8.6.76) corresponding to the fixed initial condition x_0 and any control $u \in \mathcal{U}_{ad}$, and

$$\mathcal{X} \equiv \{x \in AC(I,R^n) : x = x(u) \text{ for some } u \in \mathcal{U}_{ad}\}$$

the set of admissible trajectories. Define

$$\Gamma \equiv \{(u,x) \in \mathcal{U}_{ad} \times \mathcal{X} : x = x(u)\} \tag{8.6.79}$$

to be the set of admissible control-state pairs (trajectories) and

$$m_0 \equiv \inf\{C(u,x) : (u,x) \in \Gamma\}. \tag{8.6.80}$$

Our ultimate objective is to show that there exists a pair $(u^o,x^o) \in \Gamma$ such that $C(u^o,x^o) = m_0$. By virtue of the linear growth assumption on f, and the boundedness of the set U, the set

$$\{x(u)(T), u \in \mathcal{U}_{ad}\}$$

is contained in a bounded subset of R^n. Since φ is continuous and bounded on bounded sets and $g \in L_1^+(I)$, it follows from this that $-\infty < m_0 \leq +\infty$.

If $m_0 \equiv +\infty$ (identically), there is nothing to prove. So let $m_0 < +\infty$ and $\{(u_n, x_n)\} \in \Gamma$ a minimizing sequence. One may be tempted to use weak star convergence of the control sequence $\{u_n\}$ and uniform convergence of the accompanying trajectories directly on the system equation and the objective functional to arrive at the conclusion. But this is impossible since, in general, weak convergence is incompatible with nonlinearity. So the direct approach fails miserably unless u appears linearly in the system equation and ℓ is weak star lower semi continuous. So we must seek an indirect method. Note that $\mathcal{X}$ is a bounded set in $C(I, R^n)$. Further, it follows from the growth assumption on f that this set is also equicontinuous and hence by Ascoli-Arzela theorem (see Theorem 1.5.7 and [98]) it is relatively compact. Hence the sequence $\{x_n\}$ has a convergent subsequence and there exists an $x^o \in C(I, R^n)$ to which this subsequence converges. Define the sequence of functions $\{f_n\}$ by setting $f_n(t) \equiv \dot{x}_n(t) = f(t, x_n(t), u_n(t)), t \in I$. Again it follows from the linear growth assumption on f, that the sequence $\{f_n\}$ is contained in a bounded subset of $L_1(I, R^n)$ and that it is uniformly integrable. Thus by Dunford-Pettis theorem (see Theorem A.7.12) it has a subsequence that converges weakly in $L_1(I, R^n)$ to an element f^o. Relabeling the subsequences as their original counterparts, we have

$$x_n \xrightarrow{s} x^o \text{ in } C(I, R^n) \tag{8.6.81}$$

$$f_n \xrightarrow{w} f^o \text{ in } L_1(I, R^n). \tag{8.6.82}$$

Corresponding to this sequence we define

$$\ell_n(t) \equiv \ell(t, x_n(t), u_n(t)), t \in I. \tag{8.6.83}$$

Clearly, by definition of the contingent function Q we have

$$(\ell_n(t), f_n(t)) \in Q(t, x_n(t)), t \in I. \tag{8.6.84}$$

It follows from Mazur's theorem [see Appendix A.7.4] that, given any weakly convergent sequence, one can construct a suitable convex combination of the original sequence that converges strongly to the same (weak) limit. Let $j \in N \equiv \{0, 1, 2, \cdots\}$ and $n(j), m(j) \in N$ increasing with j, and $n(j), m(j) \to +\infty$ as $j \to +\infty$, and define

$$\alpha_{j,i} \geq 0, \sum_{i=1}^{m(j)} \alpha_{j,i} = 1, \quad \text{for all} \quad j \in N. \tag{8.6.85}$$

Define the sequence pair (q_j, g_j) as follows:

$$q_j(t) \equiv \sum_{i=1}^{m(j)} \alpha_{j,i} \ell_{n(j)+i}(t), \quad g_j(t) \equiv \sum_{i=1}^{m(j)} \alpha_{j,i} f_{n(j)+i}(t), \quad t \in I. \tag{8.6.86}$$

By Banach-Sacks-Mazur theorem, one can choose a suitable convex combination so that

$$g_j(t) \xrightarrow{s} f^o(t) \quad \text{in} \quad L_1(I, R^n). \tag{8.6.87}$$

Clearly, we also have

$$x_{n(j)+i}(t) \xrightarrow{s} x^o(t) \quad \text{in} \quad C(I, R^n). \tag{8.6.88}$$

Define

$$\ell^o(t) \equiv \liminf q_j(t), \quad t \in I. \tag{8.6.89}$$

Since the set $\mathcal{X}$ is bounded there is a finite positive number b such that $x(t) \in B_b$, for all $t \in I$ and $x \in \mathcal{X}$. Hence there exists an $h_b \in L_1(I, R)$ such that $q_j(t) \geq h_b(t)$. Thus the function $\ell^o(t)$ is well defined and $\ell^o(t) \geq h_b(t)$ for $t \in I$. We show that $\ell^o \in L_1(I, R)$. Indeed, it follows from Fatou's lemma [see A.6.3] that

$$\int_I \ell^o(t)dt \leq \liminf \int_I q_j(t)dt. \tag{8.6.90}$$

Since $x_n(t)$ converges to $x^o(t)$ for each $t \in I$, and φ is continuous and bounded on bounded sets, it is easy to verify that

$$\lim_{j\to\infty} \sum_{i=1}^{m(j)} \alpha_{j,i}\varphi(x_{n(j)+i}(T)) = \varphi(x^o(T)). \tag{8.6.91}$$

Clearly it follows from the definition of $C(u, x)$ and the expressions (8.6.90) and (8.6.91) that

$$\varphi(x^o(T)) + \int_I \ell^o(t)dt \leq \liminf_{j\to\infty} \sum_{i=1}^{m(j)} \alpha_{j,i}C(u_{n(j)+i}, x_{n(j)+i}). \tag{8.6.92}$$

Since $\{u_n, x_n\}$ is a minimizing sequence, the righthand expression equals m_0. Thus we have

$$\varphi(x^o(T)) + \int_I \ell^o(t)dt \leq m_0 < +\infty. \tag{8.6.93}$$

Since φ is continuous and bounded on bounded sets, and $h_b \in L_1(I, R)$, it follows from the above inequality that $\ell^o \in L_1(I, R)$. Now we must show that

$$(\ell^o(t), f^o(t)) \in Q(t, x^o(t)) \;\; a.e \;\; t \in I. \tag{8.6.94}$$

Define the sets

$$I_1 \equiv \{t \in I : |\ell^o(t)| = \infty\}, \quad I_2 \equiv \left\{t \in I : \lim_{j\to\infty} \| g_j(t) - f^o(t) \| \neq 0\right\}$$

$$I_3 \equiv \bigcup_{j\in N} \{t \in I : u_j(t) \notin U\}.$$

By definition of admissible controls, (8.6.87) and the fact that $\ell^o \in L_1(I, R)$, all the above sets have Lebesgue measure zero. Choosing a subsequence of the original sequence if necessary, we note that for each $t \in \tilde{I} = I \setminus \bigcup I_i$,

$$q_j(t) \longrightarrow \ell^o(t), \quad g_j(t) \longrightarrow f^o(t), \quad x_{n(j)+i}(t) \longrightarrow x^o(t).$$

Clearly, for every $\varepsilon > 0$, there exists a $j_0 \in N$ such that $x_{n(j)+i}(t) \in N_\varepsilon(x^o(t))$ for all $j > j_0$ and any $i \in N$. Since both ℓ and f are continuous in x on R^n, for $j > j_0$,

$$Q(t, x_{n(j)+i}(t)) \subset Q(t, N_\varepsilon(x^o(t))), t \in I.$$

It follows from the definition of the contingent function Q that

$$(\ell_{n(j)+i}(t), f_{n(j)+i}(t)) \in Q(t, x_{n(j)+i}(t)) \quad a.e \quad t \in I.$$

Thus, for sufficiently large $j \in N$ and for any $\varepsilon > 0$, we have

$$(\ell_{n(j)+i}(t), f_{n(j)+i}(t)) \in Q(t, N_\varepsilon(x^o(t))) \quad a.e \quad t \in I. \tag{8.6.95}$$

Hence it follows from the definition of the sequence $\{q_j, g_j\}$ given by (8.6.86) and the above inclusion, that

$$(q_j(t), g_j(t)) \in coQ(t, N_\varepsilon(x^o(t))) \quad a.e \quad t \in I. \tag{8.6.96}$$

Therefore, for each $t \in \tilde{I}$, we have

$$(\ell^o(t), f^o(t)) \in c\ell coQ(t, N_\varepsilon(x^o(t))) \quad \text{for any} \quad \varepsilon > 0,$$

and consequently,

$$(\ell^o(t), f^o(t)) \in \bigcap_{\varepsilon > 0} c\ell coQ(t, N_\varepsilon(x^o(t))), \quad t \in \tilde{I}.$$

Since Q satisfies the weak Cesari property, it follows from this that

$$(\ell^o(t), f^o(t)) \in Q(t, x^o(t)), \quad t \in \tilde{I}. \tag{8.6.97}$$

Define the multifunction G given by

$$G(t) \equiv \{v \in U : \ell^o(t) \geq \ell(t, x^o(t), v) \text{ and } f^o(t) = f(t, x^o(t), v)\}. \tag{8.6.98}$$

This is a measurable multifunction with nonempty closed subsets as values. Thus, by virtue of Kuratowski-Ryll Nardzewski selection theorem [see Appendix A.7.9], G has measurable selections. Let u^o be such a selection of G. Clearly $u^o \in \mathcal{U}_{ad}$. Corresponding to this control, it follows from the inequality defining G and the expression (8.6.93) that

$$\varphi(x^o(T)) + \int_I \ell(t, x^o(t), u^o(t)))dt \leq \varphi(x^o(T)) + \int_I \ell^o(t)dt \leq m_0 < +\infty. \tag{8.6.99}$$

From weak convergence of f_n to f^o and uniform convergence of x_n to x^o (along a subsequence if necessary) and the fact that u^o is a measurable selection of the multifunction G as defined above, we have

$$x^o(t) = x_0 + \int_0^t f^o(s)ds = \int_0^t f(s, x^o(s), u^o(s))ds, t \in I.$$

Thus $(u^o, x^o) \in \Gamma$ where Γ is as defined in (8.6.79). By definition,

$$C(u^o, x^o) = \int_I \ell(t, x^o(t), u^o(t))dt + \varphi(x^o(T)).$$

Hence it follows from the inequality (8.6.99) that $C(u^o, x^o) \leq m_0$, while the admissibility of the pair (u^o, x^o) and the fact that m_0 is the infimum of C on the set Γ, imply that $m_0 \leq C(u^o, x^o)$. Hence we have $C(u^o, x^o) = m_0$. This completes the proof. □

8.6.2 Relaxed Controls

The question of relaxed controls arises when the contingent function $Q(t, x)$ is non convex. For example, if the set U is not convex, optimal control may not exist as seen in the example of Section 8.3. There we observed that use of measure valued controls resolves the problem. There are two equivalent approaches to this problem. The contingent set Q is relaxed by introducing the set $\hat{Q}(t, x) \equiv clcoQ(t, x)$. Again assuming weak Cesari property for this set we obtain existence results as in Theorem 8.6.2 with relaxed controls replacing ordinary controls. The more direct approach, however, is to introduce measure valued controls and define

$$Q(t, x) \equiv \{(\zeta, \eta) \in R \times R^n : \zeta \geq \hat{\ell}(t, x, \mu), \eta = \hat{f}(t, x, \mu) \text{ for some } \mu \in M_1(U)\}, \tag{8.6.100}$$

where $M_1(U)$ is the space of probability measures on U. In case ℓ and f are linear in μ we can write

$$\hat{\ell}(t, x, \mu) \equiv \int_U \ell(t, x, \zeta)\mu(d\zeta), \hat{f}(t, x, \mu) \equiv \int_U f(t, x, \zeta)\mu(\zeta). \tag{8.6.101}$$

Note that in this case the contingent function is convex without requiring convexity of the set U. This problem can be solved directly by use of lower semi continuity and compactness arguments. Let ν be an admissible control policy, that is, $\nu \in \mathcal{U}_{ad}^r$. Define

$$J(\nu) \equiv C(\nu, x) \equiv \int_I \hat{\ell}(t, x(t), \nu_t)dt + \varphi(x(T)), \tag{8.6.102}$$

where $x \equiv x(\nu)$ is the solution of equation

$$\dot{x} = \hat{f}(t, x(t), \nu_t), x(\nu)(0) = x_0, t \in I. \tag{8.6.103}$$

Theorem 8.6.3. Suppose f and ℓ satisfy the assumptions of theorem 8.6.2. Then there exists an optimal control $\nu^o \in \mathcal{U}_{ad}^r$ that minimizes the functional $J(\nu)$.

Proof. Note that the set $\mathcal{U}_{ad}^r$ is a weak star compact subset of $L_\infty^w(I, M(U))$. Thus it suffices to prove that $\nu \longrightarrow J(\nu)$ is weak star lower semi continuous. Let $\nu^m \xrightarrow{w*} \nu^o$ in $\mathcal{U}_{ad}^r$ and $x^m \in C(I, R^n)$ the corresponding sequence of solutions of equation (8.6.103). Let x^o denote the unique solution of equation (8.6.103) corresponding to the control ν^o. Since f is locally Lipschitz, uniqueness is guaranteed. We show that $x^m \xrightarrow{u} x^o$ in $C(I, R^n)$. It follows from the growth assumption (see Theorem 8.3.4) that the set of admissible trajectories

$$\mathcal{X} \equiv \{x \in C(I, R^n) : x(\nu)(0) = x_0 \ \text{ and } \ x = x(\nu), \nu \in \mathcal{U}_{ad}^r\},$$

is bounded. Thus there exists a finite positive number r such that $x(t) \in B_r$ for all $t \in I$ and for all $x \in \mathcal{X}$. Using this fact, one can easily verify that

$$\psi_m(t) \leq \eta_m(t) + \int_0^t K_r(s)\psi_m(s)ds, t \in I, \tag{8.6.104}$$

where $\psi_m(t) \equiv\parallel x^m(t) - x^o(t) \parallel$ and

$$\eta_m(t) \equiv\parallel \int_0^t \left(\int_U f(s, x^o(s), \zeta)(\nu_s^m(d\zeta) - \nu_s^o(d\zeta))\right)ds \parallel . \tag{8.6.105}$$

Then by virtue of Gronwall inequality we have

$$\psi_m(t) \leq \eta_m(t) + \int_0^t \left\{\exp \int_s^t K_r(\theta)d\theta\right\}K_r(s)\eta_m(s)ds. \tag{8.6.106}$$

Since $\nu^m \xrightarrow{w*} \nu^o$, it is clear that $\eta_m(t) \longrightarrow 0$ uniformly with respect to $t \in I$. Thus it follows from the above inequality that $x^m \xrightarrow{u} x^o$ in $C(I, R^n)$.

Now we prove the continuity. Note that $J(\nu^m)$ and $J(\nu^o)$ are related by the following expression,

$$J(\nu^m) = J(\nu^o) + T_{1,m} + T_{2,m} \tag{8.6.107}$$

where

$$T_{1,m} = \int_{I\times U} \ell(t, x^o(t), \xi)[\nu_t^m(d\xi) - \nu_t^o(d\xi)]dt \tag{8.6.108}$$

$$T_{2,m} \equiv \int_{I\times U} [\ell(t, x^m(t), \xi) - \ell(t, x^o(t), \xi)]\nu_t^m(d\xi)dt. \tag{8.6.109}$$

Since ν^m converges in the weak star topology to ν^o and the set U is compact and $\xi \longrightarrow \ell(t, x, \xi)$ is continuous,

$$\lim_{m\to\infty} T_{1,m} = 0.$$

Similarly, since $x^m \longrightarrow x^o$ uniformly and ν^m is weak star convergent, we have

$$\lim_{m\to\infty} T_{2,m} = 0.$$

From these facts we conclude that $\nu \longrightarrow J(\nu)$ is weak star continuous. Since $\mathcal{U}_{ad}^r$ is weak star compact, J attains its minimum on $\mathcal{U}_{ad}^r$ proving existence of optimal relaxed controls. This completes the proof. □

Remark. Note that the proof of Theorem 8.6.3 is much simpler than that of Theorem 8.6.2. This is because we assume U to be compact and $\hat{f}$ and $\hat{\ell}$ linear with respect to the control ν. However this is natural for the original problem, though one may conceive of f as well as ℓ to be nonlinear functions of the measure valued controls. For relaxed controls even compactness of U is not essential. For noncompact U, the proof is slightly more difficult. For more general problems involving differential inclusions driven by relaxed controls see [151, 35] and references there in.

Practical Realizability: One important question is, how can one practically generate relaxed controls which at any given time must be distributed over the set U. Clearly here there is a fundamental practical limitation. However by virtue of Krein-Milman theorem, we know that any relaxed control can be approximated by chattering controls. For example if U is compact then $M_1(U)$ is weak star compact and so

$$M_1(U) = \ell co(ext M_1(U))$$

and similarly

$$L_\infty^w(I, M_1(U)) = \ell co(ext L_\infty^w(I, M_1(U))).$$

Given any relaxed control $\nu \in L_\infty^w(I, M_1(U))$, letting U_0 denote a countable dense subset of U, one may construct an approximating sequence of controls of the form

$$\nu_t^n(dv) \equiv \sum_i \alpha_{n,i}(t)\delta_{u_i}(dv), t \in I, u_i \in U_0,$$

where $\{\alpha_{n,i}(\cdot)\}$ are nonnegative bounded measurable functions satisfying

$$\sum_i \alpha_{n,i}(t) = 1 \quad \forall\, t \in I, n \in N.$$

Let $\mathcal{X}_r$ denote the class of relaxed trajectories (solutions corresponding to relaxed controls) and $\mathcal{X}_c$ the class of trajectories corresponding to chattering controls as defined above. Then by use of Theorem 8.6.3, one can verify that the set of chattering trajectories is dense in the set of relaxed trajectories, that is, $\overline{\mathcal{X}}_c = \mathcal{X}_r$.

In case the set U consists of a finite set of points (hence nonconvex) one has to choose only a finite number of switching functions $\{\alpha_{n,i}\}$ from the class of measurable functions satisfying the constraint indicated above. This is much simpler.

Remark. In view of the above comments, it is evident that development of ultra high speed switching devices (optical/quantum) in future may make it feasible to construct approximate relaxed controls.

8.6.3 Uncertain Systems/Differential Inclusions

Using similar technique, one can prove existence of optimal policies for the so called uncertain systems described by controlled differential inclusions of the form

$$\dot{x}(t) \in F(t, x(t), u(t)), x(0) = x_0, t \in I, \tag{8.6.110}$$

where $F(t, \xi, v)$ is a multivalued map. Letting $\mathcal{X}(u)$ denote the solution set corresponding to any control $u \in \mathcal{U}_{ad}$, the problem is to find a control policy u^o such that

$$J(u^o) = \inf\{J(u), u \in \mathcal{U}_{ad}\}$$

where

$$J(u) \equiv \sup\{C(u, x), x \in \mathcal{X}(u)\}$$

with the functional C given by (8.6.77). This problem is equivalent to minimizing the maximum risk. We call it the min-max problem. This kind of uncertainty arises from incomplete knowledge of the dynamics or unknown system parameters with known bounds as discussed in Chapter 3. Here we present one such result. Let $cc(R^n)$ denote the class of nonempty closed convex subsets of R^n. Define the contingent set as

$$Q(t,x) \equiv \{(\zeta,\eta) \in R \times R^n : \zeta \geq \ell(t,x,u), \eta \in F(t,x,u) \text{ for some } u \in U\}. \tag{8.6.111}$$

We need the following assumptions:

(F1): $F : I \times R^n \times U \longrightarrow cc(R^n)$

(F2): $t \longrightarrow F(t,x,u)$ is measurable for each $x \in R^n$ and $u \in U$

(F3): $u \longrightarrow F(t,x,u)$ is continuous with respect to the Hausdorff metric d_H

(F4): $x \longrightarrow F(t,x,u)$ is upper semi continuous uniformly with respect to $(t,u) \in I \times U$ and there exists a $K \in L_1^+(I)$ such that

$$d_H(F(t,x,u), F(t,y,u)) \leq K(t)[\| x - y \|] \; \forall \; u \in U$$

(F5): there exists an $h \in L_1^+(I)$ such that

$$\| z \| \leq h(t)\{1 + \| x \|\}, \; \forall \; z \in F(t,x,u) \; \forall \; u \in U.$$

Theorem 8.6.4. Suppose the multifunction F satisfy the assumptions (F1)-(F5), the set U is a closed convex bounded subset of R^m and the functions ℓ and φ, determining the cost functional C, satisfy the assumptions of Theorem 8.6.2. Then the min-max problem as stated above has a solution.

Proof. The proof is very similar to that of Theorem 8.6.2. We present only a brief outline. It follows from the assumptions (F4) and (F5) that for each control policy $u \in \mathcal{U}_{ad}$, the system

$$\dot{x}(t) \in F(t, x(t), u(t)), t \in I, x(0) = x_0$$

has a nonempty set of solutions denoted by $\mathcal{X}(u)$. Further this set is a bounded equicontinuous subset of $C(I, R^n)$ and hence by Ascoli-Arzela theorem is relatively compact. That the set $\mathcal{X}(u)$ is closed follows from the fact that F is closed convex valued. Thus for each $u \in \mathcal{U}_{ad}$ the solution set $\mathcal{X}(u)$ is compact. Using the continuity of ℓ and φ, one can then verify that for any fixed $u \in \mathcal{U}_{ad}$, the functional $x \longrightarrow C(x,u)$ attains its maximum on $\mathcal{X}(u)$. Thus the functional

$$J(u) \equiv \sup\{C(x,u), x \in \mathcal{X}(u)\}$$

is well defined. One must now prove that J attains its infimum on $\mathcal{U}_{ad}$. Let $\{u_n\}$ be a minimizing sequence for J and $\{x_n\}$ the associated sequence at which $J(u_n) = C(x_n, u_n)$. Using this sequence we then construct the sequence

$$\ell_n(t) \equiv \ell(t, x_n(t), u_n(t)), t \in I$$
$$f_n(t) \in F(t, x_n(t), u_n(t)), t \in I.$$

and note that

$$(\ell_n(t), f_n(t)) \in Q(t, x_n(t)), t \in I.$$

From here on we follow the same steps as in Theorem 8.6.2 starting from equation (8.6.84) provided Q satisfies the weak Cesari property. But this follows from the assumptions (F1)-(F4). This completes the brief outline of our proof. □

Remark. In the preceding theorem we gave sufficient conditions guaranteeing the multifunction Q to satisfy the weak Cesari property. However, these conditions are not necessary. For more on uncertain systems see Xiang and Ahmed [201].

Some Examples(Differential Games):

(E1): Consider the system with two sets of controls given by

$$\dot{x} = f(t, x, u, v), x(0) = x_0$$

where $u \in \mathcal{U}_{ad}$ and $v \in \mathcal{V}_{ad}$. Here player P1 can choose his controls from $\mathcal{U}_{ad}$ and player P2 can choose his strategies from the set $\mathcal{V}_{ad}$. Player P1 wants to reach the target D, a closed bounded convex subset of R^n, and player $P2$ tries to prevent this. This can be formulated as the min-max problem as follows:

$$\inf_{u \in \mathcal{U}_{ad}} \sup_{v \in \mathcal{V}_{ad}} \{J(u, v) \equiv d(x^{u,v}(T), D)\}$$

where d is any metric on R^n. We show that this is equivalent to the original problem treated in Theorem 8.6.4. Define the multifunction $F(t, x, u) \equiv f(t, x, u, V)$ with V denoting the control constraints of the player P2. Then the system is governed by the differential inclusion $\dot{x} \in F(t, x, u), x(0) = x_0$, with $\mathcal{X}(u)$ denoting the family of solutions corresponding to the control $u \in \mathcal{U}_{ad}$. The objective functional is given by $C(u, x) = d(x(T), D)$, for $x \in \mathcal{X}(u)$. The problem is to find a control $u^o \in \mathcal{U}_{ad}$ so that $J(u^o) \leq J(u) \equiv \sup\{C(u, x), x \in \mathcal{X}(u)\}$.

(E2): Another classical example comes from differential games. Here there are two competing players with dynamics given by

$$P1: \quad \dot{x} = f(t,x,u), \text{in } R^n, \; u \in \mathcal{U}_{ad} \tag{8.6.112}$$

$$P2: \quad \dot{y} = g(t,y,v), \text{in } R^n, \; v \in \mathcal{V}_{ad} \tag{8.6.113}$$

where the control strategies of player P1 consists of measurable functions with values in a closed bounded set $U \subset R^{m_1}$ and those of player P2 are given by measurable functions with values in a closed bounded set $V \subset R^{m_2}$. Player P1 wants to pursue P2 and player P2 wants to evade P1. Here one simple objective functional is given by

$$J(u,v) \equiv \varphi(\| x^u(T) - y^v(T) \|)$$

where φ is any continuous nonnegative nondecreasing function of its argument. The problem is to find a pair of policies $(u^o, v^o) \in \mathcal{U}_{ad} \times \mathcal{V}_{ad}$ so that

$$J(u^o, v^o) = \inf_{u \in \mathcal{U}_{ad}} \sup_{v \in \mathcal{V}_{ad}} J(u,v).$$

A pair $(u^o, v^o) \in \mathcal{U}_{ad} \times \mathcal{V}_{ad}$ is said to be a saddle point if it satisfies the following inequalities

$$J(u^o, v) \leq J(u^o, v^o) \leq J(u, v^o).$$

This raises the question of existence of saddle points. If U and V are closed bounded convex sets, the corresponding class of admissible controls $\mathcal{U}_{ad}$ and $\mathcal{V}_{ad}$ are weak star compact subsets of $L_\infty(I, R^{m_1})$ and $L_\infty(I, R^{m_2})$ respectively. Then, under standard regularity assumptions for $\{f, g\}$, the attainable sets $\{\mathcal{A}_1(T), \mathcal{A}_2(T)\}$ of the players P1 and P2 are compact. Thus the games problem stated above is equivalent to

$$\inf_{u \in \mathcal{U}_{ad}} \sup_{\mathcal{V}_{ad}} J(u,v) = \inf_{\xi \in \mathcal{A}_1(T)} \sup_{\eta \in \mathcal{A}_2(T)} \varphi(\| \xi - \eta \|).$$

From the compactness of the attainable sets and the continuity of φ follows the existence of saddle points.

For detailed study of games theory see Basar and Olsder [16] and the references therein. Further discussion on this topic is outside the scope of this book. Interested readers may refer to [151, 133, 134]. Here one will find detailed analysis of much broader class of systems in infinite dimensional Banach spaces covering systems governed by partial differential equations, stochastic partial differential equations and more.

8.6.4 Impulsive Controls

Systems controlled by impulsive forces arise in many applications such as engineering, economics, management and social sciences. These can be considered as special cases of systems driven by vector measures. For more detailed study of such problems ambitious readers are referred to [113, 114, 118, 129, 130]. Here we present only a brief introduction to optimal control of such systems. Consider the semilinear system

$$dx = f(t,x)dt + C(t)u(dt), x(0) = x_0, \tag{8.6.114}$$

and let $\mathcal{U}_{ad} \subset \mathcal{M}_c(I, R^d)$ denote the class of admissible controls which consists of countably additive bounded vector measures. The problem is to find an admissible control that minimizes the cost functional given by

$$J(u) \equiv \int_I \ell(t,x)dt + \Psi(x(T)) + \varphi(u), \tag{8.6.115}$$

where the first two terms are standard and the third term $\varphi(u)$ represents the cost associated with control policy u which may include the cost of switching.

Theorem 8.6.5. Consider the system (8.6.114) and suppose $f : I \times R^n \longrightarrow R^n$ is Borel measurable satisfying the following assumptions; there exist $K \in L_1^+(I)$ and, for each $r > 0$, $K_r \in L_1^+(I)$ such that

$$\begin{aligned} &(1) : \| f(t,x) \| \leq K(t)[1 + \| x \|], x \in R^n \quad \| f(t,x) - f(t,y) \| \\ &\leq K_r(t) \| x - y \|, x, y \in B_r(R^n) \end{aligned}$$

and (2): $C \in C(I, M(n \times d))$ bounded continuous. (3): The set of admissible controls $\mathcal{U}_{ad}$ is a weakly compact subset of $\mathcal{M}_c(I, R^d)$ and the function ℓ is measurable in t on I and lower semi continuous on R^n satisfying $\ell(t,x) \geq h(t)$ for an integrable function h, Ψ is lower semi continuous on R^n satisfying $\Psi(x) \geq c$ for $c \in R$ and φ is a nonnegative weakly lower semi continuous functional on $\mathcal{U}_{ad}$. Then the optimal control problem has a solution.

Proof. First we prove that $u \longrightarrow J(u)$ is weakly lower semi continuous. We have seen in Chapter 3, that under the assumptions (1) and (2), for each $x_0 \in R^n$ and $u \in \mathcal{U}_{ad}$, the system (8.6.114) has a unique solution $x \in B(I, R^n)$. Let u^k be a sequence from $\mathcal{U}_{ad}$ converging weakly to u^o and let $x^k, x^o \in B(I, R^n)$ denote the solutions of equation (8.6.114) corresponding to the controls u^k, u^o respectively. Since $\mathcal{U}_{ad}$ is weakly compact it is bounded and therefore it follows from the growth assumption on f and the boundedness of $C(\cdot)$ that there exists a finite positive number r such that $x^k(t), x^o(t) \in$

$B_r(R^n)$ for all $t \in I$ and all $k \in N$. Clearly it follows from equation (8.6.114) corresponding to the controls u^o and u^k that

$$x^o(t) - x^k(t) = \int_0^t \{f(s, x^o(s)) - f(s, x^k(s))\}ds + \int_0^t C(s)(u^o(ds) - u^k(ds)). \tag{8.6.116}$$

Defining

$$e_k(t) \equiv x^o(t) - x^k(t), \quad R_k(t) \equiv \int_0^t C(s)(u^o(ds)) - u^k(ds))$$

and using the local Lipschitz property of f, we have

$$\| e_k(t) \| \leq \| R_k(t) \| + \int_0^t K_r(s) \| e_k(s) \| ds.$$

Using Gronwall Lemma 2.4.1, in particular inequality (2.4.66), it follows from the above inequality that

$$\| e_k(t) \| \;\; \leq \;\; \| R_k(t) \| + \int_0^t e^{\{\int_s^t K_r(\theta)d\theta\}} K_r(s) \| R_k(s) \| ds, t \in I. \tag{8.6.117}$$

Since

$$u^k \xrightarrow{w} u^o \quad \text{in} \quad \mathcal{M}_c(I, R^d)$$

it is clear that $R_k(t) \to 0$ in R^n for each $t \in I$ and hence $\lim_{k\to\infty} \| R_k(t) \| K_r(t) \longrightarrow 0$ for almost all $t \in I$. Further, it follows from the boundedness of $C(\cdot)$ and the set $\mathcal{U}_{ad}$ that there exists a finite positive number b such that $\sup_{k\in N}\{\sup\{\| R_k(t) \|, t \in I\}\} \leq b$. Thus by dominated convergence theorem, it follows from the above expression that

$$\lim_{k\to\infty} e_k(t) \to 0$$

for each $t \in I$. In other words $x^k(t) \longrightarrow x^o(t)$ point wise in $t \in I$ whenever $u^k \xrightarrow{w} u^o$. By assumption (3), it follows from the above results that

$$h(t) \leq \ell(t, x^o(t)) \leq \liminf \ell(t, x^k(t)), \;\; \Psi(x^o(T)) \;\leq\; \liminf \Psi(x^k(T))$$

and $\varphi(u^o) \leq \liminf \varphi(u^k)$. Since $h \in L_1(I)$, by extended Fatou's Lemma we obtain

$$\int_I \ell(t, x^o(t))dt \leq \liminf \int_I \ell(t, x^k(t))dt.$$

Hence

$$\begin{aligned} J(u^o) &= \int_I \ell(t, x^o(t))dt + \Psi(x^o(T)) + \varphi(u^o) \\ &\leq \int_I \liminf_k \ell(t, x^k(t))dt + \liminf_k \Psi(x^k(T)) + \liminf_k \varphi(u^k) \\ &\leq \liminf_k \Big\{ \int_I \ell(t, x^k(t))dt + \Psi(x^k(T)) + \varphi(u^k) \Big\} = \liminf_k J(u^k). \end{aligned}$$

This proves that J is weakly lower semi continuous on $\mathcal{M}_c(I, R^d)$. Now we prove the existence of an optimal control. If $J(u) \equiv +\infty$ there is nothing to prove. So we may assume the contrary. Since $h \in L_1(I), c \in R$ and $\varphi \geq 0$, it is clear that $J(u) > -\infty$ for all $u \in \mathcal{U}_{ad}$. Let $\{u^k\} \subset \mathcal{U}_{ad}$ be a minimizing sequence so that

$$\lim_k J(u^k) = m_o = \inf\{J(u), u \in \mathcal{U}_{ad}\}.$$

Since $\mathcal{U}_{ad}$ is weakly compact, there exists a subsequence of the sequence u^k, relabeled as the original sequence, and an element $u^o \in \mathcal{U}_{ad}$ so that $u^k \xrightarrow{w} u^o$. Then by weak lower semi continuity of J we have

$$J(u^o) \leq \liminf_k J(u^k) = \lim_k J(u^k) = m_0.$$

On the other hand since $u^o \in \mathcal{U}_{ad}$ and m_0 is the infimum on this set, $J(u^o) \geq m_0$. Hence we conclude that u^o is the optimal control. This completes the proof. □

8.7 Exercises

(Q1):Consider the semi linear system,

$$\begin{aligned} &\dot{x} = f(t, x) + G(t, x)u, x(0) = x_0, t \in I \equiv [0, T], \\ &\mathcal{U}_{ad} = L_2(I, U), U \;\; \text{compact convex subset of } R^d, \end{aligned}$$

where $f : I \times R^n \longrightarrow R^n$ and $G : I \times R^n \longrightarrow M(n \times d)$. The problem is to find a control that minimizes the cost functional,

$$J(u) \equiv \int_I \{\ell(t, x) + (Ru, u)\}dt.$$

Assume that f, G are locally Lipschitz, that is, for each finite $r > 0$, there exist $K_r, L_r \in L_1^+(I)$ so that for all $x, y \in B_r$, closed ball of radius r around the origin,

$$\| f(t, x) - f(t, y) \| \leq K_r(t) \| x - y \|; \;\; \| G(t, x) - G(t, y) \| \leq L_r(t) \| x - y \|$$

and that they satisfy linear growth condition, that is, there exist $K, L \in L_1^+(I)$ such that

$$\| f(t,x) \| \leq K(t) \| [1+ \| x \|]; \| G(t,x) \| \leq L(t) \| [1+ \| x \|].$$

Prove that for the given $x_0 \in R^n$ and every $u \in \mathcal{U}_{ad}$ the system has a unique solution $x = x^u \in AC(I, R^n)$.
(Q2): For the control problem of (Q1), prove that $u \longrightarrow x^u$ is continuous on $\mathcal{U}_{ad}$ with respect to the weak star topology in $L_\infty(I, R^d)$ and point wise topology in $C(I, R^n)$.
(Q3): Assume that (a1): $x \longrightarrow \ell(t,x)$ is continuous for all $t \in I$ and bounded on bounded sets, $t \longrightarrow \ell(t,x)$ is Lebesgue integrable for all x in bounded subsets of R^n and there exists an $h \in L_1(I)$ such that $\ell(t,x) \geq h(t), t \in I$ for all $x \in R^n$. (a2): $R(t) \in M_s^+(d \times d)$ and bounded on I. Prove the existence of an optimal control for the problem as stated in (Q1).

Hints: Show that $u \longrightarrow J(u)$ is weak star lower semi continuous and note that $\mathcal{U}_{ad}$ is a weak star compact subset of $L_\infty(I, R^d)$.
(Q4): Consider the control system of (Q1) and suppose the given assumptions hold. Prove that the attainable set

$$\mathcal{A}(T) \equiv \{\xi \in R^n : \xi = x^u(T), u \in \mathcal{U}_{ad}\}$$

is compact. Using this result and assuming that Ψ is a lower semi continuous real valued function on R^n, prove that the terminal control problem

$$J(u) \equiv \Psi(x(T)) \longrightarrow \inf .$$

has a solution.
(Q5): Under the assumptions of (Q3) and (Q4), prove that the optimal control problem

$$J(u) \equiv \int_I \{\ell(t,x) + (R(t)u, u)\}dt + \Psi(x(T)) \longrightarrow \inf .$$

has a solution.
(Q6): Following Theorem 8.6.5 and using suitable assumptions for G prove the existence of an optimal control for the nonlinear impulsive system

$$dx(t) = f(t, x(t))dt + G(t, x(t-))u(dt), x(0) = x_0, u \in \mathcal{U}_{ad}$$
$$J(u) \equiv \int_I \ell(t,x)dt + \Psi(x(T)) + \varphi(u) \longrightarrow \inf .$$

(Q7): Consider the control system described in question (Q7) of Chapter 3. Impose sufficiently relaxed conditions on the regularity of the cost integrand

ℓ, the vector field f and the multifunction G that guarantee the existence of a solution of the extremal problem

$$\Phi(x) \equiv \int_0^T \ell(t,x)dt \longrightarrow \inf$$

and the existence of an optimal control.

8.8 Bibliographical Notes

Control theory and its applications have been the most active field since the early part of the 20th century. Even today it continues to grow both in depth and diversity in applications. There are three major breakthroughs in the field. The first is the introduction of feedback, the second is the invention of dynamic programming due to Bellman and the third is the maximum (or minimum) principle due to Pontryagin. Pontryagin maximum principle with calculus of variations being its precursor brought about a revolution in the field of optimal control. There has been enormous progress in the field since these foundations were laid and many outstanding contributions have been made leading to the present level of depth both in theory and applications. There are many outstanding contributors whose names may have been missed in the list presented here: Bellman [18], Pontriagin and his coworkers Boltyanski, Gamkrelidze, Mischenko [77], Cesari [24], Neustadt [71], La Salle and Hermes [43], Berkovitz [20], Boltyanski [21], Warga [95], Gamkrelidze [38], Oguztoreli [73], Clark [26], Fleming [36], Ahmed [2], Teo [89], Luenberger [65], Petrov [76]. For infinite dimensional systems see Fattorini [35], Lions [63], Balakrishnan [13], Ahmed and Teo [4], Ahmed [5, 6], Lasiecka and Triggiani [60], Zabczyk [99], Barbu and Da Prato [15], Denkowski, Migorski and Papageorgiou [29], Vrabie [94]. In this chapter we have presented both existence of optimal controls and necessary conditions of optimality only for finite dimensional systems. This has been done for both ordinary controls and relaxed controls. The notion of relaxed controls was first introduced by Warga [95] in his early papers which was later extended by Ahmed and Teo [2], Ahmed [101, 133, 138, 139, 151] and Fattorini [35] to infinite dimensional problems dealing with control of systems governed by partial differential equations and abstract differential equations on Banach spaces. For the sake of completeness we have also presented briefly some results on min-max problems and differential games. For more on games theory see Oguztoreli [73] and Basar [16]. For optimal control of impulsive systems driven by vector measures see Ahmed [113, 114, 118, 129, 130].

Chapter 9

Linear Quadratic Regulator Theory

9.1 Introduction

In Chapter 8 we considered general nonlinear control problems. The necessary conditions developed there are used to find optimal controls or more precisely controls satisfying the necessary conditions for optimality. If convexity conditions hold those necessary conditions are also sufficient. In any case by using this theory we obtain open loop controls. For many engineering problems it is absolutely essential to construct optimal feedback controls, in particular, state feedback or (even better) output feedback. For nonlinear problems this is not easy. As we shall see later this requires solving a nonlinear first order partial differential equation on R^n known as the Hamilton-Jacobi-Bellman equation. Thus very often engineers prefer to linearize the system around some operating point (normal state trajectory corresponding to normal input) and construct linear feedback regulators for the linearized problem. In the absence of large disturbances this is fairly satisfactory for many applications. Let the normal (or nominal) operating system be given by

$$\dot{z}^o = f(t, z^o, v^o), z^o(0) = z_0, t \geq 0, \tag{9.1.1}$$

where v^o denotes the normal control (input) and z^o the corresponding trajectory. The control v^o is also known as the nominal control that keeps the system running. Due to many uncertainties or disturbances, which may not have been accounted for in the system dynamics, the control may deviate from the nominal control causing deviation of the state trajectory from its nominal path. This is described by the same system equation

$$\dot{z} = f(t, z, v), z(0) = x_0, t \geq 0. \tag{9.1.2}$$

The error vector is given by

$$e(t) \equiv z(t) - z^o(t), t \geq 0$$

which corresponds to the perturbed control $w(t) = v(t) - v^o(t), t \geq 0$. Engineers would like to design a feed back control mechanism driven by the error signal so that it tends to keep the error within certain bounds or even drive it to zero. Assuming that the vector field is C^1 in both the state and the control variables with second derivatives bounded, this problem can be formulated as follows:

$$\begin{aligned} \dot{e} &= f_z(t, z^o(t), v^o(t))e + f_v(t, z^o(t), v^o(t))w(t) + h(t) \\ &= A(t)e(t) + B(t)w(t) + h(t) \end{aligned} \tag{9.1.3}$$

where A and B are the matrices of first partials as indicated and h represents all the higher order terms which can be neglected. One may use Taylor expansion and put all the higher order terms, starting from the second order ones, in the expression for h. The problem is to find a feedback control law that minimizes the functional

$$J(w) \equiv (1/2) \int_I \{(Q(t)e, e) + (R(t)w, w)\}dt + (Me(T), e(T)). \tag{9.1.4}$$

Here the first term is a measure of penalty for any deviation from the normal state trajectory, the second is a measure of deviation of control energy and the last term stands for terminal error. Due to its enormous popularity, simplicity and importance in industry, this is the problem we consider here under many different situations.

9.2 Linear Quadratic Regulator

In view of the preceding discussion, we have good reasons to study the optimal control problem for any linear system described by the following differential equation

$$\dot{x}(t) = A(t)x(t) + B(t)u(t), x(0) = x_0, \; t \in I \equiv [0, T], T < \infty. \tag{9.2.5}$$

The cost functional is given by

$$J(u) = (1/2) \int_I \{(Q(t)x, x) + (R(t)u, u)\}dt + (1/2)(Mx(T), x(T)), \tag{9.2.6}$$

where the matrices $\{Q(t), R(t), M\}$ are symmetric, real and positive semidefinite. In particular, $R(t)$ is positive definite. These assumptions are quite natural. Positive definiteness of R implies that no control is free. The problem is to find a linear state feedback control law that minimizes the quadratic cost functional as presented above. Here the control constraint set is $U = R^m$, and the admissible controls are measurable functions with values in U. Applying the Corollary 8.3.2 to the Hamiltonian,

$$H(t,x,\psi,v) = (A(t)x+B(t)u,\psi)+(1/2)(Q(t)x,x)+(1/2)(R(t)u,u), \quad (9.2.7)$$

and noting that U is the whole space, it follows from the inequality (8.3.26) that for optimality the derivative of this function with respect to the control must vanish. This leads to the form of the control

$$u(t) = -R^{-1}(t)B^*(t)\psi(t) \quad (9.2.8)$$

where ψ is the solution of the adjoint equation given by

$$\dot{\psi} = -A^*(t)\psi - Q(t)x, \psi(T) = Mx(T). \quad (9.2.9)$$

The reader can verify that the solution of the adjoint equation is given by the following expression

$$\psi(t) = \Phi^*(T,t)Mx(T) + \int_t^T \Phi^*(s,t)Q(s)x(s)ds. \quad (9.2.10)$$

where $\Phi(s,t), t \leq s < \infty$ is the transition operator corresponding to the system matrix $A(\cdot)$ and Φ^* is its adjoint. Since for any given control policy, the future values $\{x(s), s \geq t\}$ of the state x is uniquely determined by the system equation (9.2.5) and its current value $x(t)$, it follows from the above expression that ψ can be expressed by the simple relation

$$\psi(t) = K(t)x(t) \quad (9.2.11)$$

where, clearly, K determines an operator equivalent to the one expressed by the equation (9.2.10). Using this in the expression (9.2.8) we obtain the feedback control law

$$u \equiv -R^{-1}(t)B^*(t)K(t)x. \quad (9.2.12)$$

Now it remains to determine a convenient way for finding the matrix valued function K instead of going through the difficulties of determining the transition operator Φ. This can be easily accomplished by differentiating the expression (9.2.11) on both sides and using the system dynamics (9.2.5), the adjoint equation (9.2.9), and the control law (9.2.12). This leads to the following identity

$$(\dot{K}+A^*(t)K+KA(t)-KB(t)R^{-1}(t)B^*(t)K+Q(t))x(t) = 0, t \in I. \quad (9.2.13)$$

Using the feedback control given by (9.2.12) in the system dynamics (9.2.5) we obtain the state feedback system

$$\dot{x} = (A(t) - B(t)R^{-1}(t)B^*(t)K(t))x(t), x(0) = x_0. \tag{9.2.14}$$

Letting Ψ denote the transition operator corresponding to $A - BR^{-1}B^*K$, equation (9.2.13) turns into

$$(\dot{K} + A^*(t)K + KA(t) - KB(t)R^{-1}(t)B^*(t)K + Q(t))\Psi(t,0)x_0 = 0, t \in I. \tag{9.2.15}$$

Since a transition operator is always nonsingular and $x_0 \in R^n$ is arbitrary, it follows from the above equation that K must satisfy the differential equation

$$\dot{K} + A^*(t)K + KA(t) - KB(t)R^{-1}(t)B^*(t)K + Q(t) = 0, t \in I, \tag{9.2.16}$$

$$K(0) = M \tag{9.2.17}$$

called the differential matrix Riccati equation with the terminal condition following from the terminal condition for the adjoint state ψ. Thus we have solved the optimal feedback regulator problem as stated below.

Theorem 9.2.1. Consider the linear system (9.2.5) with the quadratic cost functional given by (9.2.6). The optimal state feedback control law is given by

$$u \equiv -R^{-1}(t)B^*(t)K(t)x \tag{9.2.18}$$

where K is the solution of the differential matrix Riccati equation

$$\dot{K} + A^*(t)K + KA(t) - KB(t)R^{-1}(t)B^*(t)K + Q(t) = 0, t \in I, \tag{9.2.19}$$

$$K(0) = M. \tag{9.2.20}$$

In view of this result it is clear that once the solution of the Riccati equation is available, the feedback control law given by (9.2.18) can be formally constructed. We shall discuss further on this question later.

It is interesting to observe that the optimal (linear) regulator is also stable in the Lyapunov sense. In other words optimality leads to stability. This follows as a corollary of the previous theorem.

Corollary 9.2.2 (Optimality to Stability) The closed loop (or feedback) system given by (9.2.14) is stable in the Lyapunov sense with respect to the Lyapunov function given by

$$V(t,x) = (1/2)(K(t)x, x). \tag{9.2.21}$$

Further, the system is asymptotically stable with respect to the zero state if any one of the following conditions hold (c1): Q is positive definite (c2): $Q(t) = C^*(t)C(t)$, and $(A(t), C(t))$ is observable.

Proof. First we must verify that the function V given by (9.2.21) satisfies the basic properties of a Lyapunov function. Clearly $V(t, 0) = 0$, and V is C^1 in $x \in R^n$. We must show that it is positive, that is, $V(t, x) \geq 0$. It suffices to show that $K(t)$ is at least positive semidefinite. There are several ways to verify this. Define

$$z(t) \equiv (K(t)x(t), x(t)) \equiv (\psi(t), x(t)), t \in I,$$

where we have used the expression (9.2.11). Differentiating this we have

$$\begin{aligned}
\dot{z}(t) &= (\dot{\psi}(t), x(t)) + (\psi(t), \dot{x}(t)) \\
&= -(Q(t)x, x) - (B^*(t)K(t)x, R^{-1}B^*(t)K(t)x) \\
&= -(Q(t)x, x) - (R(t)R^{-1}(t)B^*(t)K(t)x, R^{-1}(t)B^*(t)K(t)x) \\
&= -(Q(t)x, x) - (R(t)u(t), u(t)) \leq 0. \qquad (9.2.22)
\end{aligned}$$

Now note that

$$z(T) = (\psi(T), x(T)) = (Mx(T), x(T)) \geq 0.$$

Integrating and using the above expression, we find that

$$\begin{aligned}
z(t) &= (Mx(T), x(T)) - \int_t^T \dot{z}(s)ds \\
&= (Mx(T), x(T)) - \int_t^T \dot{z}(s)ds \\
&= (Mx(T), x(T)) + \int_t^T \{(Q(s)x(s), x(s)) + (R(s)u(s), u(s)\}ds. \qquad (9.2.23)
\end{aligned}$$

Since $M \geq 0, Q(t) \geq 0$ and $R(t) > 0$ and all are real symmetric matrices, it follows from this expression that

$$z(t) \equiv (K(t)x(t), x(t)) \geq 0, \ \forall\, t \geq 0. \qquad (9.2.24)$$

Clearly it follows from the equations (9.2.23) and (9.2.24) that the solution $K(t), t \geq 0$, of the differential Riccati equation is real, symmetric and at least positive semi definite. Thus V satisfies the basic properties. For stability we must show that its time derivative along any solution of the feedback or closed loop system given by (9.2.14) is at least negative semi definite. Let

$x_0 \in R^n$ be arbitrary and $x(t) \equiv x(t, x_0)$ be the corresponding solution of the closed loop system (9.2.14). Differentiating V along this trajectory we obtain

$$\begin{aligned}\dot{V} &= (1/2)\{(\dot{K}x, x) + (K\dot{x}, x) + (Kx, \dot{x})\} \\ &= (1/2)\Big\{((\dot{K} + A^*K + KA - KBR^{-1}B^*K)x, x) - (R^{-1}B^*Kx, B^*Kx)\Big\}.\end{aligned}$$

Since K is a solution of the differential Riccati equation (9.2.20), it follows from the above expression that

$$\dot{V} = -(1/2)\Big\{(Qx, x) + (R^{-1}B^*Kx, B^*Kx)\Big\}. \tag{9.2.25}$$

As $Q(t) \geq 0$ and $R^{-1}(t) > 0$, it is evident that $\dot{V} \leq 0$ along any solution of the closed loop system (9.2.14). This proves that the system is stable in the Lyapunov sense. For the statement (c1), note that if $Q(t) > 0$, $\dot{V}(t, x(t)) < 0$ till $x(t) \to 0$. Hence in this case the system is asymptotically stable with respect to the zero state. For the statement (c2), note that

$$\dot{V} = -(1/2)\Big\{\| C(t)x(t, x_0) \|^2 + (R(R^{-1}B^*Kx(t, x_0)), R^{-1}B^*Kx(t, x_0))\Big\}$$

for any initial state $x_0 \in R^n$. If the system is not asymptotically stable, there exists an $x_0 (\neq 0) \in R^n$ for which $\dot{V}(t, x(t, x_0)) \equiv 0$. Since R is positive definite this requires that $R^{-1}B^*Kx = -u \equiv 0$. Then for $\dot{V}$ to be identically zero, it is necessary that $C(t)x(t, x_0) = 0$ for all $t \geq 0$. Thus, for every finite T

$$\begin{aligned}\int_0^T \| C(t)x(t, x_0) \|^2 \, dt &= \int_0^T (\Phi^*(t, 0)C^*(t)C(t)\Phi(t, 0)x_0, x_0)dt \\ &= (Q_T(0)x_0, x_0) = 0.\end{aligned}$$

Since the system is observable, the observability matrix $Q_T(0)$ given by

$$Q_T(0) \equiv \int_0^T \Phi^*(t, 0)C^*(t)C(t)\Phi(t, 0)dt$$

is positive definite for some $T < \infty$. Thus x_0 must equal zero, which is a contradiction. Hence $\dot{V} < 0$ along any solution trajectory of the closed loop system and consequently the system is asymptotically stable. □

Remark (optimal cost). It follows from the expression for the cost functional (9.2.6) and the expressions for z given by (9.2.23) and (9.2.24) that the cost corresponding to the optimal feedback control is given by

$$J(u^o) = (1/2)z(0) = (1/2)(K(0)x_0, x_0). \tag{9.2.26}$$

In fact this relation holds for any initial state $x(0) = \xi \in R^n$ and not just for x_0 as indicated. Thus by solving the differential Riccati equation backward in time till $t = 0$, we also obtain the optimal cost.

Direct Approach.

The equations solving the linear quadratic regulator problem, that is, the differential Riccati equation for K and the feedback operator $\Gamma(t) \equiv -R^{-1}(t)B^*(t)K(t)$ can also be derived directly without the use of minimum principle. This is done by direct substitution of the expression for the state trajectory into the cost functional and then setting the functional derivative to zero. This action gives an expression for the feedback control law which after some computation leads to the Riccati equation. Letting $\Phi(t,s), s \leq t < \infty$, denote the transition operator corresponding to system matrix $A(\cdot)$, the solution of equation (9.2.5) is given by

$$x(t) = \Phi(t,0)x_0 + \int_0^t \Phi(t,s)B(s)u(s)ds, t \in I = [0,T].$$

Define the function ϕ and the operators L and L_T as follows:

$$\phi(t) \equiv \Phi(t,0)x_0, \; Lu(t) \equiv \int_0^t \Phi(t,s)B(s)u(s)ds, t \in I, \quad (9.2.27)$$

$$L_T u \equiv \int_0^T \Phi(T,s)B(s)u(s)ds. \quad (9.2.28)$$

Since our control signals are required to have finite energy, we may consider the Lebesgue space $L_2(I, R^m)$ as the appropriate choice for the admissible controls. Then L maps $L_2(I, R^m)$ to $L_2(I, R^n)$ and the operator L_T maps $L_2(I, R^m)$ to R^n. Using these notations the cost functional (9.2.6) can be rewritten as

$$\begin{aligned} J(u) = (1/2) \int_I \Big\{ (Q\phi, \phi) + 2(L^*Q\phi, u) + (L^*QLu, u) + (Ru, u) \Big\} dt \\ + (1/2)\{(M\phi(T), \phi(T)) + 2(L_T^* M\phi(T), u(T)) \\ + (L_T^* M L_T u(T), u(T))\}, \quad (9.2.29) \end{aligned}$$

where L^* and L_T^* are the adjoint operators corresponding to the operators L and L_T respectively. These are given by

$$(L^*\xi)(t) \equiv \int_t^T B^*(t)\Phi^*(s,t)\xi(s)ds, t \in I \quad (9.2.30)$$

$$L_T^*\eta(t) \equiv B^*(t)\Phi^*(T,t)\eta, t \in I. \quad (9.2.31)$$

Clearly the adjoint operator L^* maps $L_2(I, R^n)$ to $L_2(I, R^m)$, and the operator L_T^* maps R^n to $L_2(I, R^m)$. All these implications hold once the elements $\{b_{i,j}(t), t \in I\}$ of the matrix $B(\cdot)$ are in $L_2(I, R)$. Taking the functional derivative (Frechet or Gateaux) with respect to u of the functional J given by (9.2.29) and setting it to zero, we arrive at the following expression

$$L^*Q(\phi + Lu) + Ru + L_T^*M(\phi(T) + L_T u) = 0. \tag{9.2.32}$$

Recalling that $x = \phi + Lu$, the optimal control u must have the representation

$$\begin{aligned} u(t) &\equiv -R^{-1}(t)\{(L^*Qx)(t) + (L_T^*Mx(T))(t)\} \\ &\equiv -R^{-1}(t)\left\{\int_t^T B^*(t)\Phi^*(s,t)Q(s)x(s)ds + B^*(t)\Phi^*(T,t)Mx(T)\right\} \\ &\equiv -R^{-1}(t)B^*(t)\left\{\int_t^T \Phi^*(s,t)Q(s)x(s)ds + \Phi^*(T,t)Mx(T)\right\}. \end{aligned} \tag{9.2.33}$$

Defining

$$\psi(t) \equiv \int_t^T \Phi^*(s,t)Q(s)x(s)ds + \Phi^*(T,t)Mx(T), t \in I,$$

we obtain

$$u(t) \equiv -R^{-1}(t)B^*(t)\psi(t).$$

Differentiating $\psi(t)$ as given above and using the properties of the transition matrix Φ, we find that it satisfies the adjoint equation

$$-\dot{\psi}(t) = A^*(t)\psi(t) + Q(t)x(t), \psi(T) = Mx(T), \tag{9.2.34}$$

as expected. Now defining

$$\psi(t) \equiv K(t)x(t), t \in I, \tag{9.2.35}$$

we arrive at the differential Riccati equation

$$\dot{K} + A^*(t)K + KA(t) - KB(t)R^{-1}(t)B^*(t)K + Q(t) = 0, K(T) = M.$$

Thus we have obtained all the necessary conditions that we derived using the minimum principle. The control law is given by a linear operator

$$u(t) \equiv \Gamma(t)x = -R^{-1}(t)B^*(t)K(t)x.$$

Before we conclude this section, we present a result on the question of existence of solution of the differential Riccati equation,

$$\dot{K} + A^*(t)K + KA(t) - KB(t)R^{-1}(t)B^*(t)K + Q(t) = 0, t \in I, \quad (9.2.36)$$
$$K(T) = M. \quad (9.2.37)$$

Theorem 9.2.3. Suppose the matrix valued functions A, Q have locally integrable entries, B has locally square integrable elements and the elements of R and R^{-1} are essentially bounded measurable functions. Then for every $M \in M(n \times n)$ equation (9.4.58) has a unique solution $K \in AC(I, M(n \times n)$.

Proof. There are several ways to prove this result [2, 3]. These methods are indirect. Here we give a direct and simple proof [31]. The nonlinear problem is transformed into an equivalent linear problem which is then solved by use of the transition operator corresponding the linear problem. Let $M(2n \times n)$ denote the linear vector space of $2n \times n$ matrices with entries from the real line. Consider the linear system

$$\dot{Z} = \mathcal{A}(t)Z, \; Z(T) = \Gamma \equiv \begin{pmatrix} I \\ M \end{pmatrix}$$

where the $2n \times 2n$ matrix valued function $\mathcal{A}$ is given by

$$\mathcal{A}(t) \equiv \begin{pmatrix} A(t) & -B(t)R^{-1}(t)B^*(t) \\ -Q(t) & -A^*(t) \end{pmatrix}.$$

This is a linear differential equation on the vector space $M(2n \times n)$. Under our assumptions, the matrix valued function $\mathcal{A}$ has all its elements locally integrable. Hence there exists a unique transition operator $\Psi(t, s)$ that maps the linear space $M(2n \times n)$ to itself. In other words this equation has a unique solution given by

$$Z(t) = \Psi^{-1}(T, t)\Gamma, t \in [0, T].$$

Let $X(t)$ and $Y(t)$ denote the upper and lower block of $n \times n$ matrices so that

$$Z(t) \equiv \begin{pmatrix} X(t) \\ Y(t) \end{pmatrix}.$$

We show that the solution of equation (9.4.58) is given by

$$K(t) = Y(t)X^{-1}(t).$$

Indeed, taking the time derivative and using the matrix $\mathcal{A}$ and elementary matrix multiplications, we obtain

$$\begin{aligned}\dot{K} &= \dot{Y}X^{-1} + Y\dot{X}^{-1} \\ &= (-QX - A^*Y)X^{-1} - YX^{-1}\dot{X}X^{-1} \\ &= -Q - A^*K - K(AX - BR^{-1}B^*Y)X^{-1} \\ &= -Q - A^*K - KA + KBR^{-1}B^*K. \end{aligned} \tag{9.2.38}$$

This is precisely the equation (9.4.58). Since $Y(T) = M$ and $X(T) = I$, it is easy to see that the terminal condition, $K(T) = M$, holds. □

Remark. Note that in the above existence result we did not require symmetry or positivity of the matrices $\{Q(t), R(t), M\}$. However, we have already seen in the corollary 9.2.2 that if these matrices are symmetric and positive and $R(t)$ is positive definite, then the solution $K(t), t \geq 0$, is symmetric and positive. Further, $K(t)$ is positive definite if either $Q(t) > 0$ or the pair $(A(t), C(t))$ is observable.

9.3 Perturbed Regulators

P1: Here we consider a linear system subject to known perturbation, $h(t)$, $t \in I$,

$$\dot{x} = A(t)x + B(t)u + h(t), t \in I, x(0) = x_0 \in R^n. \tag{9.3.39}$$

Again the problem is to design a linear feedback control law that minimizes the same quadratic cost functional as given by (9.2.6). The fundamental problem here is that the future values of $x(s), s \geq t$, is no more determined by $x(t)$ and the control alone. The perturbation h must come into play. Thus the representation of ψ given by the expression (9.2.35) does not hold in this case. So we must introduce a compensating term to represent ψ such as

$$\psi(t) = K(t)x(t) + r(t), t \in I. \tag{9.3.40}$$

Our problem now reduces to finding K and r. Differentiating this expression and using the closed loop system equation

$$\dot{x} = Ax - BR^{-1}B^*(Kx + r) = (A - BR^{-1}B^*K)x - BR^{-1}B^*r$$

and the adjoint equation

$$-\dot{\psi} = A^*\psi + Qx,$$

we now obtain two sets of differential equations.

$$\dot{K} + A^*(t)K + KA(t) - KB(t)R^{-1}(t)B^*(t)K + Q(t) = 0, K(T) = M, \tag{9.3.41}$$

$$\dot{r} + (A^*(t) - K(t)B(t)R^{-1}(t)B^*(t))r + K(t)h(t) = 0, r(T) = 0. \tag{9.3.42}$$

The first one is the usual differential Riccati equation and the second is the compensator equation. In this case the optimal feedback control is an affine (linear) map of the state with a compensator,

$$u(t) = -R^{-1}(t)B^*(t)(K(t)x + r(t)), t \in I.$$

Note that if $h(t) \equiv 0$, then $r(t) \equiv 0$ and as a consequence we arrive at the linear law which is the standard result.

P2: Here it is the cost functional that is perturbed, in the sense that the cost functional is given by

$$J(u) = (1/2)\int_I \{(Q(t)[x - x_d], [x - x_d]) + (R(t)u, u)\}dt + (1/2)(Mx(T), x(T)), \tag{9.3.43}$$

and the system is given by

$$\dot{x} = A(t)x + B(t)u, t \in I, x(0) = x_0 \in R^n. \tag{9.3.44}$$

The objective here is to follow a given trajectory $x_d(t), t \in I$, as closely as possible. In this case the adjoint equation is given by

$$-\dot{\psi} = A^*(t)\psi + Q(x - x_d), \psi(T) = Mx(T). \tag{9.3.45}$$

Again choosing $\psi(t) \equiv K(t)x(t) + r(t)$ and carrying out similar steps as in the first case we find that K must satisfy the same Riccati equation and r must satisfy the differential equation,

$$\dot{r} + (A^*(t) - K(t)B(t)R^{-1}(t)B^*(t))r - Qx_d = 0, r(T) = 0. \tag{9.3.46}$$

This reduces to the standard case if $x_d(t) \equiv 0$.

P3: In (P2) if the system is also perturbed, the Riccati equation remains intact, but the compensator equation has an additional term. The reader can easily verify that it is given by

$$\dot{r} + (A^*(t) - K(t)B(t)R^{-1}(t)B^*(t))r + Kh - Qx_d = 0, r(T) = 0. \tag{9.3.47}$$

Remark. The closed loop system and the compensator dynamics are determined by the matrix valued function

$$\tilde{A}(t) \equiv A(t) - B(t)R^{-1}(t)B^*(t)K(t)$$

and its adjoint (or transpose) $\tilde{A}(t)^*$. Thus in case $Q(t)$ is strictly positive definite, both the systems are asymptotically stable in the absence of disturbances. Thus if the disturbance persists only for a finite interval of time, the error converges to zero asymptotically.

9.4 Constrained Regulators.

Consider the linear system (9.2.5) with the quadratic cost (9.2.6) and admissible controls

$$\mathcal{U}_{ad} \equiv \{u : I \longrightarrow R^m \text{ measurable} : u(t) \in B_r \ a.e\} \tag{9.4.48}$$

where $B_r \equiv \{v \in R^m : \| v \| \leq r\}$. Clearly the minimum principle holds, and hence the Hamiltonian must be minimized over the closed ball B_r. That is, we must minimize the functional

$$H(t, x, \psi, u) = (A(t)x + B(t)u, \psi) + (1/2)(Qx, x) + (1/2)(Ru, u)$$

on the closed ball $B_r \subset R^m$. Define the retract

$$F_r(\xi) \equiv \begin{cases} \xi, & \text{if } \xi \in B_r; \\ (r/\| \xi \|)\xi, & \text{otherwise.} \end{cases}$$

Now minimizing the Hamiltonian on the ball we obtain the form of the optimal control law which is given by

$$u^o \equiv -F_r(R^{-1}(t)B^*(t)\psi(t)). \tag{9.4.49}$$

Using this control law and substituting in the state equation we obtain

$$\dot{x} = A(t)x - B(t)F_r(R^{-1}B^*\psi) \tag{9.4.50}$$

with the adjoint equation remaining unchanged,

$$-\dot{\psi} = A^*(t)\psi + Q(t)x, \psi(T) = Mx(T). \tag{9.4.51}$$

In view of the fact that the adjoint equation (9.4.51) has remained unchanged, there exists a real symmetric positive matrix valued function K

so that $\psi(t) = K(t)x(t)$. Substituting this in (9.4.50) and (9.4.51) we arrive at the following system of equations for the regulator

$$(\dot{K} + A^*K + KA + Q)x - KBF_r(R^{-1}B^*Kx) = 0, K(T) = M, \quad (9.4.52)$$
$$\dot{x} = Ax - BF_r(R^{-1}B^*Kx), x(0) = x_0. \quad (9.4.53)$$

This is a system of coupled nonlinear differential equations. Since K is symmetric, this is a system of $n(n+3)/2$ equations with the same number of boundary conditions.

Similarly one may consider imposing state constraints through the cost functional. Suppose it is required that the state never leaves a closed bounded convex set $E \subset R^n$. In this case one can choose the cost functional as

$$J(u) = (1/2)\int_I \{(Q(t)x, x) + \chi_E(x(t)) + (R(t)u, u)\}dt + (1/2)(Mx(T), x(T)), \quad (9.4.54)$$

where χ_E is the indicator function of the set E defined by

$$\chi_E(z) \equiv \begin{cases} 0 & \text{if } z \in E; \\ \infty & \text{otherwise.} \end{cases}$$

In this situation the adjoint equation (9.4.51) turns into a differential inclusion as follows:

$$-\dot{\psi} - A^*(t)\psi \in Q(t)x(t) + (1/2)\partial\chi_E(x(t)), \psi(T) = Mx(T). \quad (9.4.55)$$

Here $\partial\chi_E(z)$ denotes the subdifferential of the indicator function at $z \in E$ and it is given by

$$\partial\chi_E(z) \equiv \{w \in R^n : (w, y - z) \le 0 \; \forall \; y \in E\}.$$

The control is given by

$$u = -R^{-1}(t)B^*(t)\psi(t) \quad (9.4.56)$$

and the state equation is given by

$$\dot{x} = Ax - BR^{-1}B^*\psi, x(0) = x_0. \quad (9.4.57)$$

In view of the adjoint inclusion (9.4.55), it is clear that ψ is not linear in the state and so a closed form Riccati equation can not be written.

Another interesting case is requiring hard terminal constraint. This can be formulated by replacing the quadratic terminal cost by an indicator function as follows:

$$J(u) = (1/2) \int_I \{(Q(t)x, x) + (R(t)u, u)\}dt + \chi_E(x(T)). \tag{9.4.58}$$

In this case the regulator equations can be derived from

$$\begin{aligned} \dot{x} &= Ax - BR^{-1}B^*\psi, x(0) = x_0 \\ -\dot{\psi} &= A^*\psi + Qx, \psi(T) \in \partial\chi_E(x(T)). \end{aligned}$$

and they are given by

$$-(\dot{K} + A^*K + KA + Q - KBR^{-1}B^*K)x = 0, K(T) = N,$$

$$Nx(T) \in \partial\chi_E(x(T)) \tag{9.4.59}$$

$$\dot{x} = (A - BR^{-1}B^*K)x, x(0) = x_0. \tag{9.4.60}$$

It is interesting to note that if $E = R^n$ then $\partial\chi_E(\xi) = \{0\}$ for all $\xi \in E$ and hence $N = 0$. This also means that $\psi(T) = 0$ which agrees with the standard transversality condition. Clearly the regulator equations given above reduce to the classical case without terminal cost. On the other extreme if $E = \{x^*\}$ is a singleton, one can verify that $\partial\chi_E(\xi) = R^n$, the entire state space. This means that $Nx(T) = \psi(T) \in R^n$. In other words the terminal condition for the co-state is free, again in agreement with transversality conditions. Linear filtering with similar constraints were treated in [3, 149]. For numerical results see [149].

9.5 Algebraic Riccati Equation: Steady State

For practical applications, the time varying problem is difficult. The feedback control law is time varying and requires the knowledge of the time varying system matrices for the entire period of optimization. For time invariant systems the problem becomes much simpler and one may wish to find a time invariant (stationary) control law (not the control) to obtain the closed loop system. Suppose all the matrices $\{A, B, Q, R\}$ are time invariant. The problem is to find an optimal control that minimizes the cost functional

$$J(u) = (1/2) \int_0^\infty \{(Qx, x) + (Ru, u)\}dt \tag{9.5.61}$$

subject to the dynamic constraint

$$\dot{x} = Ax + Bu, x(0) = x_0 \in R^n. \tag{9.5.62}$$

Again using the minimum principle one obtains the control

$$u = -R^{-1}B^*\psi$$

where ψ is the solution of the adjoint equation

$$-\dot{\psi} = A^*\psi + Qx, \psi(\infty) = 0.$$

The solution of this equation is given by

$$\psi(t) = \int_t^\infty e^{(s-t)A^*} Qx(s)ds.$$

Again setting $\psi(t) = Kx(t)$ and using the system equation (9.5.62) and the adjoint equation one can easily verify that K must be a solution of the algebraic Riccati equation (ARE) given by

$$A^*K + KA - KBR^{-1}B^*K + Q = 0. \tag{9.5.63}$$

This is a nonlinear system of algebraic equations and one must verify that it has a solution. First let us verify that if it has a solution, it must be symmetric and positive, that is $K \in M_s^+$ where we use M_s^+ to denote the class of symmetric positive square matrices. Differentiating the scalar product $(\psi(t), x(t))$ and using the state and the adjoint equations one can easily verify that

$$(d/dt)(\psi(t), x(t)) = -\Big\{(Qx, x) + (KBR^{-1}B^*Kx, x)\Big\}. \tag{9.5.64}$$

Integrating this by parts using the end conditions one arrives at the following expression,

$$\begin{aligned} &(1/2)(Kx(t), x(t)) \\ &\quad = (1/2)(\psi(t), x(t)) \\ &\quad = (1/2)\int_t^\infty \Big\{(Qx(s), x(s)) + (R^{-1}B^*Kx(s), B^*Kx(s))\Big\}ds \\ &\quad = (1/2)\int_t^\infty \Big\{(Qx(s), x(s)) + (Ru, u)\Big\}ds \end{aligned} \tag{9.5.65}$$

where u is the feedback control given by $u = -R^{-1}B^*Kx$. Setting $t = 0$ we obtain

$$\begin{aligned} (1/2)(Kx_0, x_0) &= (1/2)\int_0^\infty \Big\{(Qx(s), x(s)) + (R^{-1}B^*Kx(s), B^*Kx(s))\Big\}ds \\ &= (1/2)\int_0^\infty \Big\{(Qx(s), x(s)) + (Ru, u)\Big\}ds = J(u). \end{aligned} \tag{9.5.66}$$

Clearly if both Q and R are symmetric and positive, K is also symmetric and positive. For positive definiteness one must introduce additional conditions as in the case of differential Riccati equation. This is stated in the following result.

Theorem 9.5.1. Suppose the system matrices $\{A, B, Q, R\}$ are all constant with $Q \in M_s^+(n \times n)$, $R \in M_s^+(m \times m)$ and $R > 0$. Then $K \in M_s^+(n \times n)$ and it is positive definite if any one of the following conditions hold: (c1): Q is positive definite, (c2): Q can be written as $Q = C^*C$ and (A, C) is observable.

Proof. We prove positive definiteness. Under the assumption (c1), it is clear from the expression 9.5.66 that if $Q > 0$ then $K > 0$. Suppose now (c2) holds and K is not positive definite. Then there exists an initial state $x_0(\neq 0) \in R^n$ for which $(1/2)(Kx_0.x_0) = 0$. Since $R > 0$, this requires that $u(t) = 0$ almost every where. Thus the second term of (9.5.66) is zero and so also the first term. This later is possible if

$$Cx(t, x_0) \equiv 0.$$

Since the pair (A, C) is observable, this implies that x_0 must equal zero. This is a contradiction thereby proving the result. □

Next we prove that the algebraic Riccati equation (ARE) (9.5.63) has a solution.

Theorem 9.5.2. Suppose the assumptions of Theorem 9.5.1 hold and that the pair (A, B) is stabilizable. Then the algebraic Riccati equation has a unique positive definite solution.

Proof. If equation (9.5.63) has a solution, by Theorem 9.5.1 it is positive definite. We use Corollary 9.2.2 and Theorem 9.5.1 to prove the existence. Focusing our attention on the expression for $K(t)$ given by (9.2.23), it is clear that it depends on the data at the final time and hence the time T. So we may write this as $K^T(t), t \in [0, T]$. For $M = 0$, it follows from (9.2.23) that

$$(K^T(t)x(t), x(t)) = \int_t^T \{(Qx(s), x(s) + (Ru(s), u(s))\}ds, \qquad (9.5.67)$$

which holds for $0 \leq t \leq T$ for any finite T. Setting $t = 0$ and taking any arbitrary initial state ξ and defining $x(s) = x(s, \xi)$ and $K^T(0) \equiv K^T$, we rewrite (9.5.67) as

$$(K^T\xi, \xi) = \int_0^T \{(Qx(s), x(s) + (Ru(s), u(s))\}ds. \qquad (9.5.68)$$

Comparing this expression with (9.5.66) we notice that the solution of the ARE is the limit $\lim_{T\to\infty} K^T$ provided such a limit exists. We also note that along any solution $x(t,\xi)$ of the system (9.5.62) corresponding to any control u and any initial state $\xi \in R^n$, $T \longrightarrow K^T$ is a monotone increasing function of T, in the sense that

$$(K^{T_1}\xi,\xi) \leq (K^{T_2}\xi,\xi) \;\; \forall \;\; T_1 \leq T_2 \;\; \forall \;\; \xi \in R^n.$$

Since by assumption the system is stabilizable, there exists a $\Gamma \in M(m \times n)$ such that $(A - B\Gamma)$ is a stability matrix. Thus the system (9.5.62) with the feedback control $u(t) = -\Gamma x(t)$ is asymptotically stable, that is,

$$\lim_{t\to\infty} \{x(t) = x(t,\xi)\} = 0.$$

However, since this control is not necessarily optimal, use of this control in equation (9.5.68) leads to the following inequality,

$$(K^T\xi,\xi) \leq \int_0^T \{(Qx(s),x(s) + (R\Gamma x(s),\Gamma x(s))\}ds. \tag{9.5.69}$$

Since $x(t) \to 0$ exponentially, and $u = -\Gamma x$, u also converges to zero exponentially. This implies that there exists a positive definite matrix $\hat{K}$ such that

$$(K^T\xi,\xi) \leq \int_0^\infty \{(Qx(s),x(s) + (R\Gamma x(s),\Gamma x(s))\}ds = (\hat{K}\xi,\xi) < \infty. \tag{9.5.70}$$

To summarize, we have a family of positive definite matrices $\{K^T, T \geq 0\}$ which is monotone increasing and bounded above. Hence by monotone convergence theorem, there exists a symmetric positive definite matrix $\tilde{K}$ such that

$$\lim_{T\to\infty} K^T = \tilde{K}.$$

Thus the algebraic Riccati equation has a solution and we have shown that it is the unique limit of the solution of the differential Riccati equation. We prove that this is the only solution. Suppose $\tilde{K}$ and K are two such solutions of (9.5.63). Defining $E \equiv \tilde{K} - K$ and subtracting one equation from the other one can easily verify that E satisfies the matrix equation

$$A_1E + EA_2 = 0, \tag{9.5.71}$$

where

$$A_1 \equiv (A - BR^{-1}B^*K)^* \;\text{ and }\; A_2 \equiv (A - BR^{-1}B^*\tilde{K}). \tag{9.5.72}$$

It is well known that the solution of a matrix equation like

$$A_1E + EA_2 = L$$

is given by

$$E = \int_0^\infty e^{tA_1} L e^{tA_2} dt.$$

Since both A_1 and A_2 are stability matrices as seen earlier, the integral is well defined. In case of equation (9.5.71), $L = 0$ and hence $E = 0$ thereby proving uniqueness. □

Remark. In view of the above results, given that the assumptions hold, one can build a feedback system using the solution of the algebraic Riccati equation (ARE)

$$\dot{x} = (A - BR^{-1}B^*K)x, \; t \geq 0,$$

which is asymptotically stable with respect to the zero state. The algebraic Riccati equation can be solved using Newton-Raphson technique or more efficient versions of it [31].

9.6 Some Nonstandard Regulator Problems

(NRP1) Problems with Intermediate costs: Define the index set

$$D \equiv \{t_k : t_k \in (0,T] \; t_k < t_{k+1} \;, k = 1, 2, \cdots p, \text{ and } t_0 = 0, t_p = T\}$$

and consider the linear system

$$\dot{x}(t) = A(t)x(t) + B(t)u(t), x(0) = x_0, \; t \in I \equiv [0,T], T < \infty, \tag{9.6.73}$$

with the cost functional given by

$$J(u) = (1/2) \int_I \{(Q(t)x, x) + (R(t)u, u)\}dt + (1/2) \sum_{k=1}^{p} (M_k x(t_k), x(t_k)). \tag{9.6.74}$$

The matrices $\{Q(t), R(t)\}$ satisfy standard assumptions of section 2 and all the matrices $M_k \in M_s^+$. The problem is to find a linear feedback control law that minimizes the cost functional. Here in addition to the running cost determined by Q, special importance is placed on the cost (error) at certain intermediate points of time which may be considered important for system performance. Using the minimum principle of Chapter 8, we have the adjoint equation given by

$$\begin{aligned} -\dot{\psi} &= A^*(t)\psi + Q(t)x \\ \psi(t_k - 0) &= M_k x(t_k), k = 1, 2, \cdots p. \end{aligned} \tag{9.6.75}$$

This is a special class of impulsive differential systems. The solution has simple discontinuities on the set D. It is piecewise continuous, more precisely, left continuous having right hand limits. Starting with the final time $t_p = T$ one must solve this backward in time till t_{p-1} yielding the value of the adjoint state $\psi(t_{p-1})$ prior to the jump. Then the jump takes place and the adjoint takes the value

$$\psi(t_{p-1} - 0) = \psi(t_{p-1}) + \{M_{p-1}x(t_{p-1}) - \psi(t_{p-1})\} = M_{p-1}x(t_{p-1}).$$

Hence the jump at t_{p-1} is given by

$$\mathcal{J}(\psi(t_{p-1})) \equiv \{M_{p-1}x(t_{p-1}) - \psi(t_{p-1})\}.$$

This process is carried out for all the subintervals $(t_{k-1}, t_k]$ till $t_0 = 0$ is reached. As a consequence the differential Riccati equation is also an impulsive system given by

$$\begin{aligned} &\dot{K} + A^*K + KA - KBR^{-1}B^*K) + Q = 0 \\ &K(t_k - 0) = M_k, k = 1, 2, \cdots p, \end{aligned} \tag{9.6.76}$$

and must be solved backward. Again the Riccati equation has piece wise continuous solution (left continuous with right limits). The control law has no change

$$u = -B(t)R^{-1}(t)B^*(t)K(t)x(t)$$

except for the fact that it has distinct discontinuities exactly at the points of discontinuities of K.

NRP2 Free Controls: Consider the system (9.6.73) with the cost functional given by

$$J(u) = (1/2)\int_0^T (Qx(t), x(t))dt + (1/2)(Mx(T), x(T)) \tag{9.6.77}$$

where $M \in M_s^+$. The set of admissible controls is

$$\mathcal{U}_{ad} \equiv L_2(I, R^m).$$

Since there is no strict bounds on the controls and apparently the cost functional may be noncoercive, Pontryagin minimum principle does not hold. It is also apparent from the fact that the minimum of the Hamiltonian

$$H = (A(t)x + B(t)u, \psi) + (1/2)(Qx, x)$$

over $U = R^m$ does not make sense. However the integral minimum principle may hold, provided sufficient conditions are imposed on the data $\{A(t), B(t), M\}$ so that the functional $J(u)$ is coercive on the Hilbert space $L_2(I, R^m)$. Note that if U is a compact set this problem does not arise. For $U = R^m$, the existence problem is crucial. First we prove that this problem has a solution. For simplicity we consider only linear time invariant system.

Lemma 9.6.1. Consider the linear time invariant system

$$\dot{x} = Ax + Bu, x(0) = x_0 \in R^n, t \in I \equiv [0, T]; \mathcal{U}_{ad} = L_2(I, R^m),$$

with the cost functional (9.6.77) as defined above. Suppose the pair (A, B) is controllable, M is symmetric and positive definite and $Q \in M_s^+$, not necessarily positive definite. Then there exists an optimal control $u^o \in L_2(I, R^m)$.

Proof. Using the closed form expression for the solution one can express the cost functional (9.6.77) in the form

$$J(u) \equiv c + \ell(u) + (Gu, u) \tag{9.6.78}$$

where

$$c \equiv (1/2)(\Phi^*(T)M\Phi(T)x_0, x_0) + (1/2)\int_0^T (\Phi^*(t)Q\Phi(t)x_0, x_0)dt \tag{9.6.79}$$

$$\ell(u) \equiv \int_0^T ([B^*\Phi^*(t)M\Phi(T) + (L^*\sqrt{Q}\Phi)(t)]x_0, u(t))dt \tag{9.6.80}$$

$$(Gu, u) \equiv (1/2) \parallel L_T u \parallel^2 + (1/2)\int_0^T \parallel (Lu)(t) \parallel^2 dt \tag{9.6.81}$$

and the operators L and L_T are linear integral operators given by

$$(Lu)(t) \equiv \int_0^t \sqrt{Q}\Phi(t-s)Bu(s)ds, t \in I, \tag{9.6.82}$$

$$L_T u \equiv \int_0^T \sqrt{M}\Phi(T-s)Bu(s)ds. \tag{9.6.83}$$

The operator L maps $L_2(I, R^m)$ to $L_2(I, R^n)$ and L_T maps $L_2(I, R^m)$ to R^n and they are bounded linear. Since (A, B) is controllable and controllability implies stabilizability there exists a $\Gamma \in M(m \times n)$ such that $u = \Gamma x$ stabilizes the system, that is, the feedback system $\dot{x} = (A - B\Gamma)x$ is asymptotically stable. Thus along this solution trajectory,

$$\begin{aligned} J(u) &= c + \ell(u) + (Gu, u) \\ &= (1/2) \int_0^T (Qx, x)dt + (1/2)(Mx(T), x(T)) \leq J(0) = c. \end{aligned}$$

Since $J(u) \geq 0$, this shows that there exists a minimizing sequence $\{u_n\} \in \mathcal{U}_{ad}$ such that

$$\lim_{n \to \infty} J(u_n) = \inf\{J(u), u \in \mathcal{U}_{ad}\} \equiv m^o \leq J(0) = c.$$

Let N denote the null space

$$N \equiv Ker L_T \cap Ker L \equiv \{u \in L_2(I, R^m) : L_T u = 0, Lu = 0\}$$

and $N^\perp$ the complementary subspace of the Hilbert space $L_2(I, R^m)$. Clearly a minimizing sequence $\{u_n\}$ must be elements of $N^\perp$ since otherwise $J(u_n) \equiv J(0) = c$ for all n and so can not be a minimizing sequence. Thus $\{u_n\} \in N^\perp$. By our assumption, $M \in M_s^+$ is positive definite and (A, B) is controllable. This implies that the range of L_T, denoted by RangeL_T, is the whole space R^n. Hence L_T, restricted to $N^\perp$ is invertible. In other words

$$\left(L_T|_{N^\perp}\right)^{-1}$$

is a continuous (bounded) linear operator from R^n to $N^\perp \subset L_2(I, R^m)$. Hence there exists a $\gamma > 0$ such that

$$\| L_T u \| \geq \gamma \| u \| \;\; \forall\, u \in N^\perp.$$

Thus

$$(Gu, u) \geq (1/2)\gamma^2 \| u \|^2 \;\; \forall\, u \in N^\perp.$$

Since c is finite and $\ell(u)$ is a bounded linear functional on $L_2(I, R^m)$, it follows from the above analysis that $J(u)$ is coercive on $N^\perp$, that is,

$$\lim_{\|u\| \to \infty, u \in N^\perp} \Big(J(u) / \| u \| \Big) = +\infty.$$

Hence the minimizing sequence $\{u_n\}$ must be confined in a bounded subset of $N^{\perp} \subset L_2(I, R^m)$. Let $r > 0$ be the smallest number such that

$$\{u_n\} \subset B_r \cap N^{\perp}$$

where B_r is the closed ball

$$B_r \equiv \{u \in L_2(I, R^m) : \parallel u \parallel \ \leq r\}.$$

Since a closed bounded convex set of any Hilbert space is weakly compact, there exists a subsequence of the sequence $\{u_n\}$, relabeled as $\{u_n\}$, and a $u^o \in B_r \cap N^{\perp}$ such that

$$u_n \xrightarrow{w} u^o.$$

Since $\ell(u)$ is a continuous linear functional it is also weakly continuous and $\lim_{n\to\infty} \ell(u_n) = \ell(u^o)$. The reader can easily verify that any quadratic functional is weakly lower semi continuous and therefore,

$$(Gu^o, u^o) \leq \underline{\lim}(Gu_n, u_u).$$

Hence $u \longrightarrow J(u)$ is also weakly lower semi continuous. A weakly lower semi continuous functional attains its minimum on any weakly compact subset of any Hilbert space. In fact since $u^o \in B_r \cap N^{\perp}$ we have

$$m^o \leq J(u^o) \leq \underline{\lim}_{n\to\infty} J(u_n) = \lim_{n\to\infty} J(u_n) = m^o.$$

In other words the infimum is attained and it is the minimum. Hence u^o is the optimal control and it lies in the ball $B_r \cap N^{\perp}$. This completes the proof.

□

With the help of this result we can now construct linear regulators for the quadratic cost functional (9.6.77). This is stated in the following result.

Theorem 9.6.2. Consider the linear regulator problem with the cost functional (9.6.77) and suppose the assumptions of Lemma 9.6.1 hold. Then the optimal linear regulator has the form

$$u(x) \equiv -\alpha B^* K(t) x$$

for some $\alpha > 0$, where K is the solution of the differential Riccati equation,

$$\dot{K} + A^*K + KA - \alpha KBB^*K + Q = 0, \ \ K(T) = M. \tag{9.6.84}$$

Proof. The Hamiltonian for this problem is given by

$$H(\xi, v, y) \equiv (A\xi + Bv, y) + (1/2)(Q\xi, \xi), \xi \in R^n, v \in R^m, y \in R^n.$$

By Lemma 9.6.1, there exists a control policy $u^o \in B_r$ and a corresponding trajectory $x^o \in AC(I, R^n)$ and a $\psi^o \in AC(I, R^n)$ corresponding to the pair $\{u^o, x^o\}$ such that the triple $\{u^o, x^o, \psi^o\}$ satisfies the integral minimum principle

$$\int_I H(x^o(t), u^o(t), \psi^o(t))dt \leq \int_I H(x^o(t), u(t), \psi^o(t))dt \; \forall u \in B_r \subset L_2(I, R^m). \tag{9.6.85}$$

Further, the triple $\{u^o, x^o, \psi^o\}$ must also satisfy the canonical equations

$$\dot{x}^o = H_\psi = Ax^o + Bu^o, \; x^o(0) = x_0 \tag{9.6.86}$$

$$\dot{\psi}^o = -H_x = -A^*\psi^o - Qx^o, \; \psi^o(T) = Mx^o(T). \tag{9.6.87}$$

Then it follows from these necessary conditions that the following inequality must hold:

$$\int_I (u^o, B^*\psi^o)dt \leq \int_I (u, B^*\psi^o)dt \quad \forall \quad u \in B_r \cap N^\perp \subset L_2(I, R^m). \tag{9.6.88}$$

Define the functional

$$f(u) \equiv \int_I (u, B^*\psi^o)dt.$$

This is a continuous linear functional on the Hilbert space $L_2(I, R^m)$ and it satisfies

$$f(u^o) \leq f(u) \; \forall \; u \in B_r \cap N^\perp.$$

It follows from the inequality that u^o must be in the direction opposite to that of the vector $B^*\psi^o$ and hence u^o must be of the form

$$u^o = -\alpha B^*\psi^o$$

for some $\alpha > 0$ satisfying $\| u^o \| \leq r$. In view of the adjoint equation, there exists a matrix valued function $K \in AC(I, M_s^+)$ such that

$$\psi^o(t) = K(t)x^o(t), t \in I.$$

Differentiating ψ^o and using the adjoint and the state equation we arrive at the following expression

$$(\dot{K} + A^*K + KA - \alpha KBB^*K + Q)x^0 = 0 \; \forall t \in I$$

with the feedback control law

$$u^o = -\alpha B^*Kx^o$$

for some $\alpha > 0$. Using the feedback control law, the solution of the state equation is given by

$$x^o(t) = \Phi_\alpha(t,0)x_0,$$

where $\Phi_\alpha(t,s), 0 \leq s \leq t \leq T$, is the transition operator corresponding to the matrix $A - \alpha BB^*K(t)$. Since the initial state x_0 is arbitrary and the transition operator is always nonsingular, we arrive at the following differential Riccati equation

$$\dot{K} + A^*K + KA - \alpha KBB^*K + Q = 0, K(T) = M.$$

This completes the proof. □

Choosing $V \equiv (1/2)(Kx, x)$ as a candidate for Lyapunov function, the reader can easily verify that

$$\dot{V} = -(1/2)\{(Qx, x) + \alpha \parallel B^*Kx \parallel^2\}.$$

Again if $Q = C^*C$ and (A, C) is observable one can verify that $\dot{V} < 0$ for $x \neq 0$ and the system is asymptotically stable with respect to the zero state.

Remark. Since Q was assumed to be symmetric and only positive semidefinite, the results of Lemma 9.6.1 and Theorem 9.6.2 remain valid also for the terminal problem.

$$J(u) = (1/2)(Mx(T), x(T)). \tag{9.6.89}$$

The Riccati equation, in this case, is given by

$$\dot{K} + A^*K + KA - \alpha KBB^*K = 0, K(T) = M.$$

In this case also one can choose $V \equiv (1/2)(Kx, x)$ as a candidate for the Lyapunov function. Then $\dot{V} = -(1/2)\alpha \parallel B^*Kx \parallel^2$. The system is stable but not expected to be asymptotically stable. For numerical results and other details see [150].

NRP3 Impulsive Systems:

Consider the time interval $[0, T]$ and let $D \equiv \{t_i \in (0, T), i = 1, 2, 3 \cdots N\}$ be a set of N distinct points with $t_{N+1} \equiv T$ and let $\delta > 0$ be such that $t_i + \delta < t_{i+1}$ for all $i \in \{1, 2, 3, \cdots N\}$. A linear impulsive system may be described by the following system of equations

$$\dot{x} = A(t)x, x(0) = x_0, t \in I \setminus D, \quad (\Delta x)(t_i) = A_i x(t_i), \quad (\Delta x)(t_i + \delta) = B_i u_i, \tag{9.6.90}$$

where $\{A(t), A_i\}$ are the system matrices, $\{B_i\}$ are the control matrices, and $\{u_i \in R^m\}$ are the controls applied at times $\{t_i + \delta\}$ after each jump. The jump operator $\triangle$ is defined as follows

$$\triangle x(t) \equiv x(t+) - x(t-)$$

for any function x with $x(t+)$ denoting the value of x with t approaching from the right and $x(t-)$ that for t approaching from the left. If the function is left continuous $x(t-) = x(t)$ and if right continuous then $x(t+) = x(t)$. The cost functional is given by

$$J(u) = (1/2)\int_I (Qx, x)dt + (1/2)\sum_{i=1}^{N}(R_i u_i, u_i) + (1/2)(Mx(t), x(T)). \tag{9.6.91}$$

In this situation the Hamiltonian contains Dirac measures and its integral is well defined as given by

$$\begin{aligned}\int_I H dt = &\int_I \{(Ax, \psi) + (1/2)(Qx, x)\}dt \\ &+ \sum_{i=1}^{N}(A_i x(t_i), \psi(t_i)) + \sum_{i=1}^{N}(B_i u_i, \psi(t_i + \delta)) \\ &+ (1/2)\sum_{i=1}^{N}(R_i u_i, u_i).\end{aligned} \tag{9.6.92}$$

The optimal control has the usual form given by

$$u_i^o = -R_i^{-1} B_i^* \psi(t_i + \delta), \tag{9.6.93}$$

where ψ is the solution of the adjoint equation

$$\begin{aligned}-\dot{\psi} &= A^*\psi + Qx, t \in I \setminus D, \psi(T) = Mx(T) \\ -\triangle\psi(t_i) &= A_i^*\psi(t_i), t_i \in D.\end{aligned} \tag{9.6.94}$$

Using standard technique but taking care of the boundary conditions at each of the points of D and $D_\delta \equiv D + \delta$, one can derive the Riccati equations as follows:

$$\dot{K} + A^*K + KA + Q = 0, K(T) = M, t \in I \setminus (D \cup D_\delta), \tag{9.6.95}$$

$$\triangle K(t_i + \delta) = K(t_i + \delta)B_i R_i^{-1} B_i^* K(t_i + \delta), \text{ on } D_\delta, \tag{9.6.96}$$

$$\triangle K(t_i) = -K(t_i)A_i - A_i^* K(t_i), t_i \in D. \tag{9.6.97}$$

Finally the optimal feedback control is given by

$$u_i^o = -R_i^{-1} B_i^* K(t_i + \delta)x(t_i + \delta), i = 1, 2, \cdots, N. \tag{9.6.98}$$

For detailed and rigorous treatment of impulsive systems and their controls see [107, 108, 109].

A brief description of computational strategy is as follows. The Riccati equations are solved sequentially starting with the first one from $t = T$ till $t_N + \delta$ giving $K(t_N + \delta)$. This is then used in the second equation to compute the jump and the value at $t_N + \delta-$ giving $K(t_N + \delta-)$. Starting with this data one returns to the first equation and solves it (backward in time) till t_N is reached. The jump at this point is calculated using the third equation giving $K(t_N-)$. Using this data one returns to the first equation to continue the process till $t = 0$ is reached.

9.7 Disturbance Rejection Problem

In Section 2 we considered perturbed regulators with known disturbance. Here we consider a problem with unknown disturbance. The system is governed by the differential equation

$$\dot{x} = A(t)x + B(t)u + C(t)w, t \in I \equiv [0, T]. \tag{9.7.99}$$

Here w is the unknown disturbance. Our problem is to find a linear feedback control law that minimizes the quadratic cost

$$J(u, w) \equiv (1/2) \int_I \{(Qx, x) + (Ru, u) - (Sw, w)\}dt + (1/2)(Mx(T), x(T)). \tag{9.7.100}$$

This is a games problem; here control u is the first player playing against the nature w in the sense that u wants to minimize losses while the nature w wants to thwart this objective. We can also interpret this as an attempt to minimize the maximum losses. Suppose the matrices $\{B, C\}$ appearing in equation (9.7.99) take values from $M(n \times m)$ and $M(n \times d)$ respectively and that the elements of these matrices are also in L_2. We assume that $u \in L_2(I, R^m)$ and $w \in L_2(I, R^d)$. We define the Hamiltonian

$$H(x, u, \psi, w) \equiv (Ax + Bu + Cw, \psi) + (1/2)(Ru, u) - (1/2)(Sw, w). \tag{9.7.101}$$

Using the necessary conditions of optimality one maximizes the Hamiltonian with respect to w and minimizes it with respect to the control. This leads to the saddle point inequality

$$\begin{aligned}
&\int_I H(x^o(t), u^o(t), \psi^o(t), w(t))dt \\
&\quad \leq \int_I H(x^o(t), u^o(t), \psi^o(t), w^o(t))dt \\
&\quad \leq \int_I H(x^o(t), v(t), \psi^o(t), w^o(t))dt \;\forall w \in L_2(I, R^d), v \in L_2(I, R^m).
\end{aligned} \tag{9.7.102}$$

By this min max operation, one can easily verify that the saddle point is given by

$$u^o = -R^{-1}B^*\psi^o, \quad w^o = S^{-1}C^*\psi^o. \tag{9.7.103}$$

and the solutions of the state and the adjoint equations (corresponding to these processes)

$$-\psi^o = A^*\psi^o + Qx^o, \psi^o(T) = Mx^o(T), \tag{9.7.104}$$

$$\dot{x}^o = Ax^o - BR^{-1}B^*\psi^o + CS^{-1}C^*\psi^o, x^o(0) = x_0, \tag{9.7.105}$$

are given by the pair $\{x^o, \psi^o\}$ respectively. From similar arguments used often in the derivation of the Riccati equation, it follows from the adjoint equation that there exists a $K \in AC(I, M(n \times n)$ such that $\psi^o(t) = K(t)x(t)$. Substituting this in the adjoint equation and using the state equation we obtain the differential Riccati equation for K

$$\dot{K} + A^*K + KA - K(BR^{-1}B^* - CS^{-1}C^*)K + Q = 0, K(T) = M. \tag{9.7.106}$$

By a similar transformation as used in Theorem 9.2.3, one can write an equivalent linear system for $Z(t) \equiv \begin{pmatrix} X(t) \\ Y(t) \end{pmatrix} \in M(2n \times n)$,

$$\dot{Z} = \mathcal{A}(t)Z, \; Z(T) = \Gamma \equiv \begin{pmatrix} I \\ M \end{pmatrix}, \tag{9.7.107}$$

where the $2n \times 2n$ matrix valued function $\mathcal{A}$ is given by

$$\mathcal{A}(t) \equiv \begin{pmatrix} A(t) & -P(t) \\ -Q(t) & -A^*(t) \end{pmatrix}$$

with $P(t) \equiv B(t)R^{-1}(t)B^*(t) - C(t)S^{-1}(t)C^*(t)$. Then under similar assumptions as in Theorem 9.2.3, one can prove the existence of a unique solution $K \in AC(I, M(n \times n))$ which is given by

$$K = YX^{-1}.$$

However this time one can not guarantee the positivity of the solution K. This requires additional assumptions. One can show that if P is positive then K is also positive. This follows from the fact that the Riccati equation can be rewritten in the form

$$\dot{K} + (A - PK)^*K + K(A - PK) + KPK + Q = 0, K(T) = M. \quad (9.7.108)$$

For positivity of P we need that the following inequality holds

$$B(t)R^{-1}(t)B^*(t) \geq C(t)S^{-1}(t)C^*(t) \ \forall t \in I.$$

In the case of time invariant systems with both control and the disturbance having the same dimension, that is, $m = d$, one has a very natural interpretation of the above result. The inequality $BR^{-1}B^* \geq CS^{-1}C^*$ is equivalent to

$$Ker B^* \subset Ker C^*, S^{-1}|_{Range C^*} \leq R^{-1}|_{Range B^*}.$$

This means that the weight given to disturbance must be relatively larger than the cost of control. In other words the control must be cheap enough so that it can be liberally used to combat the disturbance.

Remark. Choosing $V \equiv (1/2)(Kx, x)$ as a candidate for the Lyapunov function, the reader can easily verify that

$$\dot{V} = -(1/2)\{(Qx, x) + (PKx, Kx)\}$$

where $P \equiv BR^{-1}B^* - CS^{-1}C^*$. Again if Q admits a factorization $Q = D^*D$ and the pair $\{A, D\}$ is observable and $P \geq 0$, one can prove the stability of the closed loop system. The crucial problem is with the question of positivity of P. If the designer has no choice on the system matrices $\{B, C\}$, it is possible that there exists no pair of matices $\{R, S\}$ for which $P \geq 0$. For more on LQR problems without constraints see the detailed presentation in [31] by Dorato, Abdallah and Cerrone .

9.8 Structurally Perturbed Impulsive Systems

Here we consider the LQR problem for the general structurally perturbed impulsive system given by equation (2.7.138) or (2.7.156) which was studied in Chapter 2, Section 2.7. We consider regular (measurable functions) controls. In this case the system is given by

$$dx(t) = A_0(t)x(t)dt + A_1(dt)x(t) + B(t)u(t)dt, t \in I. \quad (9.8.109)$$

Recall (see Theorem 2.7.2, equation (2.7.140)) that by this equation we mean the integral equation

$$x(t) = x_0 + \int_0^t A_0(s)x(s)ds + \int_0^t A_1(ds)x(s-) + \int_0^t B(s)u(s)ds$$

or equivalently the integral equation

$$x(t) = \Phi_0(t,0)x_0 + \int_0^t \Phi_0(t,s)A_1(ds)x(s-) + \int_0^t \Phi_0(t,s)B(s)u(s)ds,\ t \geq 0,$$

where Φ_0 denotes the transition operator corresponding to the matrix $A_0(\cdot)$. The cost functional is given by the quadratic functional

$$J(u) \equiv (1/2)\int_I \{(Q(t)x,x)+(R(t)u,u)\}dt+(1/2)(Mx(T),x(T)). \quad (9.8.110)$$

The problem is to find a control that minimizes this functional.

Remark. We have seen in Chapter 2 (Section 2.7) that the solutions of impulsive systems may not be continuous. They are generally elements of the Banach $B(I,R^n)$, the space of bounded measurable functions with values in R^n and furnished with the topology of sup norm. The running cost is well defined for measurable solutions. In case the operator (matrix) valued measure A_1 has an atom at $\{T\}$, the terminal state is given by $x(T) = x(T-) + A_1(\{T\})x(T-)$ where $x(T-)$ is the left limit of $x(t)$.

We prove the following necessary conditions of optimality. Let $U \subset R^d$ be a closed convex set and let

$$\mathcal{U}_{ad} \equiv \{u \in L_\infty(I,R^d) : u(t) \in U \ \ a.e. \}$$

denote the class of admissible controls.

Theorem 9.8.1. Suppose $\{A_0, A_1\}$ satisfy the assumptions of Theorem 2.7.2 ; $B \in L_1(I, M(n \times d))$, $Q(t) \in M(n \times n)$ is a real symmetric positive semidefinite matrix valued function, M is also symmetric positive semidefinite and $R(t) \in M(d \times d)$ is a positive definite symmetric matrix valued function. Then the necessary conditions for optimality of the pair $\{u^o, x^o\} \in \mathcal{U}_{ad} \times B(I,R^n)$ are that there exists a $\psi \in B(I,R^n)$ which together satisfy the following equations and inequalities:

$$(1) : dx^o = A_0(t)x^o dt + A_1(dt)x^o + B(t)u^o(t)dt, x^o(0+) = \xi \quad (9.8.111)$$

$$(2) : d\psi = -A_0^*(t)\psi dt - A_1^*(dt)\psi - Q(t)x^o dt, \psi(T) = Mx^o(T), \quad (9.8.112)$$

$$(3) : \int_I \Big\{(B^*(t)\psi(t) + R(t)u^o(t), v(t) - u^o(t))\Big\}dt \geq 0,\ \forall\ v \in \mathcal{U}_{ad} \quad (9.8.113)$$

Proof. Since the proof is very similar to that of the classical case we give a brief outline. Using variational principle, one can easily demonstrate that

$$dJ(u^o, v - u^o) = \int_I \Big\{(Q(t)x^o, y) + (R(t)u^o, v - u^o)\Big\}dt \\ + (Mx^o(T), y(T)) \geq 0, \quad \forall v \in \mathcal{U}_{ad}, \tag{9.8.114}$$

where y is the unique solution of the variational equation

$$dy(t) = A_0(t)y(t)dt + A_1(dt)y(t) + B(t)(v(t) - u^o(t))dt, t \in I. \tag{9.8.115}$$

Note that the map $B(v - u^o) \longrightarrow y$ is continuous and linear from $L_1(I, R^n)$ to $B(I, R^n)$. Thus the functional

$$B(v-u^o) \longrightarrow \ell(B(v-u^o)) \equiv \int_I \{(Q(t)x^o, y)\}dt + (Mx^o(T), y(T)) \tag{9.8.116}$$

is a bounded linear functional on $L_1(I, R^n)$. Hence by the classical representation theorem, there exists a $\psi \in L_\infty(I, R^n)$ such that

$$\ell(B(v - u^o)) = \int_I (\psi(t), B(t)(v(t) - u^o(t)))dt. \tag{9.8.117}$$

Using the identity (9.8.117) it follows from inequality (9.8.114) that

$$dJ(u^o, v - u^o) \\ = \int_I \Big\{(\psi(t), B(t)(v(t) - u^o(t))) + (R(t)u^o(t), v(t) - u^o(t))\Big\}dt \\ \geq 0, \ \forall \, v \in \mathcal{U}_{ad}. \tag{9.8.118}$$

From this follows the necessary condition (9.8.113). Now using the variational equation (9.8.115) into the above expression for dJ we obtain

$$dJ(u^o, v - u^o) = \int_I < \psi, dy - A_0(t)y(t)dt - A_1(dt)y(t) > \\ + \int_I (R(t)u^o, v - u^o)dt. \tag{9.8.119}$$

Integration by parts yields

$$\int_I < \psi, dy - A_0(t)y(t)dt - A_1(dt)y(t) > \\ = (y(T), \psi(T)) - \int_0^T < d\psi + A_0^*(t)\psi dt + A_1^*(dt)\psi, y > . \tag{9.8.120}$$

Using this in the expression (9.8.119) and comparing with the expression for dJ given in (9.8.114) one obtains the adjoint equation

$$d\psi = -A_0^*(t)\psi dt - A_1^*(dt)\psi - Qx^o dt, \psi(T) = Mx^o(T). \qquad (9.8.121)$$

The adjoint equation is entirely similar to the state equation (9.8.109) only with time running backward. Thus it follows from existence (and uniqueness) result of Theorem 2.7.2 that $\psi \in B(I, R^n) \cap BV(I, R^n)$. This concludes the derivation of all the necessary conditions of optimality. □

Corollary 9.8.2. Suppose the assumptions of Theorem 9.8.1 hold with the exception that $U = R^d$. Then the optimal control is given by

$$u^o(t) = -R^{-1}(t)B^*(t)\psi(t) = -R^{-1}(t)B^*(t)K(t)x^o(t), t \in I,$$

where K satisfies the impulsive differential Riccati equation

$$\begin{aligned} &dK + (KA_0(t) + A_0^*(t)K)dt + (KA_1(dt) + A_1^*(dt)K) \\ &\quad +Q(t) - KB(t)R^{-1}(t)B^*(t)K = 0 \\ &K(T) = M. \end{aligned} \qquad (9.8.122)$$

Remark. Note that if the operator (matrix) valued measure $A_1(\cdot)$ is absolutely continuous with respect to Lebesgue measure, we obtain the standard Riccati equation.

Remark. The reader is encouraged to develop an LQR theory for the linear system

$$dx = A(t)xdt + B(t)u(dt), u \in \mathcal{M}_c(I, R^d), \qquad (9.8.123)$$

driven by a vector measure u which may be atomic.

Here we consider the following regulator problem

$$J(u) = (1/2)\int_I (Q(t)x, x)dt + \varphi(u) \longrightarrow \inf., \qquad (9.8.124)$$

where the infimum is taken over the class of admissible controls $\mathcal{U}_{ad} \subset \mathcal{M}_c(I, R^d)$. It is not difficult to derive the following necessary conditions of optimality. We leave this as an exercise for the reader.

Theorem 9.8.3. Consider the system (9.8.123) starting from $x(0) = x_o$ and suppose A is locally integrable and B is continuous and bounded on the interval I and $\mathcal{U}_{ad}$ is a closed convex subset of $\mathcal{M}_c(I, R^d)$ and φ is subdifferentiable. In order for the pair $(x^o, u^o) \in B(I, R^n) \times \mathcal{U}_{ad}$ to be optimal for the problem (9.8.123) and (9.8.124) it is necessary that there exists a $\psi \in B(I, R^n)$ satisfying the following necessary conditions:

$$(1): \quad \int_I < B^*(t)\psi(t) + z^o(t), (u - u^o)(dt) > \; \geq 0 \; \forall \; u \in \mathcal{U}_{ad}$$

$$\text{and } \forall \; z^o \in \partial\varphi(u^o) \tag{9.8.125}$$

$$(2): \quad dx^0 = A(t)x^o dt + B(t)u^o(dt), x(0) = x_0 \tag{9.8.126}$$

$$(3): \quad \dot{\psi} = -A^*(t)\psi - Qx^o, \psi(T) = 0. \tag{9.8.127}$$

Remark. It is interesting to observe that ψ actually belongs to the smaller space $AC(I, R^n)$. See also exercise (Q8).

An Example. A simple example of a cost functional φ appearing in equation (9.8.124) is given by

$$\varphi(u) \equiv (1/2) \sum_{i=1}^{\infty} (< \zeta_i, u >)^2,$$

where $\{\zeta_i\}$ is dense in the unit ball $B_1(C(I, R^d))$. In case $\mathcal{U}_{ad} = \mathcal{M}_c(I, R^d)$, the necessary condition for u^o to be optimal is that

$$B^*(t)\psi(t) + (Lu^o)(t) = 0, t \in I,$$

where the operator L is given by

$$(Lu)(t) \equiv \sum_{i=1}^{\infty} \Big(\int_I < \zeta_i(s), u(ds) > \Big) \zeta_i(t), t \in I.$$

Clearly if image of B^* is contained in the image of L, that is, $Im(B^*) \subset Im(L)$, then

$$u^o = -L^{-1} B^* \psi.$$

Using this expression one can formally derive the differential Riccati equation

$$dK + (KA + A^*K)dt + Qdt - KBL^{-1}B^*K = 0, K(T) = 0.$$

The last term in this equation containing L^{-1} is a measure and L is a non-local operator. Hence this is a functional differential equation on the space $BV(I, M(n \times n))$. This problem is untouched in the literature.

Remark. An interesting open problem is to develop a theory of linear quadratic Gaussian regulator for structurally perturbed systems of the form

$$dx = A_0(t)xdt + A_1(dt)x + B(t)udt + C(t)dW(t), t \geq 0,$$

where W is the Wiener process (see Chapter 11).

9.9 Exercises

(Q1): Consider the problem 9.4.55 and suppose it is required to find a control that steers the system to the ball

$$E = B_r(x^*) \equiv \{x \in R^n : \| x - x^* \| \leq r\}$$

and that the problem has a solution. (a): Show that $\psi(T)$ nmust satisfy the boundary condition

$$(\psi(T), y) \leq (\psi(T), x(T)) \; \forall \; y \in B_r(x^*).$$

(b): Verify that if $x(T) \in int(B_r(x^*))$, then $\psi(T) = 0$. (c): If $x(T) \in \partial B_r(x^*)$, then $\psi(T)$ determines a supporting hyperplane passing through $x(T)$.

(Q2): Derive the algebraic Riccati equation (9.5.60) from the system equation (9.5.59) with control $u = -R^{-1}B^*\psi$ and the corresponding adjoint equation.

(Q3): Verify the expression (9.5.61) and (9.5.69).

(Q4): Verify the expressions (9.6.90)-(9.6.94).

(Q5): Using the saddle point expression (9.7.99), verify that the saddle point is attained by the expressions (9.7.100).

(Q6): Consider the structurally perturbed system with $A_1(dt) \equiv C_1\delta_{t_1}(dt) + C_2\delta_{t_2}dt)$ where C_1, C_2 are any two square matrices of the same dimension as A_0. Analyze and interpret the necessary conditions of optimality given by Theorem 9.8.1.

(Q7): Verify the first remark following the Corollary 9.8.2.

(Q8): Consider the necessary conditions of optimality given by Theorem 9.8.3. Verify that the functional $\varphi(u)$ given by $\varphi(u) \equiv (1/2) \| u \|_v^2$ where $\| u \|_v$ denotes the variation norm is admissible.

(Q9): Consider the optimality conditions of Theorem 9.8.3 and suppose $\mathcal{U}_{ad} = \mathcal{M}_c(I, R^d)$. Verify that in this case the necessary condition 9.8.122 is equivalent to $-B^*\psi \in \partial\varphi(u^o)$.

(Q10): Find sufficient conditions under which the linear operator L, given in the example following Theorem 9.8.3, is bounded from $\mathcal{M}_c(I, R^d)$ to $C(I, R^d)$. Verify that it is a linear positive and symmetric operator and that if $ImL \supseteq ImB^*$ then the optimal control is given by $u^o = -L^{-1}B^*\psi$.

9.10 Bibliographical Notes

The original contribution to linear quadratic regulator theory is due to Kalman. Because of its simplicity, this is the most popular technique used by engineers for design of feedback controls. A good account of the classical theory and its applications can be found in Dorato, Abdallah and Cerrone [31]. For related topics see also Brogan [23], Lasdon [59], Mosca [68], Aris [10]. One of the earlier extensions of this theory due to Wonham [96] covers linear quadratic regulator problems in the presence of additive white Gaussian noise. For lack of space we could not include this here. However it is useful to mention that inclusion of Gaussian white noise does not change the control law; it is only the optimal cost that has a different expression. In fact the (LQR) theory and linear quadratic regulator theory subject to Gaussian noise (LQG) have been extended by many workers in the field in many different directions. For infinite dimensional problems, involving partial differential equations and abstract differential equations on Banach spaces, see Lions [63], Balakrisnan [13], Ahmed and Teo [2], and Ahmed [4]. For LQG regulator problems on infinite dimensional spaces see [6]. In this chapter some nonstandard regulator problems have been included. These are: regulator theory without control cost ($R = 0$) Ahmed and Mouadeb [150], regulators with control constraints (bounded) Ahmed and Li [156]. Regulator theory for linear impulsive systems, and structurally perturbed impulsive systems are presented here for the first time.

Chapter 10

Time Optimal Control

10.1 Introduction

In many social, economic, management and physical problems it is desirable to achieve certain pre specified goals in the shortest time possible. This may apply to national economy (or any industry) where certain national goals (industrial production goal) are specified and the government (management) wishes to achieve that goal in the shortest possible time under given resource constraints. In other words mathematically the system dynamics is given, the initial and the target states are specified. The set of admissible controls or decision policies are given. The problem is to find a control policy from the admissible set that takes the system from the given initial state to the target state in minimum time. The reader can easily visualize that the minimum time problem is of great importance in production line (time is money).

10.2 General Problem

Let us consider the control system along with the specified initial state x_i and target state x_T and the class of admissible controls $\mathcal{U}_{ad}$:

$$\dot{x} = f(t, x, u), t \geq 0, \tag{10.2.1}$$

$$x_i = \xi \ \in R^n, x_T = \eta \in R^n \tag{10.2.2}$$

$$u \in \mathcal{U}_{ad} \equiv \{u(t), t \geq 0, u \text{ measurable and } u(t) \in U a.e\}. \tag{10.2.3}$$

where $U \subset R^m$ is generally a closed convex and possibly bounded set.
The very first question that one must settle is: does the problem have a solution. Clearly it involves the question of controllability. Given that the system is controllable, that is, there is a control policy $\tilde{u} \in \mathcal{U}_{ad}$ that transfers the system from the initial state ξ to the target state η in finite time, there

arises the next question if there is one that does it in minimum time. Naturally the third question is, if a solution does exist, how do we find it. The first question has been dealt with in Chapter 6 where the problems of controllability were discussed. Here we are concerned with the last two questions, that is the questions of time optimality and the associated computational issues.

10.3 Attainable and Reachable Sets

In words, given the time t and the initial state $x(0) = \xi$, the attainable set for any system like (10.2.1-10.2.3) is the set of states that can be attained by the system at time t, starting from the state ξ. We may denote this by

$$\mathcal{A}(t,\xi) \equiv \{z \in R^n : z = x(t,\xi,u), u \in \mathcal{U}_{ad}\} \tag{10.3.4}$$

where $x(t,\xi,u), t \geq 0$, denotes the solution of equation (10.2.1) corresponding to the initial state ξ and the admissible control u. Clearly this is a set valued map from $\{t \geq 0\} \times R^n$ to $2^{R^n} \setminus \emptyset$ where 2^{R^n} denotes the set of all subsets of R^n called the power set. For a fixed initial state $\xi \in R^n$ we may simplify the notation and use $\mathcal{A}(t)$ to denote the attainable set. In general then the time optimal control problem may be formulated as follows. Let the target be given by $\{z(t), t \geq 0\}$, a trajectory in R^n. The objective is to hit the target in minimum time. Define the set

$$\mathcal{T} \equiv \{t \geq 0 : d(z(t), \mathcal{A}(t)) = 0\}$$

where

$$d(y,K) \equiv \inf\{\| y - k \|, k \in K\}$$

denotes the distance of y from the set K. The optimal time is given by

$$\tau \equiv \inf\{\mathcal{T}\}$$

with the understanding that for any empty set $\inf(\emptyset) = \infty$. Clearly if the system is not controllable, $\mathcal{T} = \emptyset$, and if controllable the set is nonempty. Our problem is not only to find the infimum of the set $\mathcal{T}$ but also find the corresponding control.

10.4 Linear Systems

We consider linear systems of the form

$$\dot{x} = A(t)x + B(t)u, t \geq 0, \tag{10.4.5}$$

$$x(0) = x_0 \in R^n \tag{10.4.6}$$

$$u \in \mathcal{U}_{ad} \equiv \{u(t), t \geq 0, u \text{ measurable and } u(t) \in U a.e\}. \tag{10.4.7}$$

For simplicity we shall choose for U the hypercube

$$U \equiv \{v \in R^m : |v_i| \leq 1, i = 1, 2, \cdots m\}.$$

In the linear case the attainable set can be constructed explicitly. Assuming that Φ is the transition operator corresponding to the matrix valued function $A(t), t \geq 0$, we can express the attainable set as follows:

$$\mathcal{A}(t) \equiv \left\{\xi \in R^n : \xi = \Phi(t,0)x_0 + \int_0^t \Phi(t,s)B(s)u(s)ds, u \in \mathcal{U}_{ad}\right\}.$$

Let $\mathcal{R}(t)$ denote set valued function given by

$$\mathcal{R}(t) \equiv \left\{\xi \in R^n : \xi = \int_0^t X^{-1}(s)B(s)u(s)ds, \right.$$
$$\left. \equiv \int_0^t Y(s)u(s)ds, u \in \mathcal{U}_{ad}\right\}$$

where $\Phi(t,s) = X(t)X^{-1}(s)$, with X being the fundamental solution of the differential equation

$$\dot{X}(t) = A(t)X(t), X(0) = I,$$

and $Y(t) = X^{-1}(t)B(t)$. The set valued function $\mathcal{R}(t)$ is some times called the reachable set, since it is this part that contains controls and hence determines the set of attainable states. In fact, we have

$$\mathcal{A}(t) = X(t)x_0 + X(t)\mathcal{R}(t), t \geq 0,$$

and since X is nonsingular the reachable set is given by

$$\mathcal{R}(t) = x_0 + X^{-1}(t)\mathcal{A}(t).$$

So far we have not defined precisely what topology the control space carries. This is crucial in the study of existence of optimal controls and properties of attainable sets. However, note that the class of admissible control policies $\mathcal{U}_{ad}$ is a proper subset of the Banach space $L_\infty([0,\infty), R^m)$. This is a distinguished Banach space; it is the dual of $L_1([0,\infty), R^m)$ space. Consider the interval $I \equiv [0,T]$, $T < \infty$. We assume that $L_\infty(I, R^m)$ is equipped with the weak star topology, that is the topology induced by $L_1(I, R^m)$.

Thus if U is a bounded set then $\mathcal{U}_{ad}$ is a relatively weak star compact subset of $L_\infty(I, R^m)$. In other words every sequence from $\mathcal{U}_{ad}$ has a weak star convergent subsequence with the limit contained in $c\ell(\mathcal{U}_{ad})$. This essentially follows from Alaoglu's theorem [see appendix A.7.6]. Thus if U is a closed bounded convex set then $\mathcal{U}_{ad}$ is w^* closed and hence w^* compact.

Following result is fundamental for the study of time optimal control.

Lemma 10.4.1. Suppose the elements of the matrices $\{A(t), B(t)\}$ are locally integrable and the set U is a closed bounded convex subset of R^m. Then the attainable and the reachable sets $\{\mathcal{A}(t), \mathcal{R}(t)\}$ respectively are also closed bounded convex subsets of R^n.

Proof. First we prove that these sets are bounded. Let $\xi \in \mathcal{A}(t)$. Then by the very definition there is a control u from the set $\mathcal{U}_{ad}$ so that

$$\xi = x(t, x_0, u) = \Phi(t,0)x_0 + \int_0^t \Phi(t,s)B(s)u(s)ds, t < \infty. \tag{10.4.8}$$

The boundedness follows from this identity as shown below. Taking norms on either side and using the standard triangle inequality one finds that

$$\| \xi \| \leq \| \Phi(t,0) \| \| x_0 \| + \int_0^t \| \Phi(t,s) \| \| B(s) \| \| u(s) \| ds. \tag{10.4.9}$$

Since Φ is the transition operator corresponding to a locally integrable $A(\cdot)$, there exists a finite positive number a dependent on t such that

$$\sup\{\| \Phi(t,s) \|, 0 \leq s \leq t\} \leq a.$$

From boundedness of the set U it follows that there exists a finite number $b > 0$ such that

$$ess - sup\{\| u(s) \|, s \in [0,t]\} \leq b,$$

for all $u \in \mathcal{U}_{ad}$. Hence

$$\| \xi \| \leq a \| x_0 \| + ab \int_0^t \| B(s) \| ds. \tag{10.4.10}$$

The finiteness of the second term follows from local integrability of $B(\cdot)$. Since $\xi \in \mathcal{A}(t)$ is arbitrary, this proves that the attainable set is bounded for each finite t. For proof of convexity, take any two points $\xi_1, \xi_2 \in \mathcal{A}(t)$. We show that convex combination of these vectors also belongs to $\mathcal{A}(t)$. Let $u_1, u_2 \in \mathcal{U}_{ad}$ which take the system from the state x_0 to the states ξ_1, ξ_2 respectively. This means that

$$\xi_i = \Phi(t,0)x_0 + \int_0^t \Phi(t,s)B(s)u_i(s)ds, i = 1, 2.$$

Clearly for $\alpha \in [0.1]$ we have

$$(1-\alpha)\xi_1 + \alpha\xi_2 = \Phi(t,0)\xi + \int_0^t \Phi(t,s)B(s)[(1-\alpha)u_1(s) + \alpha u_2(s)]ds.$$

Since U is a closed convex set it is clear that

$$u(s) \equiv (1-\alpha)u_1(s) + \alpha u_2(s) \in U$$

for almost all $s \in [0,t]$. It is also obvious that u is measurable since the components are. Thus u so defined is also in the admissible class $\mathcal{U}_{ad}$ and consequently $\xi \in \mathcal{A}(t)$. This being true for every convex combination, we have proved the convexity of the attainable set. We show that it is closed. Take any sequence $\{\xi_n\} \in \mathcal{A}(t)$ and suppose that it converges to $\eta \in R^n$. We must verify that $\eta \in \mathcal{A}(t)$. Since $\xi_n \in \mathcal{A}(t)$ there exists $u_n \in \mathcal{U}_{ad}$ such that

$$\xi_n = \Phi(t,0)x_0 + \int_0^t \Phi(t,s)B(s)u_n(s)ds.$$

Since $\mathcal{U}_{ad}$ is a weak star compact subset of $L_\infty([0,t],R^m)$, there exists a subsequence of the sequence $\{u_n\}$, relabeled as the original sequence, and an element $u^o \in \mathcal{U}_{ad}$ such that

$$u_n \xrightarrow{w^*} u_o.$$

Using this fact we have

$$\begin{aligned}\lim_{n\to\infty} \xi_n &= \Phi(t,0)\xi + \lim_{n\to\infty} \int_0^t \Phi(t,s)B(s)u_n(s)ds \\ &= \Phi(t,0)\xi + \int_0^t \Phi(t,s)B(s)u_o(s)ds \equiv \eta. \qquad (10.4.11)\end{aligned}$$

Since $u_o \in \mathcal{U}_{ad}$, by definition $\eta \in \mathcal{A}(t)$. This proves that the attainable set is a closed bounded convex subset of R^n. Since the reachable set is just an affine transformation of the attainable set, it is also a closed bounded convex set. This completes the proof. □

Remark. Later we show that the reachable set is convex without requiring the convexity of U.

We must now prove existence. Let $z(t), t \geq 0$, denote the (moving) target, a continuous function with values in the state space R^n. Our problem is to find the minimum time τ and the associated control such that $z(\tau) \in \mathcal{A}(\tau)$. In other words τ is the first hitting time of the target, that is, $z(t) \notin \mathcal{A}(t)$

for $t < \tau$. Let x denote the solution corresponding to any control $u \in \mathcal{U}_{ad}$. Then

$$x(t) = X(t)x_0 + X(t)\int_0^t X^{-1}(s)B(s)u(s)ds, t \geq 0. \tag{10.4.12}$$

Since $X(t)$ is nonsingular this is equivalent to

$$\left(X^{-1}(t)x(t) - x_0\right) = \int_0^t X^{-1}(s)B(s)u(s)ds \equiv \int_0^t Y(s)u(s)ds, t \geq 0. \tag{10.4.13}$$

Clearly if there is a finite time $t^* > 0$, and a control $u^* \in \mathcal{U}_{ad}$ such that $x(t^*, x_0, u^*) = z(t^*)$, then

$$\tilde{z}(t^*) \equiv X^{-1}(t^*)z(t^*) - x_0 = \int_0^{t^*} Y(s)u^*(s)ds. \tag{10.4.14}$$

Thus the time optimal control problem can be restated in terms of the reachable set $\mathcal{R}(t), t \geq 0$, as follows. Find the smallest $t^o \geq 0$ such that $\tilde{z}(t^o) \in \mathcal{R}(t^o)$, that is, $\tilde{z}(t) \notin \mathcal{R}(t)$ for any $t < t^o$.

Theorem 10.4.2. If the target is reachable, there is a time optimal control u^o with the optimal time t^o.

Proof. Since the target is reachable, there is a time $\tau > 0$ such that $\tilde{z}(\tau) \in \mathcal{R}(\tau)$. By definition of the reachable set, there is a control $u^\tau \in \mathcal{U}_{ad}$ such that

$$\tilde{z}(\tau) = \int_0^\tau Y(s)u^\tau(s)ds.$$

If this is the smallest time, the proof is complete; if not, there exists a sequence of admissible controls $\{u_n\}$ and a sequence of times $\{t_n\} \leq \tau$ such that $\tilde{z}(t_n) \in \mathcal{R}(t_n)$ and

$$\tilde{z}(t_n) = \int_0^{t_n} Y(s)u_n(s)ds.$$

Since $u_n \in \mathcal{U}_{ad}$ and $\mathcal{U}_{ad}$ is a weak star compact subset of $L_\infty([0,\tau], R^m)$, there exists a subsequence $\{u_{n_k}\} \subset \{u_n\}$ and a control $u^o \in \mathcal{U}_{ad}$ and an associated decreasing sequence $t_{n_k} \leq \tau$ and a $t^o \leq \tau$ such that

$$u_{n_k} \xrightarrow{w^*} u^o \tag{10.4.15}$$

$$t_{n_k} \longrightarrow t^o. \tag{10.4.16}$$

Clearly, the sequence

$$\tilde{z}(t_{n_k}) = \int_0^{t_{n_k}} Y(s)u_{n_k}(s)ds \in \mathcal{R}(t_{n_k})$$

is a bounded sequence in R^n and hence there exists a subsequence of this sequence, relabeled as the original sequence, and an element $z^o \in R^n$ such that

$$\tilde{z}(t_{n_k}) \longrightarrow z^o.$$

We show that $z^o \in \mathcal{R}(t^o)$. Since $\{t_{n_k}\}$ is a nonincreasing sequence converging to t^o, we have

$$\tilde{z}(t_{n_k}) = \int_0^{t^o} Y(s)u_{n_k}(s)ds + \int_{t^o}^{t_{n_k}} Y(s)u_{n_k}(s)ds.$$

Noting that Y is locally integrable and $\{u_{n_k}\} \in L_\infty([0.\tau], R^m)$, it is clear that the second term converges to zero as $k \longrightarrow \infty$. Thus, letting $k \longrightarrow \infty$ in the above expression, we obtain

$$z^o = \int_0^{t^o} Y(s)u^o(s)ds.$$

Since $\tilde{z}$ is continuous $z^o = \tilde{z}(t^o)$. Thus we have proved the existence of a control u^o which is (time) optimal. This completes the proof. □

Lemma 10.4.3. Suppose $0 \in U$. Then the reachable set $\mathcal{R}(t), t \geq 0$, is an expanding set valued map (more precisely noncontracting).

Proof. Let $0 \leq t_1 \leq t_2$. Then by definition

$$\begin{aligned}\mathcal{R}(t_2) &\equiv \left\{\xi \in R^n : \xi = \int_0^{t_2} Y(s)u(s)ds, u \in \mathcal{U}_{ad}\right\} \\ &= \left\{\xi \in R^n : \xi = \int_0^{t_1} Y(s)u(s)ds + \int_{t_1}^{t_2} Y(s)u(s)ds, u \in \mathcal{U}_{ad}\right\}.\end{aligned}$$

Since elements of $L_\infty(R_0, R^m)$ are decomposable or closed under concatenation and $0 \in U$, it follows from the above expression that $\mathcal{R}(t_2) \supset \mathcal{R}(t_1)$. Since this is valid for arbitrary pair $t_1 \leq t_2$, the result follows. □

10.4.1 Bang-Bang Control

For simplicity take

$$U \equiv \{v \in R^m : |v_i| \leq 1, i = 1, 2, \cdots, m\},$$

with the boundary of U denoted by ∂U.

Lemma 10.4.4. Suppose the target $z(t), t \geq 0$, is a continuous function of time with values in R^n and define $z^o(t) \equiv X^{-1}(t)z(t) - x_0$. If $z^o(t) \in \partial\mathcal{R}(t)$

then the corresponding control must be bang-bang, that is $u^o(t) \in \partial U$ a.e $t \geq 0$ and there exists an $\eta(\neq 0) \in R^n$ such that

$$u^o(t) = \text{sign } (Y^*(t)\eta).$$

Proof. Since, for each $t \geq 0$, $\mathcal{R}(t)$ is a closed convex set and $z^o(t) \in \partial\mathcal{R}(t)$ there exists a hyperplane H_η, passing through $z^o(t)$, with unit normal $\eta \in R^n$ outwardly directed such that

$$(z^o(t), \eta) \geq (\xi, \eta) \;\; \forall \;\; \xi \in \mathcal{R}(t). \tag{10.4.17}$$

That is, $z^o(t)$ is a support point of $\mathcal{R}(t)$ and H_η is the supporting hyperplane. It follows from this inequality that

$$(z^o(t), \eta) \geq \left(\int_0^t Y(s)u(s)ds, \eta \right) \;\; \forall \;\; u \in \mathcal{U}_{ad}. \tag{10.4.18}$$

Since $z^o(t) \in \mathcal{R}(t)$, there exists a control $u^o \in \mathcal{U}_{ad}$ such that the expression (10.4.18) is equivalent to the following inequality

$$\left(\int_0^t Y(s)u^o(s)ds, \eta \right) \geq \left(\int_0^t Y(s)u(s)ds, \eta \right) \;\; \forall \;\; u \in \mathcal{U}_{ad}. \tag{10.4.19}$$

Let $\tau \in (0, t)$ and define $\sigma_\varepsilon \equiv [\tau - (\varepsilon/2, \tau + \varepsilon/2]$ for $\varepsilon > 0$ sufficiently small so that $\sigma_\varepsilon \subset [0, t]$. Clearly, as $\varepsilon \downarrow 0$, $\sigma_\varepsilon \longrightarrow \{\tau\}$. Choose

$$u(s) = \begin{cases} u^o(s), & \text{for } s \in [0, t) \setminus \sigma_\varepsilon; \\ v, & \text{for } s \in \sigma_\varepsilon, v \in U \;. \end{cases}$$

Substituting this in the expression (10.4.19) and dividing by ε and letting $\varepsilon \downarrow 0$, it follows from Lebesgue density argument that

$$(Y(\tau)u^o(\tau), \eta) \geq (Y(\tau)v, \eta) \;\; \forall \; v \in U. \tag{10.4.20}$$

Now consider the linear functional

$$\ell(v) = (Y(\tau)v, \eta) = (v, Y^*(\tau)\eta)$$

defined on U. Clearly this functional attains its unique maximum at

$$v^* = \text{sign}(Y^*(\tau)\eta) \in \partial U.$$

Since $u^o(\tau)$ maximizes the functional ℓ, it is clear that

$$u^o(\tau) = \text{sign}(Y^*(\tau)\eta).$$

This is true for almost all $\tau \in (0,t)$, and hence we have completed the proof. □

This is known as the LaSalle bang-bang principle after the name of its discoverer [58].

Lemma 10.4.5. If a control $\tilde{u}$ is given by $\tilde{u}(t) \equiv sign(Y^*(t)\nu), t \geq 0$, for some $\nu \neq 0$, then the corresponding state $\tilde{y}(t)$ given by

$$\tilde{y}(t) \equiv \int_0^t Y(s)\tilde{u}(s)ds$$

must be an element of the boundary of the reachable set, that is, $\tilde{y}(t) \in \partial\mathcal{R}(t)$.

Proof. It is clear that for the given control we have

$$\begin{aligned}(Y(s)\tilde{u}(s),\nu) &= (\tilde{u}(s), Y^*(s)\nu) = (sign(Y^*(s)\nu), Y^*(s)\nu) \\ &= |Y^*(s)\nu| \geq (u(s), Y^*(s)\nu), \forall\ s \geq 0, \quad \text{and}\ \forall\ u \in \mathcal{U}_{ad},\end{aligned}$$

where $|y| \equiv \sum_{i=1}^n |y_i|$. Integrating this over the interval $[0,t]$, we obtain

$$\int_0^t (Y(s)\tilde{u}(s),\nu)ds \geq \int_0^t (Y(s)u(s),\nu)ds, \ \forall\ u \in \mathcal{U}_{ad}.$$

This is equivalent to the following inequality:

$$(\tilde{y}(t),\nu) \geq (\xi,\nu)\ \forall\ \xi \in \mathcal{R}(t).$$

Since $\tilde{u} \in \mathcal{U}_{ad}$, $\tilde{y}(t) \in \mathcal{R}(t)$, and it follows from the above expression that it is actually on the boundary, that is, $\tilde{y}(t) \in \partial\mathcal{R}(t)$. This ends the proof. □

Theorem 10.4.6. (Bang-Bang Control) An element $z \in R^n$ is on the boundary of the reachable set $\mathcal{R}(t)$, if, and only if,

$$z = \int_0^t Y(s)u_b(s)ds$$

for some control of the form

$$u_b(t) \equiv \text{sign}(Y^*(t)\eta)$$

for a nonzero vector η. In other words, $z \in \partial\mathcal{R}(t)$ iff the control is bang-bang.

Proof. The proof follows directly from the Lemma 10.4.4 and Lemma 10.4.5. □

Remark. It is clear from the above result that the set U can be replaced by the nonconvex set ∂U, implying that the convexity of U is not essential.

Yet one can prove that the reachable set corresponding to controls taking values only from the boundary is also closed and convex. That is

$$\mathcal{R}(t) = \mathcal{R}_b(t) \equiv \left\{ \xi \in R^n : \xi = \int_0^t Y(s)u(s)ds, u \in L_\infty([0,t], \partial U) \right\}.$$

This is certainly amazing. In fact this is a special case and follows from a general theorem due to Aumann (Appendix, Theorem A.7.13).

Theorem 10.4.7. The time-optimal control is bang-bang.

Proof. Let τ^o be the optimal time and the target $z^o(\tau^o) \in \mathcal{R}(\tau^o)$. By definition, $z^o(t) \notin \mathcal{R}(t)$ for $t < \tau^o$. By virtue of the Lemmas 10.4.5-10.4.6, it suffices to show that $z^o(\tau^o) \in \partial\mathcal{R}(\tau^o)$. We prove this by establishing a contradiction. Suppose $z^* \equiv z^o(\tau^o) \in Int\mathcal{R}(\tau^o)$. By Lemma 10.4.3, $\mathcal{R}(t)$ is a continuous (in the Hausdorff metric) and expanding set valued function. Hence by continuity of the target z^o, there exists an $\varepsilon > 0$ and a $\tau < \tau^o$ such that

$$z^o(t) \in N_\varepsilon(z^*) \; \forall \; \tau \leq t < \tau^o,$$

where $N_\varepsilon(z^*) \subset \mathcal{R}(\tau^o)$ is the epsilon neighborhood of the point z^*. Since z^o is continuous, this means that there exists a $\tau^* < \tau^o$ such that

$$z^o(\tau^*) \in \mathcal{R}(\tau^*).$$

This contradicts the fact that τ^o is the minimum time. This completes the proof.

□

Now we present the necessary conditions of (time) optimality from which one can compute the optimal control. For the admissible controls $\mathcal{U}_{ad}$ we take measurable functions with values in U as described at the beginning of this subsection. Define the Hamiltonian

$$H(t, x, y, v) \equiv (A(t)x, y) + (B(t)v, y), (t, x, y, v) \in R_0 \times R^n \times R^n \times U.$$

Theorem 10.4.8. Let τ be the minimum time and u^o and x^o the optimal control and the corresponding solution trajectory. The optimality of the triple $\{u^o, x^o, \tau\}$ implies that there exists a $\psi \in AC([0,\tau], R^n)$ which satisfies the following equations and inequalities.

$$(i) : H(t, x^o(t), \psi(t), u^o(t)) \geq H(t, x^o(t), \psi(t), v), \forall \; v \in U, a.e. \quad (10.4.21)$$

$$(ii) : \dot{x}^o(t) = H_\psi = A(t)x^o(t) + B(t)u^o(t), a.a \; t \in I_\tau \equiv [0, \tau]. \quad (10.4.22)$$

$$(iii) : \dot{\psi}(t) = -H_x = -A^*(t)\psi(t), a.a \; t \in I_\tau, \quad (10.4.23)$$

including the boundary conditions,

$$x^o(0) = x_0, \psi(\tau) = \nu \equiv (X^{-1}(\tau))^*\eta, \text{ for some } \eta \neq 0. \tag{10.4.24}$$

Proof. Consider the system (10.4.5) with the initial state given by (10.4.6) and $z \in C(I_\tau, R^n)$ the target. Let $\{u^o, x^o, \tau\}$ be the optimal triple. By virtue of Theorem 10.4.7, the optimal control is necessarily bang-bang. Hence there exists an $\eta(\neq 0) \in R^n$ such that

$$u^o(t) = \text{sign } (Y^*(t)\eta), Y(t) \equiv X^{-1}(t)B(t), t \in I_\tau.$$

Let x^o denote the corresponding solution. Define

$$\psi(t) \equiv (X^{-1}(t))^*\eta, t \in I_\tau.$$

Then it follows from the properties of the fundamental solutions that

$$\dot{\psi} = -A^*(t)\psi(t), t \in I_\tau.$$

This is the adjoint equation. Let $u \in \mathcal{U}_{ad}$ be an arbitrary control. Then one can easily verify that

$$(B(t)u(t), \psi(t)) = (u(t), Y^*(t)\eta), \quad (B(t)u^o(t), \psi(t)) = (u^o(t), Y^*(t)\eta).$$

From the bang-bang principle we have

$$(u^o(t), Y^*(t)\eta) \geq (u(t), Y^*(t)\eta), t \in I_\tau$$

which is equivalent to

$$(u^o(t), B^*(t)\psi(t)) \geq (u(t), B^*(t)\psi(t)) \; \forall u \in \mathcal{U}_{ad}, t \in I_\tau.$$

Thus it follows from the definition of the Hamiltonian and bang-bang principle that

$$H(t, x^o(t), \psi(t), u^o(t)) \geq H(t, x^o(t), \psi(t), v) \; \forall \; v \in U, a.e$$

where ψ is the solution of the adjoint equation satisfying the boundary condition $\psi(\tau) = \nu \equiv (X^{-1}(\tau))^*\eta, \eta \neq 0$. This completes the proof of all the necessary conditions (1)-(iii). □

Interpretation of the Bang-Bang Principle. Defining

$$y(t) \equiv \int_0^t Y(s)u(s)ds, y(0) = 0,$$

observe that

$$\dot{y}(t) \equiv \dot{y}(t,u) = Y(t)u(t), t \in I_\tau.$$

Since τ is the optimal time and the associated control is bang-bang, there exists an $\eta \neq 0$ such that the optimal control is given by $u^o(t) \equiv \text{sign}(Y^*(t)\eta)$, $0 \leq t \leq \tau$, and it follows from the necessary condition (i) that

$$(\dot{y}(t,u^o),\eta) = (u^o(t), Y^*(t)\eta) \geq (u(t), Y^*(t)\eta) = (\dot{y}(t,u),\eta)$$

for all $u \in \mathcal{U}_{ad}$. This inequality says that to capture the target in minimum time one must maximize the speed of the pursuer towards the target at every instant of time. This can be done by use of maximum available resources (the boundary values of the control constraint set). Intuitively this is obvious.

10.5 Linear Impulsive Systems

Consider the system with impulsive control (6.6.90) as studied in section 6.6 of chapter 6 repeated here for convenience,

$$dx = A(t)xdt + B(t)u(dt), x(0) = x_0, t \geq 0. \tag{10.5.25}$$

Let $U \subset R^d$ be a closed convex bounded set containing 0 and consider the admissible controls given by the family

$$\mathcal{U}_\mu \equiv \mathcal{U}_\mu(U) \equiv \left\{ u \in \mathcal{M}^{\ell oc}(R_0, R^d) : u : \mathcal{B}_0 \longrightarrow U, \lim_{\mu(\sigma)\to 0} u(\sigma) = 0 \right\}, \tag{10.5.26}$$

where μ is any countably additive bounded positive measure defined on $\mathcal{B}_0$ having bounded variation on bounded sets. Recall that $\mathcal{M}^{\ell oc}(R_0, R^d)$ denotes the linear space of countably additive bounded vector measures with values in R^d having bounded total variation on bounded intervals. We denote by $\mathcal{R}_\mu(t)$ the reachable set. Again we can prove existence of time optimal controls given that the target is reachable.

Theorem 10.5.11. Suppose the elements of A are locally integrable and those of B are continuous and bounded on bounded intervals and let $\mathcal{U}_\mu(U)$ denote the class of admissible controls. If the system (10.5.25) is controllable then there is a time optimal control.

Proof. The reader can verify that the reachable set $\mathcal{R}_\mu(t)$ is a closed convex bounded subset of R^n for each $t \geq 0$, and that as a function of t it is a measurable set valued map. Hence the proof is quite similar to that of Theorem 10.4.2. The important point to note here is that the admissible controls $\mathcal{U}_\mu(U)$, restricted to bounded intervals, is a weakly compact subset of $\mathcal{M}^{\ell oc}(R_0, R^d)$. This is a very special case of the celebrated Bartle-Dunford-Schwartz theorem [Appendix, Theorem A.7.11] for relative weak compactness of any family of vector measures taking values in abstract Banach spaces. If the system is controllable, there exists a finite time t^* such that $\tilde{z}(t^*) \in \mathcal{R}_\mu(t^*)$. If this is the only time there is nothing to prove. If not let $\{t_n\}$ be a sequence for which $\tilde{z}(t_n) \in \mathcal{R}_\mu(t_n)$. Without loss of generality, as in Theorem 10.4.2, we may assume that $\{t_n\}$ is a decreasing sequence converging to τ and $\{u_n\}$ a corresponding sequence of admissible controls so that

$$\tilde{z}(t_n) = \int_0^{t_n} Y(s)u_n(ds). \tag{10.5.27}$$

By virtue of weak compactness of the admissible controls, there exists a subsequence, relabeled as the original sequence, and an admissible control u^o such that

$$u_n \xrightarrow{w} u^o.$$

Since the admissible controls have bounded variation over bounded intervals and Y is continuous and bounded, the sequence $\tilde{z}(t_n)$ is bounded in R^n and hence has a subsequence, relabeled as the original sequence, and an element $z^* \in R^n$ such that

$$\tilde{z}(t_n) \longrightarrow z^*.$$

Writing equation (10.5.27) along the subsequence, relabeled as n, we have

$$\tilde{z}(t_n) = \int_0^{\tau} Y(s)u_n(ds) + \int_{\tau}^{t_n} Y(s)u_n(ds). \tag{10.5.28}$$

Letting $n \to \infty$ in the above expression we have

$$z^* = \int_0^{\tau} Y(s)u^o(ds) + Y(\tau)u^o(\{\tau\}) \equiv \int_0^{\tau+} Y(s)u^o(ds). \tag{10.5.29}$$

If τ is a point of continuity of the target $\tilde{z}$, there is no jump and $u^o(\{\tau\}) = 0$. This proves the existence. □

Remark. Let $U \subset R^d$ be a closed convex bounded (so compact) set. Consider any bounded interval $I \equiv [0, T]$ and let $\mathcal{U}_{ad}$ denote the class of measurable functions on I taking values from U and $\mathcal{U}_\mu$ the class of countably additive vector measures as described by (10.5.26) but restricted to the interval I having bounded total variation on I. If $\{\mu, \nu\}$ are any two countably additive positive measures and $\mu \cong \nu$, then $\mathcal{R}_\mu(t) = \mathcal{R}_\nu(t), t \in I$. Clearly, if the Lebesgue measure, denoted by λ, is equivalent to the measure μ, then $\mathcal{R}(t) = \mathcal{R}_\mu(t)$ for all $t \geq 0$. In general, $\mathcal{R}_\mu(t) \neq \mathcal{R}_\nu(t)$ for distinct pair $\{\mu, \nu\}$.

10.6 Nonlinear Systems Including an Algorithm

Here we consider time optimal control for nonlinear systems. We prove existence of optimal controls and also present a penalty technique for computing the optimal controls. Consider the system

$$\begin{aligned} \dot{x}(t) &= f(t, x(t), u(t)), t \geq 0 \\ x(0) &= x_0 \end{aligned} \tag{10.6.30}$$

where $u \in \mathcal{U}_{ad}$ with $\mathcal{U}_{ad}$ denoting the class of measurable functions on $R_0 \equiv [0, \infty)$ with values in the set U. Let x^u denote the unique solution of the system equation (10.6.30) corresponding to the control u. Let

$$\mathcal{A}(t) \equiv \{\xi \in R^n : \xi = x^u(t), \text{ for some } u \in \mathcal{U}_{ad}\} \tag{10.6.31}$$

denote its attainable set at time $t \in R_0$. Under fairly general assumptions, we can prove some important geometric and topological properties of the attainable set $\mathcal{A}(t), t \in R_0$. Before we can prove such results we need the following definition. Let $\mathcal{P}(R^n)$ denote the power set (class of all subsets) of R^n and $c(R^n), (k(R^n)) \subset \mathcal{P}(R^n)$ the class of nonempty closed (compact) subsets of R^n. For any $C \in k(R^n)$ define

$$d(z, C) \equiv \inf\{\| z - a \|, a \in C\}.$$

Note that this is well defined since the distance function is continuous and the set C is compact. Clearly $d(z, C) = 0$ if and only if $z \in C$. Now we recall the following notion of distance between sets. For any $C, D \in k(R^n)$, define

$$d_H(C, D) \equiv \max\left\{\sup_{z \in D} d(z, C), \sup_{z \in C} d(D, z)\right\}. \tag{10.6.32}$$

This is known as the Hausdorff distance . It satisfies all the properties required of a metric. Note that $d_H(C, D) = 0$ if and only if C coincides with the set D. In general we have the following result.

Lemma 10.6.12. The spaces $\{(c(R^n), d_H), (k(R^n), d_H)\}$, furnished with the Hausdorff distance d_H, are complete metric spaces.

In the following result we study some basic topological properties of the attainable set $\mathcal{A}(t), t \geq 0$, which are useful in proving existence of time optimal controls and other useful optimization problems.

Lemma 10.6.13. Suppose (i): $f : R_0 \times R^n \times U \longrightarrow R^n$ is Lebesgue measurable in the first variable, locally Lipschitz in the second and continuous in the third argument and there exists an $h \in L_1^{\ell oc}(R_0), h \geq 0$, possibly dependent on U, such that

$$\| f(t,x,u) \| \leq h(t)\{1+ \| x \|\}, t \in R_0,$$

uniformly with respect to $u \in U$. (ii): the set U is compact and the velocity field or the contingent function $F(t,x) \equiv f(t,x,U)$ is convex with closed values for all $(t,x) \in R_0 \times R^n$. Then for each $t \in R_0$, the attainable set $\mathcal{A}(t)$ is compact and, considered as a multi function, $t(\ni R_0) \longrightarrow \mathcal{A}(t)$ is continuous in the Hausdorff metric.

Proof. Under the assumption (i), for each fixed control policy $u \in \mathcal{U}_{ad}$, it follows from standard existence results given in Chapter 3, that the system (10.6.30) has a unique solution $\xi^u \in AC(R_0, R^n)$ and it follows from Gronwall Lemma that

$$(1+ \| \xi^u(t) \|) \leq (1+ \| x_0 \|) \exp\left\{\int_0^t h(s)ds\right\}, \ \forall u \in \mathcal{U}_{ad}, \ t \in R_0. \quad (10.6.33)$$

Thus for any finite interval $I \equiv [0,T]$, the solution set, denoted by

$$\mathcal{X} \equiv \{\xi^u \in C(I,R^n), \xi^u(0) = x_0, , u \in \mathcal{U}_{ad}\},$$

is a bounded subset of $C(I,R^n)$. Hence, for each $t \in I$, $\mathcal{A}(t)$ is a bounded subset of R^n and there exists a positive number b such that $\| \xi^u(t) \| \leq b$ for all $u \in \mathcal{U}_{ad}$ and for all $t \in I$. It suffices to show that $\mathcal{A}(t)$ is closed. But this will follow as an obvious corollary if we prove that $\mathcal{X}$ is compact. For $\theta \geq 0$ and $t \in I$, we have

$$\| \xi^u(t+\theta) - \xi^u(t) \| \leq \int_t^{t+\theta} \| f(s,\xi^u(s),u(s)) \| ds \leq (1+b) \int_t^{t+\theta} h(s)ds.$$

It follows from this inequality that the set $\mathcal{X}$ is an equicontinuous subset of $C(I,R^n)$ and hence, by Ascoli-Arzela theorem, it is a relatively compact subset of $C(I,R^n)$. For compactness, we must show that it is closed. Let $\{\xi_n\} \subset \mathcal{X}$ and suppose $\xi_n \longrightarrow \xi_0$. We show that $\xi_0 \in \mathcal{X}$. Since ξ_n is a solution, there exists a control $u_n \in \mathcal{U}_{ad}$ such that

$$\xi_n(t) = x_0 + \int_0^t f(s,\xi_n(s),u_n(s))ds, t \in I.$$

Define $g_n(t) \equiv f(t,\xi_n(t),u_n(t)), t \in I$. Since the set $\mathcal{X}$ is bounded with bound b, it follows from the growth assumption on f that $\| g_n(t) \| \leq h(t)(1+b)$. Thus the sequence $g_n \in L_1(I,R^n)$ and by definition, $g_n(t) \in F(t,\xi_n(t))$.

It is also clear from this bound that $\{g_n\}$ is uniformly integrable in the sense that

$$\lim_{\lambda(\sigma)\to 0} \int_\sigma g_n(t)dt \longrightarrow 0,$$

with λ denoting the Lebesgue measure. Thus by Dunford-Pettis theorem (see appendix A.7.12) the set $\{g_n\}$ is relatively weakly sequentially compact and hence there exists a subsequence, relabeled as the original sequence, and an element $g_0 \in L_1(I, R^n)$ such that $g_n \xrightarrow{w} g_0$. In summary we have $g_n(t) \in F(t, \xi_n(t)), t \in I$ and that

$$\begin{aligned} \xi_n &\xrightarrow{u} \xi_0 \quad \text{in} \quad C(I, R^n) \\ g_n &\xrightarrow{w} g_0 \quad \text{in} \quad L_1(I, R^n). \end{aligned}$$

Since $x \to F(t, x)$ is upper semi continuous with closed convex values, by virtue of lower closure theorem [see Appendix Theorem A.7.10], it follows from the above results that

$$g_0(t) \in F(t, \xi_0(t)) \;\; a.e \;\; t \in I.$$

Define the set valued function

$$\Gamma(t) \equiv \{v \in U : g_0(t) = f(t, \xi_0(t), v)\}, \;\; a.e.$$

Since f is measurable in t and continuous in the state and control variables, the multi function Γ is measurable with closed values. Hence by the theory of measurable selections [see Appendix, Theorem A.7.9], there exists a measurable selection $u_0(t) \in \Gamma(t)$ $a.e.$ Clearly $u_0 \in \mathcal{U}_{ad}$. This implies that

$$g_0(t) = f(t, \xi_0(t, u_0(t)) \;\; a.e,$$

and that

$$\xi_0(t) = x_0 + \int_0^t f(s, \xi_0(s), u_0(s)) \, ds, \; t \in I.$$

This proves that $\xi_0 \in \mathcal{X}$. Thus we have proved that the set of attainable trajectories $\mathcal{X}$ is compact and hence any t section of it is a compact set in R^n and consequently the attainable set $\mathcal{A}(t)$ is compact for each $t \in I$. For continuity of the map $t \longrightarrow \mathcal{A}(t)$, take any pair of numbers $0 \leq t_1 \leq t_2 < \infty$. Note that for any admissible control u, we have

$$\| \xi^u(t_2) - \xi^u(t_1) \| \leq \int_{t_1}^{t_2} h(s)\{1 + \| \xi^u(s) \|\} ds.$$

This is true for all $u \in \mathcal{U}_{ad}$. Since the solution set $\mathcal{X}$ is bounded there exists a finite positive number $\beta = (1 + b)$ such that

$$d_H(\mathcal{A}(t_1), \mathcal{A}(t_2)) \leq \beta \int_{t_1}^{t_2} h(s)ds. \tag{10.6.34}$$

Since h is nonnegative and locally integrable, continuity follows from this inequality. This completes the proof. □

Since time optimal control problems are related to terminal control problems, we consider the terminal control problem. Given the system

$$\dot{x} = f(t, x, u), x(0) = x_0, t \geq 0$$

find a control that minimizes the terminal cost $\Psi(x(T))$. This problem has a solution as stated in the following theorem.

Theorem 10.6.14. Consider the terminal control problem as stated above and suppose f satisfy the assumptions of Lemma 10.6.13 and the function $\Psi : R^n \longrightarrow R$ is lower semi continuous and bounded away from $-\infty$. Then there exists an optimal control minimizing the functional Ψ.

Proof. By Lemma 10.6.13, the attainable set $\mathcal{A}(T)$ is compact for each $T \geq 0$. Let $\xi_n \in \mathcal{A}(T)$ be a minimizing sequence, that is,

$$\lim_{n \to \infty} \Psi(\xi_n) = \inf\{\Psi(z), z \in \mathcal{A}(T)\} \equiv M.$$

Since Ψ is bounded away from $-\infty$, $M > -\infty$. By virtue of compactness of $\mathcal{A}(T)$, there exist a subsequence of the sequence $\{\xi_n\}$, relabeled for convenience as the original sequence, and an element $\xi_o \in \mathcal{A}(T)$ such that $\xi_n \longrightarrow \xi_o$. It follows from lower semi continuity of Ψ, that $\Psi(\xi_o) \leq \underline{\lim}_{n\to\infty} \Psi(\xi_n)$. Thus we have

$$\Psi(\xi_o) \leq \underline{\lim}_{n\to\infty} \Psi(\xi_n) \leq \lim_{n\to\infty} \Psi(\xi_n) = M.$$

Since $\xi_o \in \mathcal{A}(T)$ and M is the infimum of Ψ on $\mathcal{A}(T)$, $M \leq \Psi(\xi_o)$ and hence it follows from the above inequality that $\Psi(\xi_o) = M$. From this we conclude that there exists an admissible control $u_o \in \mathcal{U}_{ad}$ and a corresponding trajectory x_o such that $x_o(T) = \xi_o$. This completes the proof. □

We use this result to study time optimal controls. Let $x^* \in R^n$ be a target which must be reached in minimum time. Define the function

$$\varphi(t) \equiv d(x^*, \mathcal{A}(t)), t \geq 0. \tag{10.6.35}$$

Since $\mathcal{A}(t)$ is compact for each $t \geq 0$, this function is well defined. From continuity of the map $t \longrightarrow \mathcal{A}(t)$, follows the continuity of φ. In fact one can verify that

$$|\varphi(t) - \varphi(s)| \leq d_H(\mathcal{A}(t), \mathcal{A}(s)), s, t \geq 0.$$

Clearly $\varphi(t) \geq 0$, and it is zero only if $x^* \in \mathcal{A}(t)$. Thus if $\varphi(t) > 0$ for all $t \geq 0$, the target is not reachable (system not controllable). But there exist controls that can drive the system closest to the target. Thus all the minima of the function $\varphi(t), t \geq 0$, are the time instants at which the target is nearest to the attainable set. If the system is controllable in finite time, the set

$$\{t \geq 0 : \varphi(t) = 0\}$$

is nonempty and hence by continuity of the function φ

$$\tau_m \equiv \inf\{t \geq 0 : \varphi(t) = 0\}$$

exists implying the existence of a time optimal control. For each finite $T > 0$, let $m_0(T)$ denote the minimum of φ over the interval $[0, T]$. Even though the function φ may not be monotone, the function $m_0(T), T \geq 0$, is monotone non-increasing. If the system is controllable in time $T > 0$, then $m_0(T) = 0$. Using the function φ and the minimum principle for a family of terminal control problems,

$$\begin{aligned} \dot{x} &= f(t, x, u), x(0) = x_0, t \in [0, \tau], J(\tau, u) \\ &\equiv (1/2) \parallel x^* - x(\tau, u) \parallel^2 \longrightarrow \inf., \tau \geq 0, \end{aligned}$$

an algorithm for computing time optimal controls was developed in [194]. The technique was successfully applied to some robotics and ecological problems. Unfortunately due to lack of space we can not include the details. For this the interested reader is referred to the original paper [194]. This technique is computationally intensive. In particular, if the attainable set is monotonically expanding and the system is controllable, computation time can be reduced. On the other hand if the attainable set is oscillatory computation time can be large.

Another more direct and interesting technique developed in [Teo et al.[89]] essentially introduces time as another control variable transforming the original problem into a sequence of penalized terminal control problems over a fixed time interval $[0, 1]$. We present here a constructive existence theorem from which the interested reader can easily develop a computer code for solving time optimal controls for nonlinear systems.

Theorem 10.6.15. Suppose the assumptions of Theorem 10.6.14 hold and that the system is controllable. Then the time optimal control problem has a solution which can be computed by solving a sequence of penalized terminal control problems.

Proof. Define $t = Ks, K > 0, s \in [0,1]$, and $y(s) \equiv x(Ks)$ and $v(s) \equiv u(Ks)$. Using this transformation, we have an equivalent representation of the system given by

$$\dot{y}(s) = Kf(Ks, y(s), v(s)), y(0) = x_0, s \in [0,1]. \tag{10.6.36}$$

Clearly, under the standard assumptions, for every choice of $v \in \mathcal{U}_{ad}$ and $K > 0$, equation (10.6.36) has a unique solution $y \in AC([0,1], R^n)$. For any given $K > 0$ and $v \in \mathcal{U}_{ad}$, we introduce the terminal control problem

$$J_\varepsilon(v, K) \equiv (1/2\varepsilon) \parallel y(v,K)(1) - x^* \parallel^2 + K \longrightarrow \inf, \tag{10.6.37}$$

where $y(v,K)(\cdot)$ denotes the solution of equation (10.6.36) corresponding to the parameter K and the control v. By Theorem 10.6.14, this problem has a solution. Let $(v_\varepsilon, K_\varepsilon)$ denote the pair at which the infimum is attained. Denote by $\mathcal{U}_0 \subset \mathcal{U}_{ad}$ the class of controls for which the system is controllable from x_0 to x^*. By our assumption, this set is nonempty. Then there exists a $v_0 \in \mathcal{U}_{ad}$ and a $K_0 > 0$ such that $y(v_0, K_0)(1) = x^*$ and

$$J_\varepsilon(v_0, K_0) \equiv (1/2\varepsilon) \parallel y(v_0, K_0)(1) - x^* \parallel^2 + K_0 = K_0. \tag{10.6.38}$$

In this case the transition time is given by $\tau_0 = K_0$. In view of this we have

$$J_\varepsilon(v_0, K_0) = K_0 \geq J_\varepsilon(v_\varepsilon, K_\varepsilon) = (1/2\varepsilon) \parallel y(v_\varepsilon, K_\varepsilon)(1) - x^* \parallel^2 + K_\varepsilon. \tag{10.6.39}$$

Clearly it follows from this inequality that for any $\varepsilon > 0$,

$$(1/2\varepsilon) \parallel y(v_\varepsilon, K_\varepsilon)(1) - x^* \parallel^2 \leq K_0,$$
$$K_\varepsilon \leq K_0.$$

Thus one can choose a subsequence of the sequence, $\{\varepsilon_n\} \downarrow 0$, so that along the subsequence, relabeled as the original sequence,

$$y(v_{\varepsilon_n}, K_{\varepsilon_n})(1) \longrightarrow x^* \text{ and } K_{\varepsilon_n} \longrightarrow K^*.$$

Since, for any fixed $v \in \mathcal{U}_{ad}$, the cost functional is a monotone increasing function of K, the optimal transition time is $\tau^* = K^*$. This proves that, under the assumptions of Theorem 10.6.14, and controllability, the time optimal control problem has a solution. From this we conclude that the time optimal control can be computed by solving a sequence of fixed time terminal control problems as presented above. This completes the proof. □

Remark. Note that this result also holds for time varying target. Let $x^*(t), t \geq 0$, be a continuous R^n valued function. The problem is to hit this target in minimum time. In this case the functional 10.6.37 is replaced by

$$J_\varepsilon(v, K) \equiv (1/2\varepsilon) \parallel y(v, K)(1) - x^*(K) \parallel^2 + K \longrightarrow \inf. \tag{10.6.40}$$

For a set valued target $\mathcal{T}(t), t \geq 0$, which is continuous in the Hausdorff metric with values in $c(R^n)$, this functional is taken as

$$J_\varepsilon(v, K) \equiv (1/2\varepsilon) d(y(v, K)(1), \mathcal{T}(K)) + K \longrightarrow \inf, \tag{10.6.41}$$

where $d(z, \Gamma) = \inf\{\parallel z - \xi \parallel, \xi \in \Gamma\}$ for any closed set Γ.

Another technique based on Newton's method was developed in [197].

Remark. In general, computation of time optimal controls is very time consuming. Currently we know of no single technique that could be labeled as the best. Computation time depends very much on the particular system under consideration and the distance between initial and the target state.

10.7 Exercises

(Q1): Prove Lemma 10.6.12.

(Q2): Remark following Theorem 10.4.6 states that the reachable set $\mathcal{R}(t) = \mathcal{R}_b(t)$. The former is based on L_∞ functions with values from the closed convex set U while the later is based on L_∞ functions with values from the non convex set, the boundary ∂U of the set U. Though this is counter intuitive, it follows from Aumann's thoerem A.7.13. Provide an intuitive explanation.

(Q3): Consider the nonlinear system $\dot{x} = f(t, x, u), t \in I \equiv [0, T]$, and suppose that $f : I \times R^n \times R^d \longrightarrow R^n$ is continuous in all the variables satisfying $(x, f(t, x, u)) \leq b(1+ \parallel x \parallel^2)$ for all $(t, x) \in I \times R^n$ and all $u \in U(t)$ where $t \longrightarrow U(t)$ is a continuous set valued map with compact values in R^d. Suppose that the contingent set $F(t, x) \equiv f(t, x, U(t))$ is closed convex valued. Prove that the attainable set $\mathcal{A}(t)$ is compact for each $t \in I$ and that it is continuous in the Hausdorff metric.

(Q4): Consider the two dimensional system

$$\dot{x}_1 = (1 - x_2^2)u^2, x_1(0) = 0; \quad \dot{x}_2 = u, x_2(0) = 0$$

over the time horizon $I \equiv [0, 1]$ with controls being measurable functions satisfying $|u(t)| \leq 1$. Verify that the velocity field (or the contingent set) $F(t, x)$ is not convex. Show that the attainable set $\mathcal{A}(1)$ is not closed. Hints: Partition the time interval I into $2n$ subintervals, $I \equiv \bigcup_{j=o}^{2n-1} I_j$, all of equal length. Define $u^n(t) = +1$ for $t \in I_j$ for j odd and $u^n(t) = -1$ for $t \in I_j$, j even. Then note that $x_2(1, u^n) = 0$ and $x_1(1, u^n) = 1 - \int_0^1 x_2(t, u^n)dt < 1$. From this verify that the point $(1, 0)$ is a limit point of $\mathcal{A}(1)$ but $(1, 0) \notin \mathcal{A}(1)$.
(Q5): Justify the Remark following Theorem 10.5.11.
(Q6): Find sufficient conditions on f for which the attainable set $\mathcal{A}(t), t \geq 0$, is monotone, expanding or contracting.
(Q7): Construct an example of a nonlinear system for which the attainable set $\mathcal{A}(t), t \geq 0$, is oscillatory.

10.8 Bibliographical Notes

The importance of time optimal controls can be appreciated by the fact that there is a book on this topic written 35 years ago by Hermes and La Salle [43]. For detailed study of bang-bang principle and time optimal control the reader is referred to this book. For more on this topic see [89] and [2]. A powerful computer code developed by Teo and his school is known as MISER [89, 175]. Interested readers may use MISER and the penalty technique suggested by Theorem 10.6.15 to solve time optimal control problems for nonlinear systems.

Chapter 11

Stochastic Systems with Applications

11.1 Introduction

In this chapter we consider stochastic systems. One of the overriding reasons for considering such systems is that many of the natural phenomenon that we observe can not be fully described by deterministic evolution or deterministic differential equations. This is either due to lack of complete understanding of the physics of nature and its workings or because nature is unpredictable in the first place and can only be described in probabilistic sense. It is well known to meteorologists that long term or at times even short term prediction of weather is difficult and some time their prediction turns out to be false. However, we are comfortable with overall results based on statistical analysis and predictions. In fact statistical analysis has served science and engineering so well that without this tool it would have been impossible to come to the stage it is in today. Information theory, statistical mechanics, Quantum mechanics, stochastic control, linear and nonlinear filtering are some of the outstanding examples of success brought about by this field of knowledge.

Generally Brownian motion [45], also known as Wiener process, and Poisson process (more generally counting process), are used as basic building blocks for construction of large classes of stochastic processes covering Markov and non Markovian processes. This is done by use of stochastic differential equations (SDE) driven by those basic processes. In the following sections we will discuss these briefly.

11.2 Stochastic Systems

We need some basic notations and terminologies which are considered standard in the field of stochastic processes. Let $(\Omega, \mathcal{F}, P)$ denote a complete probability space, where Ω represents the sample space, $\mathcal{F}$ the σ algebra (Borel algebra) of subsets of the set Ω and P the probability measure on the algebra $\mathcal{F}$. Let $\mathcal{F}_t, t \geq 0$, be an increasing family of complete subsigma algebras of the σ algebra $\mathcal{F}$. For any random variable (or equivalently $\mathcal{F}$ measurable function) X, we let $EX \equiv \int_\Omega X(\omega)P(d\omega)$ denote the expected value provided it exists. It does exist if $X \in L_1(\Omega, P)$. For a subsigma algebra $\mathcal{G} \subset \mathcal{F}$, the conditional expectation of X, relative to $\mathcal{G}$, is denoted by

$$E\{X|\mathcal{G}\} \equiv Y.$$

The random variable Y is $\mathcal{G}$ measurable and it is uniquely defined through Radon Nikodym derivative (RND) [3]. Conditional expectation plays a very important role in the study of stochastic integrals and martingales. We state here some very basic properties of conditional expectations which are intuitively obvious having rigorous proofs. Let $\mathcal{G}$ and $\mathcal{G}_1 \subset \mathcal{G}_2 \subset \mathcal{F}$ be any three (complete) subsigma algebras of the sigma algebra $\mathcal{F}$.
The conditional expectation is a linear operation in the sense that
(P1): For $\alpha_1, \alpha_2 \in R$, $E\{\alpha_1 X_1 + \alpha_2 X_2|\mathcal{G}\} = \alpha_1 E\{X_1|\mathcal{G}\} + \alpha_2 E\{X_2|\mathcal{G}\}$.
For any integrable $\mathcal{F}$ measurable random variable X we have
(P2): $E\Big\{E\{X|\mathcal{G}_1\}|\mathcal{G}_2\Big\} = E\Big\{E\{X|\mathcal{G}_2\}|\mathcal{G}_1\Big\} = E\{X|\mathcal{G}_1\}$.
If Z is a bounded $\mathcal{G}$ measurable random variable with $\mathcal{G} \subset \mathcal{F}$ then
(P3): $E\{ZX|\mathcal{G}\} = ZE\{X|\mathcal{G}\}$ which is a $\mathcal{G}$ measurable random variable.
If X is a random variable independent of the sigma algebra $\mathcal{G} \subset \mathcal{F}$ then
(P4): $E\{X|\mathcal{G}\} = E\{X\}$.
(P5): For any $\mathcal{F}$ measurable and integrable random variable Z, the process

$$Z_t \equiv E\{Z|\mathcal{F}_t\}, t \geq 0,$$

is an $\mathcal{F}_t$ martingale in the sense that for any $s \leq t < \infty$,

$$E\{Z_t|\mathcal{F}_s\} = Z_s.$$

Further notations will be introduced as and when required.

11.2.1 SDE Based on Brownian Motion

We have seen that a deterministic system is governed by a differential equation of the form

$$\dot{\xi} = F(t, \xi)$$

where the vector field F is a suitable map from $I \times R^n$ to R^n, $I \equiv [0,T]$. To start with, one may consider a stochastic system to be a perturbed version of this given by

$$\dot{\xi} = F(t,\xi) + G(t,\xi)\zeta(t), t \in I, \tag{11.2.1}$$

where $G : I \times R^n \longrightarrow M(n \times m)$ is a suitable map. The process ζ is an m-dimensional white noise and is given by the derivative of the wiener process $W(t), t \geq 0$. The power spectral density of this process is flat meaning infinite energy and therefore physically meaningless. It is well known that with probability one, Brownian motion is no where differentiable and that the path length (that is, the Brownian excursion), on any interval of positive length, is infinite with probability one. However, it has generalized (distributional) derivatives of all orders. So one may interpret the white noise as the generalized derivative of the Brownian motion.

A more logical and convenient description which admits rigorous mathematical interpretation is given by the integral equation

$$\xi(t) = \xi_0 + \int_0^t F(s,\xi(s))ds + \int_0^t G(s,\xi(s))dW(s), t \in I. \tag{11.2.2}$$

The second term involves Lebesgue integration and the third term involves Ito integration. This is explained as follows. Let $\mathcal{F}_t, t \geq 0$, denote the family of increasing sub sigma algebras, as introduced earlier, and let $W(t), t \geq 0$, be a standard $\mathcal{F}_t$ Brownian motion satisfying the following basic properties:

(B1): $P\{W(0) = 0\} = 1$,

(B2): $E\{W(t)|\mathcal{F}_s\} = W(s), s \leq t$, (martingale property);

(B3): $(W(t_{i+1}) - W(t_i)) \perp (W(t_{j+1}) - W(t_j))$ whenever $[t_i, t_{i+1}) \cap [t_j, t_{j+1}) = \emptyset$.

(B4): $E\{(W(t),\xi)(W(s),\eta) = (t \wedge s)\ (\xi,\eta)$.

For any $\eta \in R^m$, it follows from the property (B4) that

$$E\{(W(t) - W(s),\eta)^2\} = \parallel \eta \parallel^2 |t-s|.$$

The last term in the integral equation (11.2.2) is interpreted as the Ito integral (see Ahmed [2, 3]) thereby resolving the interpretational issue. Indeed, the stochastic integral

$$L(K) \equiv \int_I K(t)dW(t)$$

is first defined for square integrable $\mathcal{F}_t$ adapted simple random processes with values in the space of $n \times m$ matrices $M(n \times m)$, denoted by $\mathcal{S}_e$, and then extended to $\mathcal{F}_t$ adapted random processes from the class

$$L_2^e(I, M(n \times m)) \equiv \left\{K : E \int_I \| K(t) \|^2 \, dt < \infty\right\}.$$

To start with, let K be a simple random process in the sense that there exists a finite partition of I, $I = \cup I_i \cup \{0\}$, by disjoint intervals of the form $I_i \equiv (t_i, t_{i+1}]$ such that $K(t) = K_i$ for $t \in I_i$ and that K_i is $\mathcal{F}_{t_i}$ measurable and $E \| K_i \|^2 < \infty$. For such processes the integral is given by

$$L(K) = \sum_i K_i(W(t_{i+1}) - W(t_i)).$$

Since finite sums and products of $\mathcal{F}$ measurable functions are $\mathcal{F}$ measurable, $L(K)$ as defined above is $\mathcal{F}$ measurable. By use of the properties of the conditional expectations (P1)-(P5) and the properties of the Wiener process, the reader can easily verify that $E(L(K)) = 0$ and

$$E \| L(K) \|^2 = E \int_I \| K(t) \|^2 \, dt < \infty.$$

Thus L is a bounded linear operator from $\mathcal{S}_e$ to $L_2(\mathcal{F}, R^n)$. Its extension by continuity from $\mathcal{S}_e$ to $L_2^e(I, M(n \times m))$ is the Ito integral. Indeed, since the class of simple functions $\mathcal{S}_e$ is dense in $L_2^e(I, M(n \times m))$ [2], for any $K \in L_2^e(I, M(n \times m))$ there exists a sequence of $K_n \in \mathcal{S}_e$ such that $K_n \xrightarrow{s} K$ in $L_2^e(I, M)$, and

$$L(K_n) \xrightarrow{s} L(K), \quad \text{in} \quad L_2^e(\mathcal{F}, R^n) \quad \text{as } n \to \infty.$$

The limit $L(K)$ is called the Ito integral of K. Thus under certain assumptions on G as a function of $t \in I$ and $x \in R^n$, the stochastic integral of $K(\cdot) \equiv G(\cdot, \xi(\cdot))$ with respect to the Wiener process W appearing in the integral equation (11.2.2) is well defined. For more details on stochastic integrals see Ahmed [2, 3], Friedman [37] and Skorohod [84].

Symbolically we write the integral equation (11.2.2) as follows:

$$d\xi = F(t, \xi)dt + G(t, \xi)dW, t \geq 0, \tag{11.2.3}$$

which is known as stochastic differential equation or Ito equation after the name of the founder who laid the foundation of stochastic calculus. Given $\xi(\tau) = x \in R^n$, at time $\tau \geq 0$, it follows from (11.2.2) that

$$\xi(\tau + \Delta t) \approx x + F(\tau, x)\Delta t + G(\tau, x)\Delta W, \tag{11.2.4}$$

where $\Delta W = W(\tau + \Delta t) - W(\tau)$. Since ΔW is a Gaussian random vector with values in R^m, it is clear that $\xi(\tau + \Delta t)$ is also Gaussian random vector with values in R^n. Clearly $E\Delta W = 0$, and hence $E\xi(t+\Delta t) = x + F(\tau, x)\Delta t$. Thus given the state $\xi(\tau) = x$ at time τ, the (infinitesimal) mean direction of motion is given by the vector field $F(\tau, x)$, and the corresponding expected state (position), the system is found in, is given by $x + F(\tau, x)\Delta t$. Note that, given τ and x, this is linearly proportional to Δt. Because of the presence of the third term, which is a Gaussian random vector, the true state is expected to fluctuate around the position $x + F(\tau, x)\Delta t$ with an intensity determined by the trace of the covariance matrix $G^*(\tau, x)G(\tau, x)$. Indeed, for any $e \in R^n$, $(G(\tau, x)\Delta W, e)$ is a real valued Gaussian random variable with mean zero and variance

$$E(G\Delta W, e)^2 = E(G^*e, \Delta W)^2 = \| G^*(\tau, x)e \|^2 \Delta t. \qquad (11.2.5)$$

Thus the covariance matrix of the random vector $G(\tau, x)\Delta W$ is given by $\Delta t(GG^*)$ and it is important to observe that this is also linearly proportional to Δt. This fact is particularly important for a heuristic but transparent derivation of the Ito formula. Indeed, in view of equations (11.2.4) and (11.2.5), for any $\varphi \in C^{1,2}(I \times R^n)$ we have the approximation,

$$\varphi(t + \Delta t, \xi(t + \Delta t)) \approx \varphi(t + \Delta t, x + F(t, x)\Delta t + G(t, x)\Delta W).$$

By use of Taylor series and disregarding terms of order $o(\Delta t)$, such as $(\Delta t)(\Delta W), (\Delta t)^\kappa, \kappa > 1$, we have

$$\begin{aligned} d\varphi &= \varphi_t dt + (D\varphi, F)dt + (G^* D\varphi, dW) \\ &\quad + (1/2)\left\{P - \lim_{\Delta t \to dt} < D^2\varphi G\Delta W, G\Delta W >\right\}. \end{aligned}$$

Since W is a standard Brownian motion, the reader can easily verify that, for any $\varepsilon > 0$ and any component w_i of W,

$$\lim_{\Delta t \to 0} P\{|(\Delta w_i)^2 - \Delta t| > \varepsilon\} = 0.$$

Using this fact and some simple matrix manipulations we obtain

$$P - \lim_{\Delta t \to dt} \{< D^2\varphi G\Delta W, G\Delta W >\} = tr(D^2\varphi \; GG^*)dt.$$

Thus we have given a heuristic derivation of the famous Ito formula

$$d\varphi = \{\varphi_t + (D\varphi, F) + (1/2)tr(D^2\varphi GG^*)\}dt + (G^* D\varphi, dW). \qquad (11.2.6)$$

For rigorous derivation the reader may consult [64, 84, 87]. Returning to the expression (11.2.5), note that the trace is given by the sum of the eigen values,

$$tr(G^*(\tau,x)G(\tau,x)) = \sum \lambda_i(\tau,x) \geq 0. \tag{11.2.7}$$

The larger these eigen values are, the greater is the fluctuation. In finance this is known as volatility; the larger is its value the greater is the fluctuation of stock price and the greater is the risk for investors. In view of the expression (11.2.7), if $\lambda_k(\tau,x)$ is the largest eigen value with the corresponding eigen vector e_k one may expect maximum fluctuation in the direction e_k. As noted above $\xi(\tau+\Delta t)$ is a Gaussian random vector. If the matrix G^*G is nonsingular, the probability density of $\xi(\tau+\Delta t)$ is given by

$$g(m,Q,z) \equiv \frac{1}{(2\pi)^{n/2}\, detQ} \exp\{(1/2)(Q^{-1}(z-m),(z-m))\}$$

with the mean vector $m \equiv x + F(\tau,x)\Delta t$ and covariance matrix $Q \equiv G^*(\tau,x)G(\tau,x)\Delta t$.

In case $F(t,x) = A(t)x$ and $G(t,x) = G(t)$, the system (11.2.3) reduces to a linear stochastic differential equation

$$d\xi = A(t)\xi dt + G(t)dW(t), t \geq 0. \tag{11.2.8}$$

The solution for this equation is given by the variation of constants formula

$$\xi(t) = \Phi(t,0)\xi_0 + \int_0^t \Phi(t,s)G(s)dW(s), t \geq 0, \tag{11.2.9}$$

where Φ is the transition operator corresponding to the matrix $A(t), t \geq 0$, and ξ_0 is the initial state. If ξ_0 has a Gaussian distribution with finite second moment and $G(t), t \geq 0$, is a square integrable $M(n \times m)$ valued function, then $\xi(t), t \geq 0$, is also a second order Gaussian random process with values in R^n.

As in the deterministic case, the question of existence of solutions of the stochastic differential equation is also important. We consider its evolution over a finite interval $I \equiv [0,T]$. Under similar assumptions on the drift term F and the diffusion term G, we can prove existence of unique solutions for the nonlinear system (11.2.3). Towards this goal we introduce the following basic assumptions:

(AF): $F : I \times R^n \longrightarrow R^n$ is Borel measurable and there exists $K \in L_2(I)$ such that

$$\| F(t,x) \|^2 \leq K^2(t)[1 + \| x \|^2], \quad \| F(t,x) - F(t,y) \|^2 \leq K^2(t) \| x - y \|^2 .$$

(AG): $G : I \times R^n \longrightarrow M(n \times m)$ is Borel measurable and there exists an $L \in L_2(I)$ such that

$$\| F(t,x) \|^2 \leq L^2(t)[1+ \| x \|^2], \quad \| G(t,x) - G(t,y) \|^2 \leq L^2(t) \| x - y \|^2 .$$

Let $\mathcal{F}_0$ be a complete subsigma algebra of the sigma algebra $\mathcal{F}$ and $L_2(\mathcal{F}_0, R^n)$ the Hilbert space of $\mathcal{F}_0$ measurable R^n valued random variables having finite second moments, that is, for any $\zeta \in L_2(\mathcal{F}_0, R^n)$

$$E \| \zeta \|^2 \equiv \int_\Omega \| \zeta(\omega) \|^2 P(d\omega) < \infty.$$

Let Ξ denote the the vector space of $\mathcal{F}_t$ adapted R^n-valued random processes having bounded second moments with sample paths being continuous with probability one. Furnished with the norm topology

$$\| \xi \|_\Xi \equiv \left(E \sup\{ \| \xi(t) \|^2, t \in I\} \right)^{1/2},$$

it is a Banach space. It is easy to see that Ξ is a normed vector space. The fact that it is a Banach space follows from Borel-Cantelli Lemma ([2], Proposition 7.1.7).

Theorem 11.2.1. Consider the system (11.2.3) in its integral form

$$\xi(t) = \xi_0 + \int_0^t F(s, \xi(s))ds + \int_0^t G(s, \xi(s))dW(s), t \in I.$$

and suppose the drift and the diffusion coefficients F, G satisfy the assumptions (AF) and (AG) respectively and the σ-algebra $\mathcal{F}_0$ is independent of $\mathcal{F}_t, t > 0$ with W being an R^m-valued $\mathcal{F}_t$ adapted standard Brownian motion. Then for every $\xi_0 \in L_2(\mathcal{F}_0, R^n)$, the system (11.2.3) has a unique solution $\xi \in \Xi$ and further there exists a finite positive number C_T such that

$$\| \xi \|_\Xi^2 \ \leq \ C_T(1 + E\{\| \xi_0 \|^2\}).$$

Proof. Define the operator S by

$$(S\xi)(t) \equiv \xi_0 + \int_0^t F(s, \xi(s))ds + \int_0^t G(s, \xi(s))dW(s), t \in I. \qquad (11.2.10)$$

First we verify that S maps Ξ into Ξ and then show that S has a unique fixed point in Ξ. It is clear that with probability one,

$$\sup_{0 \leq s \leq t} \{\| (S\xi)(s) \|^2\} \leq 3\Big\{\| \xi_0 \|^2 + T \int_0^t \| F(r, \xi(r)) \|^2 dr + \sup_{0 \leq s \leq t} \| \int_0^s G(r, \xi(r))dW(r) \|^2\Big\}. \qquad (11.2.11)$$

For any continuous square integrable $\mathcal{F}_t$-adapted martingale $M_t, t \geq 0$, it follows from Doob's martingale inequality ([2], Proposition 7.1.12) that

$$E\left\{\sup_{0\leq t\leq T} \| M_t \|^2\right\} \leq 4E \| M_T \|^2 . \tag{11.2.12}$$

One can easily verify that for each $\xi \in \Xi$, it follows from assumption (AG) that

$$M_t \equiv \int_0^t G(r,\xi(r))dW(r), t \geq 0,$$

is a continuous $\mathcal{F}_t$ adapted square integrable martingale. Hence by the martingale inequality (11.2.12) we have

$$E \sup_{0\leq s\leq t} \| \int_0^s G(r,\xi(r))dW(r) \|^2 \leq 4E \| \int_0^t G(r,\xi(r))dW(r) \|^2 .$$

Using this inequality in (11.2.11) and the assumptions (AF) and (AG) and carrying out some elementary computations, we find that

$$\| S\xi \|_\Xi^2 \leq 3\Big\{E \| \xi_0 \|^2 +(T \| K \|^2 +4 \| L \|^2)(1+ \| \xi \|_\Xi^2)\Big\}. \tag{11.2.13}$$

This inequality implies that if ξ_0 has finite second moment and $\xi \in \Xi$, then $S\xi \in \Xi$ also. Thus S is a self map in the Banach space Ξ. Now we show that it has a fixed point. We prove this by demonstrating that some iterate (or power) of S is a contraction in Ξ and then use Banach fixed point theorem to complete the proof. For any pair $\xi, \eta \in \Xi$ and $t \in I$, define

$$\rho_t^2(\xi,\eta) \equiv E\left\{\sup_{0\leq s\leq t} \| \xi(s) - \eta(s) \|^2\right\}.$$

Since Ξ is a Banach space, furnished with the metric topology $\rho \equiv \rho_T$, it is a complete metric space. Note that

$$\begin{aligned}(S\xi)(t) - (S\eta)(t) &= \int_0^t [F(r,\xi(r)) - F(r,\eta(r))]dr \\ &+ \int_0^t G(r,\xi(r)) - G(r,\eta(r))]dW(r), t \in I.\end{aligned}$$

Using once again the martingale inequality, and the Lipschitz properties (AF)-(AG), it is easy to verify that

$$\rho_t(S\xi, S\eta) \leq \int_0^t \alpha^2(r)\rho_r(\xi,\eta)dr \tag{11.2.14}$$

where

$$\alpha^2(t) \equiv TK^2(t) + 4L^2(t), t \in I,$$

which, by our assumption, is integrable. Thus the function β, given by

$$\beta(t) \equiv \int_0^t \alpha^2(s)ds, t \in I,$$

is bounded and also has bounded variation. By substituting (11.2.14) into itself, it follows after n iterations, that

$$\rho_t(S^n\xi, S^n\eta) \leq \rho_t(\xi,\eta)\Big(\beta^n(t)/n!\Big).$$

Hence

$$\rho(S^n\xi, S^n\eta) \leq \rho(\xi,\eta)\Big(\beta^n(T)/n!\Big) \leq \gamma_n \rho(\xi,\eta). \tag{11.2.15}$$

Clearly for n large enough $0 \leq \gamma_n < 1$ and so S^n is a contraction in the complete metric space (Ξ, ρ) and hence by Banach fixed point theorem [Corollary 1.5.7], S^n and consequently S has a unique fixed point in Ξ. The last inequality follows from Gronwall Lemma. Indeed, let $\xi \in \Xi$ be a solution of equation (11.2.3) or equivalently the integral equation as stated in the theorem. Define

$$\varphi(t) \equiv E\{\sup \parallel \xi(s) \parallel^2, 0 \leq s \leq t\}.$$

Using the integral equation, Cauchy inequality and the martingale inequality once again, one can easily verify that

$$\varphi(t) \leq 3E \parallel \xi_0 \parallel^2 + 3T\int_0^t K^2(s)(1+\varphi(s))ds + 12\int_0^t L^2(s)(1+\varphi(s))ds, t \in I.$$

Then it follows from Gronwall inequality that

$$\varphi(t) \leq 3\Big(\exp\{3\int_I \alpha^2(s)ds\}\Big)\Big\{E \parallel \xi_0 \parallel^2 + \int_I \alpha^2(s)ds\Big\}$$

where

$$\alpha^2(t) \equiv TK^2(t) + 4L^2(t), t \in I.$$

It is clear from this that there exists a finite positive constant C_T such that

$$\varphi(t) \leq C_T(1 + E \parallel \xi_0 \parallel^2),$$

where

$$C_T = 3\exp\left\{3\int_0^T \alpha^2(t)dt\right\}\max\left\{1, \int_0^T \alpha^2(t)dt\right\}.$$

This completes the proof. □

Corollary 11.2.2. Under the assumptions of Theorem 11.2.1, the solution ξ is Lipschitz continuous with respect to the initial state. There exists a constant, $C = C(K, L, T) > 0$, such that

$$E\{\| \xi(t,x) - \xi(t,y) \|^2\} \leq C \| x - y \|^2 .$$

It is easy to see that given the value of ξ at time s, the future evolution of ξ given by the solution of the integral equation,

$$\xi(t) = \xi(s) + \int_s^t F(r, \xi(r))dr + \int_s^t G(r, \xi(r))dW(r), t \geq s,$$

is independent of the past. Thus the solutions of stochastic differential equations of the form (11.2.3) are always Markovian. For any Borel set $\Gamma \subset R^n$, the probability, that at time t $\xi(t) \in \Gamma$ given that at time $s < t$ it was in state $\xi(s) = x$, is called the transition probability. The transition probability

$$P\{\xi(t) \in \Gamma | \xi(s) = x\} \equiv P(s, x; t, \Gamma), s < t$$

satisfies the Chapman-Kolmogorov equation

$$P(s, x; t, \Gamma) = \int_{R^n} P(s, x; r, dy)P(r, y; t, \Gamma), 0 \leq s \leq r \leq t$$

where Γ is any Borel subset of R^n. We may define the corresponding transition operator (semigroup in the time invariant case) on the space $B_b(R^n)$ of bounded measurable functions furnished with the sup norm topology as follows,

$$(U(t,s)\phi)(x) \equiv E\{\phi(\xi(t))|\xi(s) = x\}.$$

Thus, given the initial distribution $\mu_0 = \nu \in \mathcal{M}_1(R^n)$ for $t = 0$, we have

$$< U(t,0)\phi, \nu > = < \phi, U^*(t,0)\nu >,$$

where by the duality product we mean the action of the measure μ on ϕ given by the expression

$$< \phi, \mu > = \int \phi(z)\mu(dz).$$

For fixed $t > 0$ and $0 \leq s \leq t$, define $\psi(s, x)$ by

$$\psi(s, x) \equiv E\{\phi(\xi(t))|\xi(s) = x\}.$$

Then it follows from Ito's formula (11.2.6) that ψ satisfies the backward Kolmogorov equation

$$\partial\psi/\partial s + \mathcal{A}(s)\psi = 0, \quad \psi(t, x) = \phi(x), \tag{11.2.16}$$

where the differential operator (infinitesimal generator) $\mathcal{A}$ is given by

$$(\mathcal{A}(s)\varphi)(x) = (D\varphi(s, x), F(s, x)) + (1/2)Tr(GG^* D^2\varphi)(s, x),$$
$$0 \leq s \leq t, x \in R^n.$$

Similarly one can justify that $\mu_t \equiv U^*(t, 0)\nu$ satisfies the forward Kolmogorov equation

$$(d/dt)\mu_t = \mathcal{A}^*(t)\mu_t, \mu_0 = \nu, \tag{11.2.17}$$

which is generally interpreted in the weak sense. That is, for any C^∞ function φ with compact support, (11.2.17) is equivalent to

$$(d/dt)\mu_t(\varphi) = \mu_t(\mathcal{A}(t)\varphi), \mu_0(\varphi) = \nu(\varphi), t \geq 0.$$

Thus given the initial probability measure ν for the random variable ξ_0, the adjoint transition operator determines the measure induced by the random variable $\xi(t)$ at time t and it is given by $\mu_t = U^*(t, 0)\nu$. For further details on Kolmogorov equations see [37].

11.2.2 SDE Based on Fractional Brownian Motion

In recent years, specially in the study of traffic in computer communication network, it has been observed that the real traffic has characteristics similar to those of fractional Brownian motion. Thus fractional Brownian motion with appropriate choice of the Hurst parameter has been used to approximate the fluid or aggregate model for network traffic. The original construction of the fractional Brownian motion is obtained by a specific linear operation (an integral operator) of the standard Brownian motion. This is of the form

$$B_H(t) \equiv \int_0^t K_H(t - s)dB(s) \tag{11.2.18}$$

where $B \equiv W$ stands for the standard Brownian motion (Wiener Process) and $H \in [0, 1]$ for the Hurst parameter. The simplest kernel K_H is of the form

$$K_H(t) = C_H \, t^{H-1/2}.$$

For more general kernels see [187, 120]. For $H < (1/2)$ we have singular kernel and in this case the integral (11.2.18) will generate generalized Gaussian processes. Some standard properties of the fractional Brownian motion are as follows:
(FB1): $P\{B_H(0) = 0\} = 1$;
(FB2): for each $t \geq 0$, $B_H(t)$ is $\mathcal{F}_t$ measurable having Gaussian distribution with $E\{B_H(t)\} = 0$;
(FB3): for $t, s \in R_0$, $E\{(B_H(t), \xi)(B_H(s), \eta)\} = (1/2)(\xi, \eta)\{t^{2H} + s^{2H} - |t - s|^{2H}\}$
(FB4): $P\{B_H \in C([0, \infty), R^m)\} = 1$ but nowhere differentiable.

It follows from property (FB3) that the covariance is given by

$$E\{(B_H(t), \xi)^2\} = \| \xi \|^2 t^{2H}.$$

In case $H = (1/2)$ we have the standard Brownian motion with covariance

$$E\{(B_{1/2}(t), \xi)^2\} = E\{(W(t), \xi)^2\} = \| \xi \|^2 t.$$

Throughout this section we take the Hurst parameter $H > (1/2)$ and introduce the function

$$\varphi_H(t) \equiv H(2H - 1)|t|^{2H-2}.$$

One can easily verify that

$$E\{(B_H(t), \xi)(B_H(s), \eta)\} = \int_0^t \int_0^s \varphi_H(u - v)(\xi, \eta) du dv.$$

The reader is encouraged to verify this. Using the function φ_H, one can introduce a Hilbert space appropriate for the problem. Let $L_2^{\varphi}(I, M(n \times m))$ denote the space of $\mathcal{F}_t$ adapted random processes with values in the space of matrices $M(n \times m)$ such that for each Φ in this class,

$$Z \equiv \int_I \Phi(s) dB_H(s)$$

is a well defined R^n valued square integrable random variable. We introduce the scalar product

$$\begin{aligned}(\Phi, \Psi)_{\varphi} &\equiv E \int_I \int_I \varphi_H(s - r) Tr(\Phi(r)\Psi^*(s)) ds dr \\ &= \int_I \int_I \varphi_H(s - r) E\{Tr(\Phi(r)\Psi^*(s))\} ds dr\end{aligned}$$

and the norm (square) by

$$\| \Phi \|_{\varphi}^{2} \equiv \int_I \int_I \varphi_H(s-r) E\{Tr(\Phi(r)\Phi^*(s))\} ds dr.$$

Completing this with respect to this inner product we have a Hilbert space. Clearly for any $\eta \in R^n$, we have

$$E\left\{\left(\int_I \Phi(r) dB_H(r), \eta\right)^2\right\} = \int_I \int_I \varphi_H(u-v) E\{(\Phi^*(u)\eta, \Phi^*(v)\eta)\} du dv. \tag{11.2.19}$$

This functional (as well as the preceding one) shows the long range interdependence property of the FB-motion. In fact if $H \downarrow (1/2)$ it follows from this functional that

$$\lim_{H \downarrow (1/2)} E\{(B_H(t), \xi)(B_H(s), \eta)\} = t \wedge s \ (\xi, \eta),$$

which corresponds to the standard Brownian motion. Its increments over non overlapping intervals of time are independent while those of FB-m are interdependent. The larger the Hurst parameter H is, the longer is the range of interdependence. Another significant property is the so called self-similarity (SS) which means some form of invariance with respect to changes in time scales [187]. One can verify that

$$B_H(\alpha t) \overset{d}{\Longleftrightarrow} \alpha^H B_H(t), \alpha > 0,$$

the equivalence being in the sense of distribution. As a result of convolution with a smooth enough kernel as presented above, one can see that the FB-motion is smoother than Brownian motion. This makes this process physically more meaning full than Brownian motion. Now we are prepared to study differential equations driven by fractional Brownian motion. This is given by

$$d\xi = F(t, \xi(t)) dt + G(t, \xi(t)) dB_H(t), t \geq 0, \tag{11.2.20}$$

replacing the Brownian motion W by the FB-motion B_H.

We present an existence result without detailed proof. The proof is quite similar to that of Theorem 11.2.1.

Theorem 11.2.3. Suppose F and G satisfy the assumptions (AF)-(AG) as in Theorem 11.2.1 and let $H > (1/2)$. Then for any initial state $\xi_0 \in L_2(\mathcal{F}_0, R^n)$, the system (11.2.20) has a unique solution $\xi \in \Xi$ having finite second moments.

Proof. (Outline) The proof follows exactly the same steps. The only point where some difference arises is in the stochastic integral now with FB-m. Using the expression (11.2.19) we have

$$E \parallel \int_0^t G(r,\xi(r))dB_H(r) \parallel^2 \\ \leq E \int_0^t \int_0^t \parallel G(s,\xi(s)) \parallel \parallel G(r,\xi(r)) \parallel \varphi_H(r-s)drds.$$

Using Fubini's theorem and interchanging the order of integration, it follows from the symmetry of the function φ_H that

$$E \int_0^t \int_0^t \parallel G(s,\xi(s)) \parallel \parallel G(r,\xi(r)) \parallel \varphi_H(r-s)drds \\ \leq \left(\int_0^t E\{\parallel G(r,\xi(r)) \parallel^2\} \left\{ \int_0^t \varphi_H(s-r)ds \right\} dr \right).$$

Now it suffices to verify that there is a finite positive number b such that

$$Z(t) \equiv \int_0^t \varphi_H(s-r)ds \leq \int_0^T \varphi_H(s-r)ds \leq b.$$

Under the assumption, $H > (1/2)$, this follows easily. In fact one can take

$$b = (2H)T^{2H-1}.$$

Thus we have

$$E \parallel \int_0^t G(r,\xi(r))dB_H(r) \parallel^2 \leq b\, E \int_0^t \parallel G(r,\xi(r)) \parallel^2 dr.$$

Using this estimate, the rest of the proof can be carried out in exactly the same way as in Theorem 11.2.2. □

For more detailed study of fractional Brownian motion and its applications, see [187, 169, 120], where one will find interesting applications of FB-m to hydrology, infinite dimensional control theory, and linear Filtering.

Remark. If one uses the Ito formula for the process ξ generated by the SDE (11.2.20) driven by FB-m, one finds that

$$< G^* D^2 \varphi G \Delta B_H, \Delta B_H >= tr(G^* D^2 \varphi G)|\Delta t|^{2H}.$$

Since $H > (1/2)$, the Ito differential given by (11.2.6) reduces to the classical differential given by

$$d\varphi = \{\varphi_t + (D\varphi, F)\}dt + (G^* D\varphi, dB_H). \qquad (11.2.21)$$

Dai and Heyde [167] introduces stochastic integrals with respect to FB-m that yields the above result.

11.3 Differential Equations Based on Poisson Process

In many applications point processes (counting processes, jump processes) are used to model systems whose state may change abruptly due to events that may occur at random time with various levels of intensity leading to varied results or consequences. One classical model is based on random Poisson measure. Consider a finite interval $I \equiv [0, T]$ and a closed (bounded or unbounded) set $D \subset R^d$. Let Σ denote the Borel algebra of subsets of the set $I \times (D \setminus \{0\})$. A random measure p defined on the sigma algebra Σ is said to be a Poisson measure if there exists a bounded countably additive positive measure Λ on the Borel algebra of subsets of the set D such that for each set $J \times \Gamma \in \Sigma$,with J being a Borel subset of I and Γ a Borel subset of D,

$$P\{p(J \times \Gamma) = n\} = \exp\{-|J|\Lambda(\Gamma)\}\frac{(|J|\Lambda(\Gamma))^n}{n!},$$

where we have used $|J|$ to denote the Lebesgue measure of the set J and $\Lambda(\Gamma)$ the expected number of events of type Γ per unit time. The measure Λ is called the Levy measure. A Physical interpretation of this measure is as follows. For any set $J \times \Gamma \in \Sigma$, the random measure $p(J \times \Gamma)$ denotes the number of events of intensity (or jumps of size) Γ that occurred during the time interval J. Suppose the random measure is $\mathcal{F}_t$ adapted in the sense that for any interval $[s,t)$, $p([s,t) \times \Gamma)$ is $\mathcal{F}_t$ measurable for all $s(< t) \in I$. Define

$$q(J \times \Gamma) \equiv p(J \times \Gamma) - Ep(J \times \Gamma) = p(J \times \Gamma) - |J|\Lambda(\Gamma).$$

The random measure q is a martingale with zero mean and for Borel sets $J, K \subset I$ with $J \cap K = \emptyset$, $E\{q(J \times \Gamma)q(K \times \Upsilon)\} = 0$ for all $\Gamma, \Upsilon \in \mathcal{B}_D$.

Remark. If the set D consists of only one point $D \equiv \{\varpi\}$ then we have

$$Ep(J \times \{\varpi\}) = |J|\Lambda(\{\varpi\}) \equiv |J|\lambda$$

where λ denotes the number of events per unit time. This is the classical homogeneous Poisson process with intensity λ. We will have occasion to use this process in Computer communication network.

Let $L_2^a(I \times D, dt \times d\Lambda)$ denote the Hilbert space of $\mathcal{F}_t$-adapted R^n-valued random processes for which

$$E\int_{I \times D} \| H(t,v) \|^2 \, dt d\Lambda < \infty.$$

Let $\mathcal{S} \subset L_2^e(I \times D, dt \times d\Lambda)$ denote the class of simple functions. For each $H \in \mathcal{S}$ there exists a finite number of disjoint sets of the form $\{J_i \times D_j\}$ such that $\cup(J_i \times D_j) = J \times D$ and the function takes constant values on these sets which are denoted by $\{h_{i,j}\}$ each of which has finite second moment. For functions from this class, the integral is given by the finite sum as indicated below:

$$\int_{I\times D} H(t,v)q(dt \times dv) \equiv \sum_{i,j} h_{i,j} q(J_i \times D_j).$$

Since $\mathcal{S}$ is dense in $L_2^e(I \times D, dt \times d\Lambda)$, for any $H \in L_2^e(I \times D, dt \times d\Lambda)$ there exists a sequence $H_n \in \mathcal{S}$ that converges to H in the L_2 sense and we have

$$\int_{I\times D} H(t,v)q(dt \times dv) = \lim_{n\to\infty} \int_{I\times D} H_n(t,v)q(dt \times dv)$$

in the mean. The limit is unique and independent of the approximating sequence.

Remark. In view of this result we note that the stochastic integrals with respect to to Poisson random measures are very similar to those with respect to Wiener processes. In fact one can use any square integrable martingale (continuous or not) to define more general stochastic integrals.

Now we are prepared to consider the SDE

$$dx(t) = \int_D H(t, x(t), v)q(dt, dv), x(0) = x_0, t \in I, \tag{11.3.22}$$

where $H : I \times R^n \times D \longrightarrow R^n$. Let $B_{\infty,2}^e(I, R^n)$ denote the vector space of $\mathcal{F}_t$ adapted R^n-valued stochastic processes defined on I. Furnished with norm topology

$$\| x \|_\infty \equiv \left(\sup\{E \| x(t) \|^2, t \in I\}\right)^{1/2},$$

this is a Banach space.

Theorem 11.3.1. Suppose $H : I \times R^n \times D \longrightarrow R^n$ is Borel measurable, and there exists an $h \in L_2(I)$ such that
(J1): $\int_D \| H(t,x,v) \|^2 \Lambda(dv) \leq h^2(t)[1+ \| x \|^2]$
(J2): $\int_D \| H(t,x,v) - H(t,y,v) \|^2 \Lambda(dv) \leq h^2(t)[\| x - y \|^2]$.
Then for every $x_0 \in L_2(\mathcal{F}_0, R^n)$, independent of the random measure q, the system (11.3.22) has a unique solution $x \in B_{\infty,2}^e(I, R^n)$.

Proof. As usual, by a solution (strong) of the SDE (11.3.22) we mean a solution of the integral equation

$$x(t) = x_0 + \int_0^t \int_D H(s, x(s), v)q(ds, dv), t \in I. \tag{11.3.23}$$

Let x be a solution of this equation. Then using the assumptions (J1), the reader can easily verify that

$$(1+E \parallel x(t) \parallel^2) \leq 2(1+E \parallel x_0 \parallel^2) + 2\int_0^t h^2(s)(1+E \parallel x(s) \parallel^2)ds, t \in I.$$

Hence by virtue of Gronwall Lemma, we have

$$(1+E \parallel x(t) \parallel^2) \leq 2(1+E \parallel x_0 \parallel^2) \exp\left\{2\int_0^t h^2(s)ds\right\}, t \in I.$$

It follows from the $\mathcal{F}_t$-measurability of x and the above inequality that $x \in B^e_{\infty,2}(I, R^n)$. Thus the operator $\mathcal{N}$ defined by

$$(\mathcal{N}x)(t) \equiv x_0 + \int_0^t \int_D H(s, x(s), v)q(ds, dv), t \in I,$$

maps $B^e_{\infty,2}(I, R^n)$ into itself. Now using the assumption (J2), one can easily verify that

$$E \parallel (\mathcal{N}x - \mathcal{N}y)(t) \parallel^2 \leq \int_0^t h(s)^2 E \parallel x(s) - y(s) \parallel^2 ds, t \in I.$$

Defining

$$\rho_t^2(x, y) \equiv E \parallel x(t) - y(t) \parallel^2,$$

it follows from the above expression that

$$\rho_t^2(\mathcal{N}x, \mathcal{N}y) \leq \int_0^t h^2(s)\rho_s^2(x, y)ds.$$

Using this inequality one can easily verify that for some finite integer n, $\mathcal{N}^n$ (the n-th iterate of the operator $\mathcal{N}$) is a contraction on the Banach space $B^e_{\infty,2}(I, R^n)$. Hence $\mathcal{N}^n$, and consequently the operator $\mathcal{N}$ itself, has a unique fixed point $x^* \in B^e_{\infty,2}(I, R^n)$. In other words x^* is the unique solution of the integral equation (11.3.23). This completes the proof. □

Remark. Note that the solution of equation (11.3.23) is purely discontinuous. Since the integrand in equation (11.3.23) has no discontinuities of the second kind, the solution x^* has only discontinuities of the first kind.

Consider the system

$$d\xi = \int_D H(t, \xi, v)p(dt, dv), \xi(0) = \xi_0, t \in I. \tag{11.3.24}$$

If $\Lambda(D) < \infty$, it follows from the previous result that equation (11.3.24) has also a unique solution.

Following exactly the same procedure we can prove the existence and uniqueness of solution of the following more general equation

$$d\xi = F(t,\xi)dt + G(t,\xi)dW + \int_D H(t,\xi,v)q(dt,dv), t \in I. \qquad (11.3.25)$$

Theorem 11.3.2. Consider the system (11.3.25) and suppose F, G satisfy the assumptions (AF),(AG) of theorem 11.2.1 and H satisfy those of theorem 11.3.1. Then for every $x_0 \in L_2(\mathcal{F}_0, R^n)$ equation (11.3.25) has a unique solution in $B^e_{\infty,2}(I, R^n)$ with no discontinuities of the second kind.

Using the above results we can derive the Kolmogorov equations. First let us consider the pure jump process. Define

$$\psi(t,x) \equiv E\{\phi(\xi(T))|\xi(t) = x\}. \qquad (11.3.26)$$

We prove the following result.

Corllary 11.3.3. Consider the system (11.3.24) and suppose the assumptions of Theorem 11.3.1 hold. Then ψ satisfies the following backward Kolmogorov integro-partial differential equation,

$$\partial\psi/\partial t + \int_D \Big(\psi(t, x + H(t,x,v)) - \psi(t,x)\Big)\Lambda(dv) = 0, t \in I, \psi(T,x) = \phi(x). \qquad (11.3.27)$$

Proof. Note that by uniqueness of solution of equation (11.3.24) we have

$$\begin{aligned}\psi(t,x) = E\{\phi(\xi(T))|\xi(t) = x\} &= E\{\phi(\xi(T))|\xi_{t,x}(t+\Delta t)\} \\ &\equiv E\psi(t+\Delta t, \xi_{t,x}(t+\Delta t)),\end{aligned} \qquad (11.3.28)$$

where $\xi_{t,x}(\cdot)$ denotes the solution of equation (11.3.24) starting at time t from state x. For Δt sufficiently small, the process may experience at most one jump or no jump at all during the interval $[t, t+\Delta t)$. Considering a jump of size dv and recalling that p is a Poisson random measure, we observe that $\psi(t+\Delta t, \xi_{t,x}(t+\Delta t))$ can assume either of two values $\psi(t+\Delta t, x+H(t,x,v))$ or $\psi(t+\Delta t, x)$ with probabilities $\Delta t\Lambda(dv)$ or $(1 - \Delta t\Lambda(dv))$ respectively. Using the properties of iterated conditional expectation, specially (P2), and summing over all possible jump sizes, we find that

$$\begin{aligned}&E\psi(t+\Delta t, \xi_{t,x}(t+\Delta t)) \\ &\quad = \psi(t+\Delta t, x) + \Delta t\Big\{\int_D \{\psi(t+\Delta t, x + H(t,x,v)) - \psi(t,x)\}\Lambda(dv)\Big\} \\ &\qquad + o(\Delta t),\end{aligned} \qquad (11.3.29)$$

where $o(\Delta t)$ is a random variable satisfying the property

$$P - \lim_{\Delta t \to 0} \{o(\Delta t)/\Delta t\} = 0.$$

Thus it follows from (11.3.28) and (11.3.29) that ψ satisfies the backward Kolmogorov equation (11.3.27). This completes the proof. □

Remark. Following the same arguments as in Corollary 11.3.3, the reader can easily prove that in the case of the system (11.3.23), with q in place of p, the function ψ satisfies the backward Kolmogorov equation

$$\partial\psi/\partial t + \int_D \Big(\psi(t, x + H(t, x, v)) - \psi(t, x) - (D\psi(t, x), H(t, x, v)) \Big) \Lambda(dv) = 0, t \in I, \psi(T, x) = \phi(x). \quad (11.3.30)$$

where $D\psi$ denotes the gradient of ψ.

Remark. For the full system (11.3.25) including the drift,diffusion and jump,ψ as defined above satisfies the following backward Kolmogorov equation

$$\partial\psi/\partial t + \mathcal{A}_c\psi + \mathcal{A}_d\psi = 0, t \in I, \psi(T, x) = \phi(x).$$

where the operators $\mathcal{A}_c$ and $\mathcal{A}_d$ are given by

$$\mathcal{A}_c\varphi \equiv (D\varphi, F) + (1/2)tr(D^2\varphi GG^*)$$
$$\mathcal{A}_d\varphi \equiv \int_D \Big(\varphi(t, x + H(t, x, v)) - \varphi(t, x) - (D\varphi(t, x), H(t, x, v)) \Big) \Lambda(dv).$$

11.3.1 Application to Auto-Insurance

Consider an automobile insurance company. Let $S(t)$ denote it's current wealth, $e(t)$ the net premium (premium minus the administration cost) collection rate, $r(t)$ the bank interest rate, $H(e, v)$ the instantaneous claims payment of size $v \in \Gamma$, possibly dependent on the premium level e, where $\Gamma \subset (0, \infty)$ denotes the possible size of claims. The claims size is a random variable with distribution denoted by the finite positive measure $\pi(dv)$ known as the Levy measure. In other words π denotes the distribution of claims size. Note that claim size may also depend on the level of premiums e received by the company. The system can be described by the following SDE

$$dS(t) = (r(t)S(t) + e(t))dt - \int_\Gamma H(e(t), v)p(dt \times dv), t \geq 0, \quad (11.3.31)$$

where p denotes the random Poisson measure (as defined earlier) with $E\{p(dt \times K)\} = dt \times \pi(K)$, for any Borel set $K \subset R_0 \equiv (0, \infty)$. The company wants an estimate of its wealth (or a function of it) at a future time T, given the current value. In general the question can be answered by solving for the conditional expectation,

$$\psi(t,x) \equiv E\{\varphi(S(T))|S(t) = x\}.$$

Using Taylor's formula along with equation (11.3.27), we arrive at the following equation for ψ,

$$\begin{aligned} &\partial\psi/\partial t + (r(t)x + e(t))D\psi(t,x) \\ &\quad + \int_\Gamma \Big(\psi(t, x - H(e(t), v)) - \psi(t,x)\Big)\pi(dv) = 0, \\ &\psi(T,x) = \varphi(x), x \in R_0. \end{aligned}$$

Solving this equation backward in time one arrives at the desired estimate $\psi(0,x)$.

As mentioned above, the variable e, which represents the rate of incoming revenue, can be controlled by improving the speed of claims settlement and also introducing ad-campaign. In other words $e(t) = f(t, u_1(t), u_2(t)) \equiv f(t, u(t))$, where f is a suitable function (non-negative) of time and the two control variables. The variable u_1 stands for the average time for claims settlement, and u_2 stands for the level of advertisements. The function f is measurable t and continuous in the control variables, and it increases with the decrease of u_1 and increase of u_2. More precisely, for a fixed t and u_1, f is a nondecreasing function of u_2. The insurance company wants to increase it's wealth by choice of a control policy satisfying the constraints such as

$$U \equiv \{(u_1, u_2) \in R_+^2 : 0 < \alpha_1 \le u_1 \le \alpha_2 < \infty, \;\; 0 \le u_2 \le \alpha_3\}.$$

The problem is to find a Markovian control policy that maximizes the expected revenue (discounted) over the time horizon $[0, T]$,

$$J(u) \equiv E\Big\{\int_0^T \ell(t, S(t), u(t))dt\Big\}$$

subject to the dynamic constraint (11.3.31). Here ℓ is a measure of the rate of discounted revenue earnings. Define the value function

$$V(t,x) \equiv \sup\{J(t,x,u), u \in \mathcal{U}_{ad}\}$$

where

$$J(t,x,u) \equiv E\Big\{\int_t^T \ell(\tau,S(\tau),u(\tau))d\tau | S(t)=x\Big\}.$$

Using Bellman's principle of optimality and the Ito formula given above, one can readily verify that V must solve the following HJB (Hamilton-Jacob-Bellman) equation

$$\partial V/\partial t + \sup_{u\in U}\Big\{\ell(t,x,u) + (r(t)x + f(t,u))DV + \int_\Gamma \{V(t,x - H(f(t,u),v)) - V(t,x)\}\pi(dv)\Big\} = 0, \qquad (11.3.32)$$

$$V(T,x) = 0, \quad t\in I, x\in(0,\infty).$$

For application of this model, one must identify the function f and the measure π by field measurements.

11.3.2 Application to Portfolio Management

Continuous Case: An investor decides to make investment in the stock market for a fixed period of time $I \equiv [0,T]$. He chooses n stocks, which he believes has better history of growth, and a bond. The stocks are risky while the bonds are risk free. At times the returns from stocks may be much larger than those from bonds though this is not always guaranteed. In fact occasionally the reverse may also be true wiping out the investors capital. Let $S \equiv \{S_0, S_1, \cdots, S_n\}$ denote the vector of bond and stock prices respectively. The dynamics of the bond market is given by

$$dS_0(t) = r(t)S_0(t)dt \qquad (11.3.33)$$

while that of the stock market is given by

$$dS_i(t) = S_i(t)\{b_i(t,S(t))dt + (\sigma_i(t,S),dW)\}, i = 1,2,\cdots n, \qquad (11.3.34)$$

where r denotes the interest rate, $b_i(t,S)$ denotes the mean short term growth rate, given the current value of the related stocks, and $\sigma_i(t,S)$ denotes the volatility vector (fluctuation) associated with i-th stock. The n stocks may be correlated and their price fluctuation may depend on the status of other closely related stocks. For example, the energy stocks are related to oil stocks, steel stocks are related to iron ore stocks etc. This is the well known Black-Schole's model [97, 93] and because of the structure chosen, the stock prices remain nonnegative which is one of the reasons for its adoption in stock price modeling. The fact that this model preserves the requirement of positivity, follows from Ito formula. Indeed, by applying Ito formula to the function $\log S_i(t)$ we have

$$d(\log S_i(t)) = \Big\{b_i(t, S(t)) - \| \sigma_i(t, S(t)) \|^2\Big\}dt + (\sigma_i(t, S(t)), dW(t)), t \geq 0,$$

and hence

$$S_i(t) = S_i(0) \exp\Big\{\int_0^t (b_i - \| \sigma_i \|^2) ds + \int_0^t (\sigma_i, dW(s))\Big\}, i = 1, 2, \cdots, n,$$

which is always nonnegative given that $S_i(0)$ is nonnegative. Let $\{u_i(t), i = 0, 1, 2, \cdots, n\}$ denote the number of bonds and stocks held by the investor at time t. Then his total wealth at time t, invested in the stock market, is given by

$$x(t) = u_0(t)S_0(t) + \sum_{i=1}^{n} u_i(t)S_i(t).$$

Suppose the stocks are traded and dividends paid continuously. Let $\{m_i, i = 1, 2, \cdots, n\}$ denote the dividend payments associated with each of the stocks and $c(t)$ the consumption rate. Including all these, the change of his wealth over the trading period $[t, t + \triangle t]$ is given by

$$\begin{aligned} & x(t + \triangle t) - x(t) \\ & \quad \approx (u_0(t + \triangle t) - u_0(t))S_0(t) + u_0(t)(S_0(t + \triangle t) - S_0(t)) \\ & \quad + \sum_{i=1}^{n} (u_i(t + \triangle t) - u_i(t))S_i(t) + \sum_{i=1}^{n} u_i(t)(S_i(t + \triangle t) - S_i(t)) \\ & \quad + \sum_{i=1}^{n} m_i(t)u_i(t)S_i(t)\triangle t - c(t)\triangle t. \end{aligned}$$

Assuming that the investment strategy is smooth, it follows from this expression that

$$dx(t) = u_0(t)dS_0(t) + \sum_{i=1}^{n} u_i(t)dS_i(t) + \sum_{i=1}^{n} m_i(t)u_i(t)S_i(t)dt - c(t)dt.$$

Substituting the increments of S_i in the above expression, one obtains the following equation for the investors wealth,

$$dx = rxdt + \Big\{\sum_{i=1}^{n} [S_i(b_i + m_i - r)]u_i(t) - c\Big\}dt + \sum_{i=1}^{n} u_i(S_i\sigma_i, dW).$$

Clearly the objective of the investor is to maximize wealth and personal consumption over the period I. Thus, the utility functional may be chosen as

$$J(u) \equiv E\Big\{\int_I e^{-\delta t} L(c(t), u(t))dt + e^{-\delta T}\Psi(x(T))\Big\},$$

where the first term represents the discounted utility of consumption including transaction and borrowing cost, while the last term gives the discounted utility of bequest. Note that $u_i < 0$ indicates short selling and $u_i > 0$ indicates buying. Let U be a compact subset of R^n defining the limits of buy and sell. The control (investment) policies must be functions taking values from U and adapted to (and measurable with respect to) the complete sigma algebra $\mathcal{F}_t$ defining the past history of the securities. We may denote these control policies by $\mathcal{U}_{ad}$. The objective is to find an investment policy that maximizes the utility functional J. Defining the vector $\xi \equiv col(S_1, S_2, \cdots, S_n, x)$ as the state variable, one can rewrite the equations in the canonical form and develop the HJB equation for the portfolio optimization problem as stated above. This is left as an exercise for the reader. For detailed study of financial markets and portfolio optimization see Yong and Zhou [97].

Discontinuous Case: The author believes that to capture the phenomenon of unusual market movements (including crash), it may be more reasonable to include in the dynamics of stock price the jump process (marked Poisson process). For simplicity of presentation, we consider one bond and one stock investment problem. The reader can easily extend the model to multiple stocks. The price dynamics of the bond is given by

$$dS_0(t) = r(t)S_0(t)dt$$

as before while that of the stock is given by

$$dS(t) = S(t)\{b(t)dt + \sigma(t)dW + \int_\Gamma h(t,v)p(dt \times dv))\},$$

where p is a Poisson random measure having expected value $Ep(dt \times dv) = dt\pi(dv)$ with π being the Levy measure representing the distribution of jump sizes. The function $h(t,v)$, defined on $R_0 \times \Gamma$, determines the impact on the stock price for each jump of size v. Through out this section we assume without further notice that the Brownian motions and the Poisson processes are statistically independent. Denoting by ξ the investors wealth (in the market) and recalling the notations for the dividend payments m, and disregarding consumption c, the wealth dynamics is given by

$$d\xi = \alpha(\tilde{u}, \xi)dt + \tilde{u}\sigma dW + \tilde{u} \int_{\Gamma} h(t, v)p(dt \times dv), t \geq 0,$$

where

$$\alpha(\tilde{u}, \xi) = \{r\xi + \tilde{u}[b + m - r]\}, \text{ and } \tilde{u} \equiv uS.$$

Note that these are functions of time though the time variable is not explicitly shown. Clearly, if $E\{b(t) + m(t)\} \leq r(t)$, there is no incentive for investment in the stock market. For investment in the market, the reverse inequality must hold most of the time. Let Φ be a nonnegative monotone increasing function defined on R_0 with $\Phi(0) = 0$. For notational convenience only, from now on we set $\tilde{u} \Longrightarrow u$ without changing its meaning. The investor's problem is to choose a policy that maximizes his expected wealth

$$E\{\Phi(\xi^u(T))\} \longrightarrow \max$$

at the terminal time. Define $J(t, x, u) \equiv E\{\Phi(\xi^u(T))|\xi(t) = x\}$ to denote the revenue earned over the time horizon $(t, T]$, given the investor's wealth at time t and the investment policy chosen during that period. The value function is then given by $V(t, x) \equiv \sup\{J(t, x, u), u \in \mathcal{U}_{ad}\}$. Here we have used the same notation for admissible controls as in the previous case. The only difference is that the set U is now a compact subset of R. Again, using the principle of optimality, one can derive the following HJB equation

$$\begin{aligned}
&\partial V/\partial t + \sup_{u \in U}\Big\{ \alpha(u, x)DV + (1/2)u^2\sigma^2 D^2 V \\
&\quad + \int_{\Gamma}\Big(V(t, x + uh(t, v)) - V(t, x)\Big)\pi(dv)\Big\} = 0, \\
&V(T, x) = \Phi(x), x \in R_+ \equiv (0, \infty).
\end{aligned} \tag{11.3.35}$$

Solving this equation backward in time, one obtains $V(0, x)$ which is the investors wealth at the time of maturity of his policy. This is a second order integro-partial differential equation on $I \times R_+$ which can be numerically solved by use of finite difference scheme.

11.3.3 Population Dynamics

It is interesting to observe that the population dynamics has almost similar form as the dynamics of the stock price. Let $\{N_i, i = 1, 2, \cdots, n\}$ denote the population of n species living in a given habitat. This system of population interacts through prey and predation (competition and cooperation) and migration(emigration) etc. The dynamics can be written as

$$dN_i(t)/N_i(t) = \left[\alpha_i + \sum_{j=1}^{n} \beta_{i,j} N_j(t) + \sum_{j=1}^{m} \gamma_{i,j} u_j\right] dt + \sigma_i(N) dW_i \quad (11.3.36)$$

where α_i denotes the intrinsic growth rate of i-th population, $\beta_{i,j}$ denotes the interaction of specie i with the specie j which is negative if specie j preys upon the specie i and positive if the pair is cooperative, and zero if neutral. The function $u = col(u_1, u_2, \cdots, u_m)$ denotes the vector of feeds, chemicals, pesticides and herbicides etc. The variable $\gamma_{i,j}, j = 1, 2, \cdots, m$ denotes the effect of the agent j on the specie i. It is positive if it promotes the growth, negative if it inhibits it and zero if neutral. The last term denotes the fluctuation of population due to migration,emigration,epidemic etc. The function $\sigma_i(N)$ denotes the migration or emigration rate of the specie i which may depend on the total population level of the habitat and availability of food supply. Note that if $\beta_{i,i}$ is positive the specie i is cooperative otherwise it is competitive. This is the stochastic version of the celebrated Lotka-Volterra population model. This system of equations has been used as a model in agricultural firms to control population of pests and herbs by use of pesticides and herbicides. We do not wish to pursue this any further, the reader interested in control of such population is referred to [2].

11.4 Applications to Computer Network

In computer communication, two main protocols (among others) used are known as TCP/IP (Transmission Control Protocol and Internet Protocol). TCP/IP is built into the UNIX operating system and is used by the internet. This is the standard system used for transmitting data over networks. TCP is a connection-oriented protocol to the upper layers that enable an application to be sure that a datagram sent out over the network was received in its entirety. IP's main task is addressing of datagrams of information between computers and managing the fragmentation process of these datagrams. IP does not have a checksum for the data contents of a datagram, only for the header, the data verification work is left for upper layers. In this section we study traffic flow dynamics and their control including optimization. We concentrate mainly on fluid approximation of aggregate data traffic. Fluid approximation is justified when considering traffic flow on national and international trunk lines carrying hundreds of mega bits of traffic per second.

For individual packet flow dynamics, the reader is referred to the works of [108, 110, 111, 142, 147].

11.4.1 Dynamic Model for Access Control

Traffic Model: Required characteristics for aggregate traffic are self-similarity (ss) and long range dependency (LRD). These are satisfied by fractional Brownian motion. We use this to construct a source model. Define,

$$I(S) = \begin{cases} 1, & \text{if the statement S is true} \\ 0, & \text{otherwise.} \end{cases}$$

Assuming the sources are independent, the aggregate packet arrival rate from the i-th source can be described by the following scalar SDE

$$da_i = h_i(t)\{I(h_i(t) > 0) + I(a_i(t) > 0, h_i(t) \leq 0)\}dt + \sigma_i I\{a_i(t) > 0\}dB^i_H(t),$$

$i = 1, 2 \cdots, m$. Each source is characterized by two parameters $\{h_i, \sigma_i\}$ which determine the infinitesimal mean rate and the volatility or fluctuation. These parameters can take both positive and negative values but the source data rates $\{a_i\}$ must be nonnegative. The model as given above guarantees the nonnegativity of these rates. For each source, the fundamental parameters $\{h_i, \sigma_i\}$ must be identified by measurement or experiment. Thus the source dynamics is governed by the following stochastic differential equation in R^m

$$da = b(t, a)dt + \sigma(t, a)dB_H \tag{11.4.37}$$

driven by FB-m. Note that for complete characterization of the source, one must also determine the Hurst parameter H by measurement studies.

Access Control Mechanism: As described above the entry of sources to the router multiplexer is controlled by token buckets. A source can access the multiplexer (MP) only if there are tokens available in the token pool. Each source is served by a dedicated token bucket. The state of a token pool is described by its contents. Thus the dynamics of a TB is given by

$$d\rho_i = \{u_i(t)I(\rho_i(t) < T_i) - a_i(t)I(\rho_i(t) > 0)\}dt, \quad i = 1, 2 \cdots, m. \tag{11.4.38}$$

This is simply a balance equation where u_i denotes the token supply rate and a_i the traffic arrival rate. The first term on the right hand side of the equation represents token acceptance (supply) rate which is zero if the bucket is full. The second term represents the consumption rate which is zero if the token pool is empty. The indicator functions provide the logic and define the boundary of excursions.

Multiplexer Queue Dynamics: The dynamics of the multiplexer is given by,

$$dq = -C(t)I(q(t) > 0)dt + \left\{\left[\sum_{i=1}^{m} a_i(t)I(\rho_i(t) > 0)\right] I(q(t) < Q)\right\}dt, \tag{11.4.39}$$

where $C(t)$ denotes the available capacity (bandwidth). Again this is a balance equation; change of MP population equals fresh traffic entry rate from all the sources minus the packet departure rate.

Remark. Note that in developing these dynamic equations we have assumed that both the TBs and the multiplexer are noiseless.

Full System Model: Considering the equations (11.4.37)-(11.4.39) and defining,

$$\begin{aligned} f_i(t, \rho_i, a_i, u_i) &\equiv \{u_i I(\rho_i < T_i) - a_i I(\rho_i > 0)\} \\ f(t, \rho, a, u) &\equiv col\{f_i(t, \rho_i, a_i, u_i), i = 1, 2 \cdots m\} \\ g(t, \rho, a, q) &\equiv \left\{-C(t)I(q > 0) + (\Sigma_{i=1}^{n} a_i I(\rho_i > 0))I(q < Q)\right\}, \end{aligned}$$

we have the system equation given by

$$\begin{aligned} da &= b(t, a)dt + \sigma(t, a)dB_H \\ d\rho &= f(t, \rho, a, u)dt \\ dq &= g(t, \rho, a, q)dt. \end{aligned} \tag{11.4.40}$$

This is a $2m + 1 \equiv n$ dimensional stochastic differential equation describing the dynamics of our system that consists of the source (traffic) dynamics and the dynamics of the access control mechanism. This model was originally developed by Ahmed and Li [102].

Compact State Space Model: Denoting the state by $\xi \equiv \text{col}\{a, \rho, q\}$ and control by u, and the drift and diffusion matrices by F and G, we have the system dynamics given by the SDE,

$$d\xi(t) = F(t, \xi(t), u(t))dt + G(t, \xi(t))dB_H, t \geq 0. \tag{11.4.41}$$

This is a controlled system similar to the one given by equation (11.2.20). To introduce control and optimization problems we must now define an appropriate objective functional and the class of controls to be used.

Performance Measures: The objective functional for a network provider proposed in [102] is given by:

$$\begin{aligned} J(u) \equiv \quad & E\Bigg\{ \int_l \lambda_1(t) \left\{ \sum_{i=0}^{n} a_i(t) I(\rho_i(t) = 0) \right\} dt \\ & + \int_l \lambda_2(t) \left\{ \left(\sum_{i-0}^{n} a_i(t) I(\rho_i(t) > 0)) I(q(t) = Q \right) \right\} dt \\ & + \int_l \lambda_3(t) q(t) dt \Bigg\}. \end{aligned} \tag{11.4.42}$$

The first and the second term represent packet losses at the TB and MP respectively. The third term is a measure of queueing delay suffered at the router. Using the notations of the full system equation (11.4.40)-(11.4.41), we have the cost functional in compact form,

$$J(u) = E\left\{ \int_0^T \ell(t, \xi(t)) dt \right\}. \tag{11.4.43}$$

One of the objectives of the network provider is to improve the system performance by using control strategies that minimize this cost functional.

Other performance measures: There are other very important performance measures such as frequency and duration of congestions.

(A) System Congestion

Definition 11.4.1 (Congestion) The system is said to be in the state of congestion whenever the multiplexer state q enters the closed interval $Q_\alpha \equiv [\alpha Q, Q]$, where $0 < \alpha \leq 1$, and Q is the multiplexer buffer size.

(B) Mean First Time to Congestion: Let C_T denote the set of time instants during which the multiplexer hits the congestion zone. In symbols, this is given by,

$$C_T \equiv \{t \in [0, T] : q(t) \in Q_\alpha\}$$

where $q(t)$ is the state of the multiplexer. The first time to congestion, denoted by τ_c, is then given by,

$$\tau_c \equiv \begin{cases} \inf\{C_T\}, & \text{if the set is nonempty} \\ T, & \text{if the set is empty} \end{cases} \tag{11.4.44}$$

We are interested in the expected value of this, denoted by

$$T_c \equiv E\{\tau_c\}. \tag{11.4.45}$$

(C)Mean Residence Time in Congestion Zone

Definition 11.4.2 (Residence Time) For any sample path $q(t), t \in I \equiv [0,T]$, representing the history of multiplexer queue, the residence time denoted by τ_r is the time spent in the set Q_α as given by

$$\tau_r \equiv \int_0^T I(q(t) \in Q_\alpha)\, dt. \tag{11.4.46}$$

Again we are interested in the expected value giving

$$T_r \equiv E\{\tau_r\} \tag{11.4.47}$$

Note that the cost functional given by (11.4.46) can be absorbed in the general functional given by (11.4.43) by adding the integrand of (11.4.46) to those of (11.4.42) with desired weight. In the paper Ahmed and Li [109] these functionals were computed using Monte Carlo technique. For detailed numerical results and their comparative analysis the reader is referred to the original paper [109].

Control Strategies: Admissible Controls: For short, let $M(\mathcal{F}_t)$ denote the class of $\mathcal{F}_t$ adapted measurable random processes and

$$\mathcal{U}_{ad} \equiv \{u : u \in M(\mathcal{F}_t), u(t) \in U \subset R_+^m\},$$

the class of admissible controls. For any $t \in I \equiv [0,T]$, let $\xi(t) = x$ denote the state of the system given at time t. Let $u \in \mathcal{U}_{ad}$ and suppose $\xi_{t,x}(s), s \in (t,T]$ is the unique strong solution of equation (11.4.41) starting from the state x at time t and driven by the control policy u. Define the functional

$$J(t,x,u) = E\int_t^T \ell(s,\xi_{t,x}(s))ds + \Psi(\xi_{t,x}(T)), t \in [0,T],$$

where Ψ is the terminal cost, possibly representing a penalty for failing to clear the traffic by the end of the observation or contract period. The value function is then given by

$$V(t,x) \equiv \inf_{u \in U_{[t,T]}} J(t,x,u), \tag{11.4.48}$$

where $\mathcal{U}_{[t,T]}$ is the restriction of $\mathcal{U}_{ad}$ over the time interval $[t,T]$. This is the optimal cost corresponding to the starting time $t \in [0,T]$ and the state $x \in R^n$. Our goal is to determine the value function V and then from this find the optimal cost $V(0,x)$ and the corresponding optimal control $u^o \in \mathcal{U}_{ad}$ if one exists.

In case $H = 1/2$, $B_H = W$ is the standard Brownian motion. Using Bellman's principle of optimality [2] and the approximate value of $\xi_{t,x}(t+\Delta t)$ obtained from equation (11.4.41) and carrying out some computation one can easily verify that, in the limit $(\Delta t \to 0)$, the value function V must satisfy the HJB equation given by

$$\frac{\partial V(t,x)}{\partial t} + \ell(t,x) + H(t,x,DV) + (1/2)Tr(D^2VGG^*)\} = 0,$$
$$V(T,x) = \Psi(x), \tag{11.4.49}$$

where

$$H(t,x,p) \equiv \inf_{v \in U}(p, F(t,x,v)).$$

Here we have a degenerate HJB equation with discontinuous coefficients. Theoretically it is a difficult problem (existence,uniqueness, regularity, verification theorem). In case of fractional Brownian motion, based on Ito differential proposed by Dai and Heyde [167], the trace term disappears and we have a first order nonlinear HJB equation,

$$\frac{\partial V(t,x)}{\partial t} + \ell(t,x) + H(t,x,DV)\} = 0, \; V(T,x) = \Psi(x). \tag{11.4.50}$$

Practically it is formidable, since, only for 100 sources (users), we must solve a 201 dimensional nonlinear PDE. Even for present day super computers this will be a tough job. However with the development of quantum computers in the future this may become practical. On the other hand, in case of aggregated sources generated by WAN (Wide Area Network) covering large geographical regions, the number is small giving low dimensional HJB equation making the optimization practically feasible.

In [109] Ahmed and Li proposed a simple feedback control law based on neural networks and compared the results with those corresponding to open loop control laws such as (OPC) with rate equal to the capacity (bandwidth), (OPM) with rate equal to the mean arrival rate, and (OPP) with rate equal to the peak arrival rate. These are the controls used in current practice. For numerical illustration, we used a system consisting of 3 users, 3 token buckets and 1 multiplexer. The control law based on neural network consists of 3 layers with the input layer having 7 inputs (user traffic rates, token bucket states and the multiplexer state), hidden layer, and the output layer giving 3 control outputs. The results are presented in rows 2-4 of the Table 11.4.1. The first row FB(RRS) presents the results corresponding to feedback control based on neural network. The network parameters (weights) are optimized by use of random recursive search technique (RRS) [202]. The cost J corresponding to the above control policies are shown for different values of link capacities. The results show that substantial improvement of performance can be obtained by fairly simple feedback laws. For further details see [109].

Control	Cost J			
Strategies	C=2 M	C=3 M	C=4 M	C=5 M
FB(RRS)	51.2574	27.7329	9.8481	0
OPC	71.4317	47.7816	13.1386	0.1486
OPM	95.4174	55.3937	13.3965	1.0167
OPP	96.4339	56.4101	14.5820	0.1465

Table 11.4.1: Cost vs. Control Strategies

In Table 11.4.2 we present the results on expected residence time and expected first hitting time of the congestion zone. Again it is clear from the results that the feedback control law does provide a superior performance. Compared to other control laws, it has the largest mean (expected) first time to congestion and smallest mean (expected) residence time in the state of congestion.

Control	First Time to Congestion (ms)			Residence Time (ms)		
Strategies	C=2 M	C=3 M	C=4 M	C=2 M	C=3 M	C=4 M
FB(RRS)	3578	3991	4000	310	10	0
OPC	34	44	1106	3871	3621	1627
OPM	16	30	1054	3985	3969	1747
OPP	16	30	1015	3985	3969	2006

Table 11.4.2: T_c, T_r vs. Control strategies

The results presented above correspond to controls having full state information $\xi = col\{a_1, a_2, a_3, \rho_1, \rho_2, \rho_3, q)$. This is very often impractical and costly. So controls based on (available) partial information requiring less costly measurements is preferable. Results based on such strategies can be found in the original paper [109]. It was found that performance based on full information is always better than that corresponding to partial information as expected. However, results based on partial information are still better than those based on open loop controls [109].

11.4.2 TCP Flow Control and Active Queue Management

A class of interesting SDE driven by intensity controlled Poisson processes arises in the study of TCP (transfer control protocol) flow control [86, 148]. Here the source windows (flow rate) are controlled by the router itself by sending congestion signals to the sources. Every time a congestion signal, equivalently an warning signal, is received, the source is bound to reduce its window size by half. To avoid congestion an active queue management system is deployed which randomly discards packets before congestion develops. This protocol is known as RED (Random Early Discard System) [172, 165, 170, 189, 191].

(A) Dynamic Model for TCP-RED System: The basic philosophy behind the TCP congestion control algorithm constitutes additive Increase and Multiplicative Decrease (AIMD) of window size [86]. For each round-trip time, the system decreases the window size by one half of the current window size if a packet is lost. For simplicity, here we ignore the slow start and retransmit timer mechanisms. Let $n-1$ denote the number of TCP flows (sources) connected to a router. The dynamics of the TCP congestion control system can then be described in terms of window size of the sources (users) and queue size of the router. Window size determines the TCP flow rate. In other words, the flow rate is proportional to the window size and this is governed by the following point process driven stochastic differential equation,

$$dw_i(t) \equiv \frac{1}{R_i(q(t))}dt - I(w_i(t) > 0)\frac{w_i(t)}{2}dN_i^{\lambda}(t), i = 1, 2 \cdots n-1, \quad (11.4.51)$$

where $I(\cdot)$ denotes the indicator function given by,

$$I(S) \equiv \begin{cases} 1 \text{ if } S \text{ is true} \\ 0 \text{ otherwise.} \end{cases}$$

The function $R_i(q(t))$ is the round trip time dependent on the router q size, $w_i(t)$ is the window size and the process $N_i^{\lambda}(t)$ is a point process which represents the number of times the source i has been forced to cut down its window size over the time interval $[0, t]$. In other words the intensity process $\{\lambda(t), t \geq 0\}$ is used to regulate TCP flows and hence we consider it as the control variable. The round-trip time is generally given by the following expression,

$$R_i(q(t)) \equiv a_i + q(t)/C,$$

where a_i denotes the round-trip propagation time between the source i and the router in case of no congestion. The second term in this expression represents an additional time required for a complete round trip if the router is not free. In summary, the first term on the right hand side of equation (11.4.51) gives the window opening rate and the second the closing rate.

The router queue dynamics can be described as follows,

$$dq \equiv -CI(q(t) > 0)dt + \left(\sum_{i=1}^{n-1} \frac{w_i(t)}{R_i(q(t))}\right) dt \tag{11.4.52}$$

where $q(t)$ is the current queue size, C is the channel (or link) capacity. The first term on the right hand side of equation (11.4.52) represents the service rate (departure rate) and the second term measures the traffic arrival rate at the router (from all the sources). The model presented above is due to [148] which is a modified version of a similar model proposed in [189].
Before we consider application of this system, we present a general theory of intensity controlled SDE.

(B): HJB Equation for Intensity Control: We consider the stochastic system

$$d\xi = F(\xi)dt + G(\xi)dN^{\lambda}(t), \xi(0) = x_0, \tag{11.4.53}$$

in R^n driven by an m-dimensional point process $\{N^{\lambda}(t), t \geq 0\}$, indexed by its intensity vector $\{\lambda(t), t \geq 0\}$. The function $F : R^n \longrightarrow R^n$ and $G : R^n \longrightarrow M(n \times m)$ are Borel measurable maps. The cost functional to be minimized is given by

$$J(\lambda) \equiv E\left\{\int_I \ell(t, \xi)dt + \Psi(\xi(T))\right\}. \tag{11.4.54}$$

The system is controlled by an appropriate choice of the intensity vector (function) of the m-variate Poisson process $\{N^{\lambda}(t), t \in I\}$. In fact this is an abstract model for the TCP flow dynamics including a router controlling the traffic. Note that we have added a terminal cost Ψ to our cost functional. This is important for the problem we have in mind. In the study of traffic control, it is required that at the end of the running period, the router queue is emptied or brought close to zero.

For admissible controls, first we note that each of the individual intensities, considered as control variables, must be physically limited because of finite propagation time between the sources and the router. Let this upper limit be denoted by $\gamma > 0$. Define the set

$$\Lambda \equiv \{\lambda \in R_+^m : 0 \leq \lambda_i \leq \gamma, i = 1, 2, \cdots, m\}.$$

The admissible set of controls is then given by the set $\mathcal{U}_{ad}$ where

$$\mathcal{U}_{ad} \equiv \{\lambda : \lambda(t) \text{ is } \mathcal{F}_t \text{ adapted and } \lambda(t) \in \Lambda : P - a.s \text{ for } a.a \ t \in I\}. \tag{11.4.55}$$

Clearly the class of Markovian controls denoted by $\mathcal{U}_M$ is a proper subset of the admissible class $\mathcal{U}_{ad}$. By a Markovian control we mean that the current value of control is a measurable function of only the current state. It is also clear from the definition that each member of the admissible class belongs to the set $L_\infty(I, \Lambda)$ with probability one.
For any $t \in I$ and $x \in R_+^n \subset R^n$, define

$$J(t, x, \lambda) = E\left\{\int_t^T \ell(s, \xi_{t,x}(s))ds + \Psi(\xi_{t,x}(T))\right\}, \tag{11.4.56}$$

where $\{\xi_{t,x}(s), s \in (t, T]\}$ is the solution of equation (11.4.53) corresponding to any given admissible intensity $\lambda \in \mathcal{U}_{ad}$. Closely associated with this functional is the value function

$$V(t, x) \equiv \inf\{J(t, x, \lambda), \lambda \in \mathcal{U}_{ad}\}. \tag{11.4.57}$$

This function represents the optimal cost for the remaining period of time $(t, T]$. We can prove the following result.

Theorem 11.4.3. Consider the system (11.4.53) with the cost functional (11.4.54) and admissible controls given by (11.4.55). Suppose that for $\lambda \in \mathcal{U}_M$ the components of the Poisson process $\{N^\lambda\}$ are mutually independent. Then the function V as defined by (11.4.57), satisfies the Hamilton-Jacobi-Bellman (HJB) equation given by

$$\begin{aligned} -\partial V/\partial t &= \ell(t, x) + (F(x), DV(t, x)) \\ &\quad + \inf_{\lambda \in \Lambda}\left\{\sum_{i=1}^m (V(t, x + G(x)e_i) - V(t, x))\lambda_i\right\}, \\ V(T, x) &= \Psi(x), \end{aligned} \tag{11.4.58}$$

where $DV \equiv$ grad V and $\{e_i \in R^m\}$ are the unit vectors.

Proof. Let $\mathcal{F}_t^\lambda$ denote the sigma algebra induced by the intensity process $\{\lambda(s), 0 \leq s \leq t\}$. This is an increasing family of right continuous sigma algebras having left hand limits. We use spike (local/needle) variation of the intensity and the resulting variation of the value function to prove the result. Indeed, for any given $\lambda \in \mathcal{U}_M$, and $h > 0$, probability that N_i^λ makes k jumps during the time interval $[t, t+h]$ is given by

$$\begin{aligned} &P\{N_i^\lambda(t+h) - N_i^\lambda(t) \\ &\quad = k | \mathcal{F}_{t+h}^\lambda\} = (1/k!)\Big(\int_t^{t+h} \lambda_i(s)ds\Big)^k exp\Big\{-\int_t^{t+h} \lambda_i(s)ds\Big\}. \end{aligned}$$

Hence for almost all $t \in I$, we have,

$$P\{N_i^\lambda(t+\Delta t) - N_i^\lambda(t) = 1\} \approx \lambda_i(t)\Delta t + o(\Delta t) \tag{11.4.59}$$

$$P\{N_i^\lambda(t+\Delta t) - N_i^\lambda(t) = 0\} \approx (1 - \lambda_i(t)\Delta t) + o(\Delta t). \tag{11.4.60}$$

Now whatever the optimal intensity may be, it follows from (11.4.59), (11.4.60) and the principle of optimality that, an arbitrary choice of $\beta \in \Lambda$ replacing the optimal intensity over the interval $[t, t+\Delta t]$, leads to the following inequality

$$V(t,x) \leq E\Big\{\int_t^{t+\Delta t} \ell(s, \xi_{t,x}^\beta(s))ds + V(t+\Delta t, \xi_{t,x}^\beta(t+\Delta t))\Big\}. \tag{11.4.61}$$

For $\Delta t > 0$ sufficiently small, it follows from the system equation (11.4.53) that

$$\xi_{t,x}^\beta(t+\Delta t) = x + F(x)\Delta t + G(x)(N^\beta(t+\Delta t) - N^\beta(t)) + o(\Delta t).$$

Substituting this in (11.4.61) and using the fact that the components of N^β are mutually independent, we obtain

$$\begin{aligned} V(t,x) &\leq \ell(t,x)\Delta t + E\Big\{V(t+\Delta t, x + F(x)\Delta t \\ &\qquad + G(x)(N^\beta(t+\Delta t) - N^\beta(t)) + o(\Delta t))\Big\} \\ &\leq \ell(t,x)\Delta t + \sum_{i=1}^m V(t+\Delta t, x + F(x)\Delta t + G(x)e_i)\beta_i\Delta t \\ &\qquad + V(t+\Delta t, x + F(x)\Delta t)\prod_{i=1}^m (1 - \beta_i\Delta t) + o(\Delta t), \end{aligned} \tag{11.4.62}$$

where $\{e_i\} \in R^m$ denotes the canonical basis of R^m. The second term involving the summation represents the expected value corresponding to jumps of the individual components and the third term involving the product gives the expected value corresponding to no jumps of any of the components. It is easy to verify that for Δt sufficiently small $\prod_{i=1}^{m}(1-\beta_i\Delta t) = (1-\sum_{i=i}^{m}\beta_i\Delta t)+o(\Delta t)$. Substituting this in the expression (11.4.62) we obtain

$$\begin{aligned}V(t,x) \leq \ell(t,x)\Delta t + \sum_{i=1}^{m}[&V(t+\Delta t, x+F(x)\Delta t + G(x)e_i)\\ &-V(t+\Delta t, x+F(x)\Delta t)]\beta_i\Delta t\\ &+V(t+\Delta t, x+F(x)\Delta t)+o(\Delta t).\end{aligned}$$

Now using the Taylor's expansion of the last term and rearranging the resulting expression and dividing by Δt, we arrive at the following inequality,

$$\begin{aligned}&\frac{-(V(t+\Delta t,x)-V(t,x))}{\Delta t} \leq \Big\{\ell(t,x)+(DV(t+\Delta t,x),F(x))\\ &+\sum_{i=1}^{m}\Big(V(t+\Delta t, x+F(x)\Delta t+G(x)e_i)-V(t+\Delta t, x+F(x)\Delta t)\Big)\beta_i\Big\}\\ &+\frac{o(\Delta t)}{\Delta t}.\end{aligned}$$

Letting $\Delta t \to 0$, we obtain

$$-\frac{\partial V(t,x)}{\partial t} \leq \Big\{\ell(t,x)+(DV(t,x),F(x))+\sum_{i=1}^{m}\Big(V(t,x+G(x)e_i)-V(t,x)\Big)\beta_i\Big\}.$$

This holds for all $\beta \in \Lambda$ and hence

$$\begin{aligned}-\frac{\partial V(t,x)}{\partial t} &= \ell(t,x)+(DV(t,x),F(x))\\ &+\inf_{\beta\in\Lambda}\Big\{\sum_{i=1}^{m}\Big(V(t,x+G(x)e_i)-V(t,x)\Big)\beta_i\Big\}.\end{aligned}$$

Thus we have the HJB equation (11.4.53) as stated. The terminal condition follows from the definition of (11.4.56) and (11.4.57). □

Using this result we can prove a verification theorem as follows.

Theorem 11.4.4 (Verification Theorem) Suppose the system (11.4.58) has a C^1 solution. Then the optimal cost is given by $V(0,x_0)$ for any given initial state $x_0 \in R^n$. If x_0 is a random vector with $\mathcal{L}(x_0)=\pi_0 \in M_1(R^n)$, then the optimal cost is given by

$$J^o = \int_{R^n} V(0,z)\pi_0(dz).$$

Proof. (Outline) Define the function H mapping $C(R^n) \times \Lambda$ to R given by

$$H(W,\lambda) \equiv \sum_{i=1}^{m}(W(x+G(x)e_i) - W(x))\lambda_i, W \in C(R^n), \lambda \in \Lambda \quad (11.4.63)$$

and

$$M(W) \equiv \inf\{H(W,\lambda), \lambda \in \Lambda\}.$$

Let V be a C^1 solution of equation (11.4.58). Define the multi function Σ by

$$\Sigma(t,x) \equiv \{\lambda \in \Lambda : H(V(t,x),\lambda) = M(V(t,x))\}, (t,x) \in I \times R^n.$$

Since the set Λ is compact and the functional within the braces of equation (11.4.63) is continuous, Σ is a well defined nonempty measurable set valued function. Thus it has measurable selections. Let λ^o be a measurable selection of Σ, that is, $\lambda^o(t,x) \in \Sigma(t,x)$ for almost all $(t,x) \in I \times R^n$. Consider the system (11.4.53) with λ replaced by λ^o and let ξ^o denote the corresponding solution. Note that λ^o is a Markovian control and that, with the substitution of λ^o, equation (11.4.53) turns into a (state) feedback system. Now computing the total derivative of V along this trajectory, that is, $V(t,\xi^o(t))$, we arrive at the identity

$$\begin{aligned}(d/dt)V(t,\xi^o(t)) &= \partial V/\partial t + (DV(t,\xi^o(t)), F(\xi^o(t))) + M(V(t,\xi^o(t)))\\ &= \partial V/\partial t + (DV(t,\xi^o(t)), F(\xi^o(t)))\\ &\quad + H(V(t,\xi^o(t)), \lambda^o(t,\xi^o(t))).\end{aligned}$$

Since V is a solution of the HJB equation (11.4.58), it follows from this equation that

$$(d/dt)V(t,\xi^o(t)) = -\ell(t,\xi^o(t)), t \in I.$$

Integrating this over the set $I \times \Omega$ with respect to the $dt \times dP$ measure, we find that

$$V(0,x_0) = E\left\{\int_0^T \ell(t,\xi^o(t))dt + \Psi(\xi^o(T))\right\} = J(\lambda^o).$$

We conclude from this identity and the definition of the value function that λ^o is an optimal control. Since the set Λ is also convex and H is linear in λ, uniqueness of the optimal control λ^o follows readily. This implies that the multi function Σ reduces to a single valued function,that is, $\Sigma(t,x) = \{\lambda^o(t,x)\}$ for all $(t,x) \in I \times R^n$. In case x_0 has the distribution π_0, using conditional expectation one can easily justify that the optimum cost is given by the expression as stated in the theorem. □

Corollary 11.4.5. Under the assumptions of Theorem 11.4.4, the intensity control is bang-bang.

Proof. It follows from the geometry of the constraint set Λ and linearity of the Hamiltonian that the minimum is attained at λ^o with the components given by

$$\lambda_i^o(t,x) = (\gamma/2)\Big(1 - \text{sign}\ (V(t,x+G(x)e_i) - V(t,x))\Big), i = 1,2\cdots,m, \tag{11.4.64}$$

where

$$\text{sign}\ z \equiv \begin{cases} 1 \text{ if } z \geq 0 \\ -1 \text{ otherwise.} \end{cases}$$

Clearly this is a measurable function and the control is bang-bang. In this case

$$M(V) = \gamma \sum_{i=1}^{m} \Big\{V(t,x+G(x)e_i) - V(t,x)\Big\}\ I\Big(V(t,x+G(x)e_i) - V(t,x) < 0\Big).$$

This completes the proof. □

Remark. Examining the control policy given by the expression (11.4.64), it is evident that the window size of the users must be cut at the maximum permissible level whenever this action improves the cost. Clearly this is in conformity with intuition. This is also the content of the bang-bang principle.

Again numerical solution of the HJB equation is formidable if the dimension is high. Further the complete state information may not be available. The only information that is readily available is the state of the router queue q.

A Simple Partially Observed Control Law: In view of the practical difficulties mentioned above, a simple feedback control law based on partial information was proposed in [109] [148]. For detailed numerical results, the reader may refer to this paper. Note that we can combine the TCP flow equations (11.4.51) and (11.4.52) into the canonical form (11.4.53). We choose the cost functional as

$$\begin{aligned} J(\lambda) &\equiv E\Big\{-\Big(\int_I \beta_1(t) C I(q(t) > 0)dt\Big) + \Big(\int_I \beta_2(t) I(q(t) \in Q_\alpha)dt\Big)\Big\} \\ &= \int_I \ell(t,\xi)dt. \end{aligned} \tag{11.4.65}$$

The first term represents the negative of throughput and the second is a measure of congestion weighted by two nonnegative measurable functions $\{\beta_1, \beta_2\}$ respectively. The objective is to minimize this functional by an appropriate choice of a feedback law f, dependent only on the router q, giving $\lambda = f(q)$. The control law f must be a nonnegative, nondecreasing function with domain, range and boundary conditions given by

$$f : [0, Q] \longrightarrow [0, \infty), f(0) = 0, f(Q) = \gamma$$

with γ being equal to the reciprocal of the smallest propagation time. The function f was chosen from the class

$$F \equiv \{f : f(q) = P_n(c, q)I(P_n(c, q) > 0), c \in R^{n+1}\}$$

satisfying the boundary conditions stated above where P_n denotes a polynomial of degree n with the coefficient vector c. The optimal values for $c = col(c_0, c_1, \cdots, c_n)$ are determined by the RRS algorithm proposed by Ye [202].

The Table 11.4.3 shows the costs $J(L)$ and $J(Q)$ corresponding to linear and quadratic control laws respectively for different values of control transmission delays. These are the time delays between control signal sent out (by the router) and received (by the source). The system consists of three TCP sources and one router with a buffer. The basic parameter values used for numerical experiments are: $C = 3MB$, $Q = 0.2MB$, propagation time a_i uniformly distributed over the interval $[0.01s, 0.20s]$, $\gamma = 100$. The weights for the cost functional are $\beta_1 = 1, \beta_2 = 100$. It is clear that the cost increases with the increase of delay. It is interesting to note that the linear control law provides a better performance compared to the quadratic law. This is probably due to the quadratic control law being more aggressive (compared to the linear law) in cutting down the window size during heavy loads and the reverse during the light loads, and thereby unnecessarily reducing the throughput.

	Delay 0	Delay 2	Delay 4	Delay 6
J(L)	-14715	-14090	-13289	-12079
J(Q)	-14235	-13376	-11912	-9919

Table 11.4.3: Cost vs. delay

One may question, if the linear law is truly optimal why then the algorithm, starting with any higher order control law, does not lead to the linear law. The answer to this question may lie in two factors. One is the lack of rigorous mathematical basis of all the random search algorithms such as RRS, GA (genetic algorithm) and SA (simulated annealing), and the other is the enormous computation time required for convergence.

Open Questions: There are several limitations with the HJB formulation. First note that the HJB equations (11.4.49),(11.4.50) and (11.4.58) are usually degenerate with discontinuous drift and diffusion coefficients F and G. Thus the question of existence, uniqueness and regularity properties are important challenges for theoretical work. The second limitation is computational one. Solving HJB equations in Euclidean spaces of high dimension such as 100 or over is a formidable task. However with the development of quantum computers this may become a reality in the future. The third set of limitations arise form the fact that the HJB equations are derived under the assumption that the entire state variable is physically measurable. When there are large number of sources in the network it is not practically possible neither economically feasible to monitor the status of all the sources and the router. Here, the most interesting open problem is to develop nonlinear filter theory for estimating the status of sources from the simple observation of the router queue which may be a very difficult task. The next step in the process is to develop optimal feedback control laws based on estimated state and the observed queue.

11.5 Real Time Control and Optimization

Optimal control theory as presented in Chapters 8,10,11 are rather difficult to apply on-line in real time. In this section we present a simple procedure whereby one can construct suboptimal controls in real time. Here the controls must be computed on-line and executed in real time as the system evolves. The technique presented here depends on the availability of sampled state information.

11.5.1 Deterministic System

Consider the system

$$\dot{x} = f(t, x(t), u(t)), t \in [0, T] \equiv I, x(0) = x_0, \tag{11.5.66}$$

with admissible controls $\mathcal{U}_{ad}$ consisting of measurable functions with values in a compact set $U \subset R^d$. It is assumed that $f : I \times R^n \times R^d \longrightarrow R^n$

is continuous and bounded in all the variables. We consider the Lagrange problem,

$$J(u) \equiv \int_I \ell(t, x(t), u(t))dt. \tag{11.5.67}$$

Suppose the interval I is given by the union of disjoint intervals $\{I_j\}$, $I = \bigcup_{j=0}^{K-1} I_j$ where $I_j = [t_j, t_{j+1}), j = 0, 1, 2. \cdots, K-1$, with $t_0 = 0$ and $t_K = T$. The state is sampled at times $\{t_0, t_1, \cdots, t_{K-1}\}$. We want the controls to be constant on each of these intervals taking values from the compact set U. Given the state $x(t_j)$ at time t_j, the objective is to choose a control for the interval $[t_j, t_{j+1})$ that minimizes the cost of running the system over this period. Considering the first interval I_0 with state information $x(t_0) = x_0$, and any control $v \in U$, the state at time t_1 is given by

$$x(t_1, v) \approx x_0 + f(t_0, x_0, v)(t_1 - t_0), \tag{11.5.68}$$

and the running cost for the interval I_0 is approximated by the expression

$$J_0(v) \equiv \ell(t_0, (1/2)[x_0 + x(t_1, v)], v)(t_1 - t_0). \tag{11.5.69}$$

For sufficiently small intervals (frequent sampling) these approximations may be acceptable in practice. Since f is continuous in the control variable, $v \to x(t_1, v)$ is continuous and hence continuity of ℓ implies continuity of $J_0(v)$. By compactness of U, J_0 attains its minimum at some $v_0^o \in U$. This gives the sub optimal control for the interval I_0. Using this control one obtains

$$x(t_1, v_0^o) \equiv x^o(t_1).$$

Thus, in general, given the sub optimal state $x^o(t_k)$ at the sampling time t_k, the cost functional to be minimized for the interval I_k is given by

$$J_k(v) \equiv \ell(t_k, x^o(t_k) + (1/2)f(t_k, x^o(t_k), v)(t_{k+1} - t_k), v)(t_{k+1} - t_k), \tag{11.5.70}$$

for $k = 0, 1, 2, \cdots, K-1$. By virtue of compactness of the set U and continuity of $J_k(v)$, there exists a $v_k^o \in U$ such that

$$J_k(v_k^o) = \inf_{v \in U} J_k(v). \tag{11.5.71}$$

This procedure gives the sequence

$$\{x^o(t_k), v_k^o, J_k(v_k^o)\}, k = 0, 1, 2, \cdots, K-1, \tag{11.5.72}$$

which may be considered suboptimal. Clearly (11.5.70) and (11.5.71)provide a very simple recursive algorithm for real time optimization and control.

Note that ℓ may contain the sampling and measurement cost. In case sampling cost is ignored, performance is expected to improve with the increase of sampling frequency. The required frequency of sampling is also dependent on the frequency of variation of the time varying system.

11.5.2 Stochastic System

Now we consider the stochastic differential equation

$$dx(t) = F(t, x(t), u(t))dt + G(t, x(t), u(t))dW(t), t \in I, x(0) = x_0 \quad (11.5.73)$$

in place of the deterministic system (11.5.66 and the cost functional

$$J(u) \equiv E \int_I \ell(t, x(t), u(t))dt \quad (11.5.74)$$

in place of (11.5.67). Let F be an n-vector and G a $M(n \times d)$ matrix valued functions continuous in all the variables. Suppose the state is monitored at each of the time instants $\{0 = t_0, t_1, t_2, t_3, \cdots, t_{K-1}\}$. Given the state $x(t_k)$ at time t_k and any control $v \in U$, the state at time t_{k+1} is given by the random vector

$$x(t_{k+1}, v, \xi) \approx x(t_k) + F(t_k, x(t_k), v)(t_{k+1} - t_k) + G(t_k, x(t_k), v)\xi, \quad (11.5.75)$$

$$\xi \equiv (W(t_{k+1}) - W(t_k)), k = 0, 1, 2, \cdots, K-1. \quad (11.5.76)$$

Clearly this is an approximation of the true state and its accuracy very much depends on the length of the interval I_k. Note that ξ is a Gaussian random vector with mean zero and covariance $(t_{k+1} - t_k)I_d$ where I_d denotes the identity matrix of dimension d. Our objective is to find a control that minimizes the average running cost for the interval I_k. Thus we must minimize

$$J_k(v) = E\{\ell(t_k, (1/2)(x(t_k) + x(t_{k+1}, v, \xi)), v)(t_{k+1} - t_k)\} \quad (11.5.77)$$

$$\equiv \int_{R^d} \{\ell(t_k, (1/2)(x(t_k) + x(t_{k+1}, v, \xi)), v)(t_{k+1} - t_k)\} P_k(d\xi) \quad (11.5.78)$$

where P_k is the Gaussian measure on R^d given by

$$P_k(\Gamma) = (1/2\pi(t_{k+1} - t_k))^{n/2} \int_\Gamma \exp -\{(1/2(t_{k+1} - t_k)) \parallel \xi \parallel^2\}d\xi, \quad (11.5.79)$$

where Γ is any Borel subset of R^n. Note that since P_k is a Gaussian measure, the integral (11.5.78) exists and is finite for a very large class of cost integrands ℓ that includes polynomial growth. Since F, G, ℓ are continuous in all the variables, it is clear from (11.5.78) that the function $v \to J_k(v)$ is continuous and since U is assumed to be compact there exists a $v_k^0 \in U$ at which $J_k(v)$ attains its minimum. This procedure generates the sequence

$$\{x(t_k), v_k^o, J_k(v_k^o)\}, k = 0, 1, 2, \cdots, K-1, \tag{11.5.80}$$

which may be considered suboptimal. More precisely this piecewise optimization is expected to improve the performance of the system measured in terms of running cost. Clearly (11.5.75)-(11.5.76) and (11.5.78) provide a very simple recursive algorithm for real time optimization and control given that the process is monitored at each of the time instants indicated. Note that the optimal control v_k^o for the interval I_k and the corresponding optimal cost $J_k(v_k^o)$ depend only on the state $x(t_k)$. Thus the control is Markovian.

Time Spent for Data Processing Here we have tacitly assumed that all the necessary computations can be done instantaneously. Clearly this is practically impossible even with ultra high-speed computers. Thus for real time applications, it is essential to invest certain amount of time for data processing during each of the sampling periods. If only a very small part of the interval I_j, $j = 0, 1, 2, \cdots K-1$, such as $I_j^\delta \equiv [t_j, t_j + \delta)$ with $t_j + \delta << t_{j+1}$ is used for data processing, the algorithm can be expected to provide improved performance though not optimal.

Remark. The recursive algorithm presented above also works for fractional Brownian motion B_H replacing the Wiener process W of the system (11.5.73). But since Gaussian measure has the maximum entropy (uncertainty), the cost estimate is conservative.

Similar on-line optimization procedure can be developed for systems driven by point processes.

Remark. If U contains only a finite number of elements, the optimization process is drastically simplified. In this situation $J_k(v)$ has to be computed only for a finite number of elements and so can be executed very rapidly.

Example 1 (TCP Flow Control) Consider the real time optimization of TCP flow control governed by the stochastic differential equation (11.4.53) with the cost functional given by (11.4.65). Again assuming the sampling times to be $\{t_k, k = 0, 1, \cdots K-1\}$, the discrete time approximation is given by

$$x(t_{k+1}) = x(t_k) + F(t_k, x(t_k))(t_{k+1} - t_k) + G(t_k, x(t_k))v, k = 0, 1, 2, \cdots, K-1$$

where $v = N(t_{k+1}) - N(t_k)$. Here the controls are commands asking for adjustment of window size: "cut or not cut". Thus controls take their values from the set $U \equiv \{1,0\}^m$ and hence on any interval of time I_k the admissible control $v \in U$. Clearly this a compact set (even finite). In this case the optimal control for the interval $I_k \equiv [t_k, t_{k+1})$ is given by v_k^o that minimizes the functional $J_k(v) \equiv J(x(t_k), v)$. This is given by

$$J_k(v_k^o) = \inf_{v \in U}\{\ell(t_k, x(t_k) + (1/2)[F(t_k, x(t_k))(t_{k+1} - t_k) + G(t_k, x(t_k))v]\}. \tag{11.5.81}$$

Example 2 (TCP Control With Random Round Trip Time) Recall that the round trip time for the i-th source is given by $R_i(q) \equiv a_i + (q/C)$, where a_i is the round trip propagation time. In practice this quantity can not be exactly determined since it depends on the length of the path traversed by the packets in the network. Thus it can be considered as a random variable taking nonnegative values. In view of this the system model (11.4.53) may be modified to

$$d\xi = F(\xi, \alpha)dt + G(\xi)dN^{\lambda}(t), \xi(0) = x_0, \tag{11.5.82}$$

where $\alpha \in R_+^{n-1}$ represents the vector of round trip delays. Let $P_{\triangle}(d\alpha)$ denote the probability measure, having possibly a compact support $K \subset R_+^{n-1}$, that describes the distribution of the random vector α. In this case the optimal control is given by v_k^o that minimizes the functional $J_k(v) \equiv J(x(t_k), v)$. In other words we have

$$\begin{aligned} J_k(v_k^o) = \inf_{v \in U} \int_K &\{\ell(t_k, x(t_k) + (1/2)(F(t_k, x(t_k), \alpha)(t_{k+1} - t_k) \\ &+ G(t_k, x(t_k))v)\}P_{\triangle}(d\alpha). \end{aligned}$$

11.6 Exercises

(Q1) Show that (a): Brownian motion is Hölder continuous exponent $0 \leq \alpha < (1/2)$, (b): it is no where differentiable and (c): the path length of the Brownian trajectory on any interval of time of positive length is $+\infty$ with probability one.
(Q2): Let $\{w(t), t \geq 0\}$ be a standard Brownian motion and define $\Delta w \equiv w(t + \Delta t) - w(t)$. Prove that for every $\varepsilon > 0$

$$\lim_{\Delta t \to 0} P\{|(\Delta w)^2 - \Delta t| > \varepsilon\} = 0.$$

(Q3): Let $\mathcal{F}_t^w \equiv \sigma\{w(s), s \leq t\}$ denote the smallest sigma algebra induced by the standard Brownian motion w up to time t. Let $L_2^e(I)$ denote the space of $\mathcal{F}_t^w$ adapted random processes $\{\varphi(t), t \in I\}$ satisfying $E\int_I |\varphi(t)|^2 dt < \infty$. Verify that

$$E\Big(\int_I \varphi(t)dw(t)\Big)^2 = E\int_I |\varphi(t)|^2 dt.$$

(Q4): Consider the scalar equation

$$dy = \alpha(t)ydt + \gamma(t)ydw(t), y(0) = y_0$$

driven by one dimensional Brownian motion w. Show that the solution of this equation is given by

$$y(t) = y_0 \exp\left\{\int_0^t (\alpha(s) - (1/2)\gamma^2(s))ds + \int_0^t \gamma(s)dw(s)\right\}.$$

(Q5): Given that $N(0) \in R_+^n$ justify that the solution of equation (11.3.36) lies in R_+^n.
(Q6): Refer to Theorem 11.2.1 and verify the estimate for the constant C_T given at the end of the proof of the theorem.
(Q7): Give the derivation of the HJB equation (11.3.30).
(Q8): Give a detailed derivation of the HJB equation (11.3.32).
(Q9): Develop an on-line optimization procedure for the stochastic system given by

$$dx = F(t, x, u)dt + G(t, x, u)dN, t \geq 0,$$

where $N(t), t \geq 0$, is a d-dimensional homogeneous Poisson process with intensity vector $\lambda = (\lambda_1, \lambda_2, \cdots, \lambda_d)$.

11.7 Bibliographical Notes

The objective of this chapter was to present a brief introduction to stochastic systems and their applications. We develop stochastic dynamic models for practical problems arising in management, finance and computer engineering. For basic theory on stochastic differential equations see [3, 36, 198, 37, 39, 42, 45, 54, 61, 64, 68, 81, 84, 87, 97]. Some of the models, such as the auto insurance dynamics and the stock market dynamics driven by Poisson random measure, are new and appear in the book for the first time. For abstract theory of financial markets see Kholodnyi [179]. In computer network, token buckets and multiplexors are widely used for access control mechanism [195, 173, 163, 108, 110, 111, 142, 147, 155].

In the section on application to computer network, there are two major dynamic models based on fractional Brownian motion and intensity controlled Poisson process. The source dynamics associated with the first one is based on fluid model driven by fractional Brownian motion. The complete model that includes the token buckets and multiplexors was proposed in [102, 109]. TCP is another widely used protocol in computer communication network. For details see [86] and the references therein. The second dynamic model mentioned above is used in modeling congestion control [148, 162, 165, 166, 170, 171, 172, 180, 189, 191, 202] in TCP flows by appropriate choice of the intensity controller. This is a new contribution. For numerical solution of HJB equations see the methods developed by Wang et al [200] and Huang et al [174].

Appendix A

Basic Results from Analysis

A.1 Introduction

Lebesgue integration stands out as one of the most monumental discovery of mathematics of the 20-th century. There are functions which are integrable in the sense of Lebesgue but their Riemann integrals are not defined. In this appendix, we present briefly the basic spirit of Lebesgue integration. For detailed study the reader is referred to the celebrated books on measure and integration as indicated in bibliographical notes. We also present some important results from analysis used in this book.

A.2 Measures and Measurable Functions

For any (abstract) set Ω let $\mathcal{B}_\Omega$ denote the class of subsets of the set Ω satisfying the following properties:

$$\begin{aligned}
&(1) : \emptyset(emptyset), \text{and } \Omega \in \mathcal{B}_\Omega \\
&(2) : \Gamma \in \mathcal{B}_\Omega \Rightarrow \Gamma' \in \mathcal{B}_\Omega, \\
&(3) : \{\Gamma_m\} \in \mathcal{B}_\Omega \Rightarrow \bigcup_{m \geq 1} \Gamma_m \in \mathcal{B}_\Omega.
\end{aligned}$$

The set $\Gamma' \equiv \Omega \setminus \Gamma$. The class $\mathcal{B}_\Omega$ is called the σ-algebra of Borel subsets of the abstract set Ω. The pair $(\Omega, \mathcal{B}_\Omega)$ is called a measurable space. Now we are prepared to define (positive) measures. A set function

$$\mu : \mathcal{B}_\Omega \longrightarrow [0, \infty]$$

is said to be a measure if for every $\Gamma \in \mathcal{B}_\Omega$, $\mu(\Gamma)$ is defined and takes values from $[0, \infty]$, and for every $\Gamma_1, \Gamma_2 \in \mathcal{B}_\Omega$ with $\Gamma_1 \subset \Gamma_2$ we have $\mu(\Gamma_1) \leq \mu(\Gamma_2)$, and $\mu(\emptyset) = 0$. The measure is said to be countably additive if for any disjoint sequence $\{\Gamma_i\}_{i\geq 1}$ of $\mathcal{B}_\Omega$ measurable sets,

$$\mu\left(\bigcup_{i\geq 1}\Gamma_i\right) = \sum_{i\geq 1}\mu(\Gamma_i).$$

Loosely speaking, if $\Omega = R^n$ and $\Gamma \in \mathcal{B}_\Omega$, the volume (area for $n = 2$, length for $n = 1$) of the set Γ is its Lebesgue measure. We denote this measure by $\lambda(\Gamma) = Vol(\Gamma) = \int_\Gamma \lambda(dx)$. This is universally denoted by $Vol(\Gamma) = \int_\Gamma dx$. In fact one can put any measure μ on the measurable space $(\Omega, \mathcal{B}_\Omega)$ turning this into a measure space which is written as $(\Omega, \mathcal{B}_\Omega, \mu)$. If $\mu(\Omega) = 1$, the measure space $(\Omega, \mathcal{B}_\Omega, \mu)$ is called a probability space. For example, consider R^3 and let $\rho(x) \equiv \rho(x_1, x_2, x_3)$ denote the mass density of matter in R^3. Define the measure μ on $\mathcal{B}_{R^3}$ by

$$\mu(A) \equiv \int_A \rho(x)\lambda(dx) \equiv \int_A \rho(x_1, x_2, x_3)dx_1dx_2dx_3, \text{ for } A \in \mathcal{B}(R^3).$$

If $\mu(A) < \infty$, the region A has finite mass and if $\mu(R^3) < \infty$ the universe has finite mass.

Let $\mathcal{Z} \equiv R^n$ and let $\mathcal{B}_\mathcal{Z}$ denote the σ-algebra of Borel subsets of the set $\mathcal{Z}$. A map (function)

$$f : \Omega \to \mathcal{Z}$$

is said to be a measurable function (Borel measurable map) if for every $K \in \mathcal{B}_\mathcal{Z}$,

$$\{\omega \in \Omega : f(\omega) \in K\} \equiv f^{-1}(K) \in \mathcal{B}_\Omega.$$

Let us denote the class of functions that satisfy this property by $M(\Omega, \mathcal{Z})$. One can show that this a real linear vector space in the sense that

$$(1) : f \in M(\Omega, \mathcal{Z}) \Rightarrow \alpha f \in M(\Omega, \mathcal{Z}) \text{ for all } \alpha \in R,$$
$$(2) : f_1, f_2 \in M(\Omega, \mathcal{Z}) \Rightarrow f_1 + f_2 \in M(\Omega, \mathcal{Z}).$$

Note that $M(\Omega, R)$ is an algebra in the sense that point wise multiplication of any two real valued measurable functions is also a real valued measurable function. Clearly the class $M(\Omega, \mathcal{Z})$ is quite general and includes continuous as well as discontinuous functions.

Indeed, if both Ω and $\mathcal{Z}$ are metric spaces, a function $f : \Omega \longrightarrow \mathcal{Z}$ is said to be continuous if the inverse image of every open set in $\mathcal{Z}$ is an open set in Ω. Thus, open sets being measurable sets, continuous functions are also measurable. Since every open or closed set and countable union and intersection of such sets are all Borel sets, by definition, continuous functions constitute a very small subclass of the class of measurable functions.

A.3 Riemann and Riemann-Stieltjes Integral

Let $I = [a, b]$ be a closed bounded interval and $f : I \longrightarrow R$ a continuous function. Partition the interval into n subintervals $a = x_0 \le x_1 \le x_2, \cdots, \le x_n = b$. Let $\{\tau_i \in [x_i, x_{i+1}]\}_{i=0}^{n-1}$ be any family of points in the intervals indicated. Since f is continuous $f(\tau_i)$ is uniquely defined for each τ_i. Then, we define Riemann integral by the limit

$$R(f) = \int_I f(x)dx \equiv \lim_{n\to\infty} \sum_{i=0}^{i=n-1} f(\tau_i)(x_{i+1} - x_i).$$

The Stieltjes integral is defined against a function of bounded variation. Let $\alpha : I \longrightarrow R$ be a function satisfying the following property

$$\sup \sum_{i=0}^{i=n-1} |\alpha(x_{i+1}) - \alpha(x_i)| < \infty,$$

where the supremum is taken over all such partitions. The class of functions satisfying this property is denoted by $BV(I, R)$ and called the space of functions of bounded variation. For $\alpha \in BV(I, R)$ define

$$\| \alpha \|_{BV} \equiv |\alpha(a)| + \sup \sum_{i=0}^{i=n-1} |\alpha(x_{i+1}) - \alpha(x_i)|.$$

The reader can easily verify that this is a norm. Furnished with this norm, $BV(I, R)$ is a Banach space. The Riemann-Stieltjes integral of a continuous function f with respect to a function α of bounded variation is defined by

$$R_s(f) = \int_I f(x)d\alpha(x) = \lim_{n\to\infty} \sum_{i=0}^{i=n-1} f(\tau_i)(\alpha(x_{i+1}) - \alpha(x_i)),$$

whenever the limit exists. Every bounded continuous function is integrable with respect to any function of bounded variation in the Riemann-Stieltjes sense. Indeed the reader can easily verify that for any $f \in C(I, R)$ and $\alpha \in BV(I, R)$ we have

$$|R_s(f)| \le \| f \|_{C(I,R)} \| \alpha \|_{BV} < \infty.$$

If α is once continuously differentiable, the integral reduces to the Riemann integral.

$$R_s(f) = \int_I f(x)d\alpha(x) = \int_I f(x)\dot{\alpha}(x)dx,$$

where $\dot{\alpha}$ denotes the derivative of α.

A.4 Lebesgue Integral

We have seen that Riemann and Riemann-Stieltjes integrals are defined for continuous functions. This is certainly a serious restriction. This was overcome by the discovery of Lebesgue integral which revolutionized 20-th century mathematics. Here we shall briefly introduce this notion for readers not familiar with the subject. First we consider an elementary introduction to Lebesgue integral and complete the discussion with a brief mention of the more general approach.

Elementary Approach: For an integer d, let $\Omega \subset R^d$ be a bounded measurable set. First we define integrals for nonnegative bounded measurable functions $f : \Omega \longrightarrow [0, b]$ with $0 < b < \infty$. Then we extend this to the case $b = \infty$ and finally indicate how this can be extended further to cover arbitrary measurable functions. Let us partition the interval $[0, b]$ by a set of $n(\in N)$ closed subintervals $\{J_i, 1 \leq i \leq n\}$ with $J_i \equiv [a_{i-1}, a_i]$ so that $a_0 = 0$ and $a_n = b$ and $[0, b] = \bigcup_{i=1}^{n} J_i$. We give a name to this partition and call it Π_n. Let $d(\Pi_n) \equiv max\{a_i - a_{i-1}, 1 \leq i \leq n\}$ denote the diameter of the partition. Define

$$\Gamma_i \equiv \{x \in \Omega : f(x) \in J_i\} \equiv f^{-1}(J_i).$$

Since f is a measurable function, these sets are measurable. Now we introduce two sums, the lower sum, and the upper sum, approximating the integral of f on Ω with respect to Lebesgue measure. These are given by

$$S_{\ell,n} \equiv \sum_{i=1}^{i=n} a_{i-1}\lambda(\Gamma_i)$$

$$S_{u,n} \equiv \sum_{i=1}^{i=n} a_i\lambda(\Gamma_i).$$

Clearly

$$S_{\ell,n} \leq S_{u,n}$$

for all $n \in N$. The sequence $\{S_{\ell,n}\}_{n \in N}$ is monotone increasing and the sequence $\{S_{u,n}\}_{n \in N}$ is monotone decreasing while maintaining the above inequality. Now it is well known that any monotone sequence of real numbers has a limit in the extended real number system. Thus both these sequences have limits, and if they converge to one and the same limit, we call it the Lebesgue integral of f and write

$$L(f) \equiv \int_\Omega f(x)\lambda(dx) \equiv \int_\Omega f(x)dx \equiv \lim_{n\to\infty} S_{\ell,n} \equiv \lim_{n\to\infty} S_{u,n}.$$

Next we consider the case where f is a bounded measurable function

$$f : \Omega \longrightarrow [-b, +b].$$

For any such function one can define two nonnegative functions as

$$f^+(x) \equiv f(x) \vee \{0\}, \ f^-(x) \equiv -f(x) \vee \{0\}$$

giving $f = f^+ - f^-$. By the previous construction, it is clear that the Lebesgue integrals $L(f^+)$ and $L(f^-)$ are well defined. Thus the Lebesgue integral of f is well defined and is given by

$$L(f) = L(f^+) - L(f^-).$$

Note that the sum makes sense only if both are not simultaneously equal to $+\infty$. Next, we consider the general case. Here f is a finite measurable function, not necessarily bounded. For this we define

$$f_k(x) \equiv \begin{cases} f(x), & \text{if } |f(x)| \leq k; \\ k \ \text{sign} f(x), & \text{otherwise} . \end{cases}$$

Clearly, for each $k \in N$, f_k is a bounded measurable function and by the previous construction the Lebesgue integral $L(f_k)$ is well defined and it is given by

$$L(f_k) = L(f_k^+) - L(f_k^-).$$

Now we consider the limits

$$\lim_{k \to \infty} L(f_k^+), \lim_{k \to \infty} L(f_k^-).$$

Clearly both $\{L(f_k^+)\}$ and $\{L(f_k^-)\}$ are monotone nondecreasing sequences and hence have limits in the (positive) extended real number system. If these limits are finite the Lebesgue integral of f is given by

$$L(f) = \lim_{k \to \infty} L(f_k).$$

If at least one of them is finite, again the integral is well defined with values possibly from the extended real number system and is given by the above limit. If both the limits are $+\infty$, the function f does not have Lebesgue integral. The reader can verify that the process of construction of Lebesgue integral as given above also holds for unbounded set Ω with very minor modification.

Universally, the class of Lebesgue integrable functions are denoted by $L_1(\Omega, \lambda)$ or simply by $L_1(\Omega)$ when it is understood that the integration is performed with respect to the Lebesgue measure. However, this procedure of construction also holds for arbitrary bounded positive measures, not just for Lebesgue measure. It is clear from the construction of the Lebesgue integral that any bounded measurable function, continuous or not, is integrable on any set of finite Lebesgue measure.

A More General Approach: Let $(\Omega, \mathcal{B}_\Omega, \mu)$ denote a complete measure space and M the space of extended real valued $\mathcal{B}_\Omega$ measurable functions defined on Ω. An element $f \in M$ is said to be a simple function if there exists an integer m and a finite set of real numbers $\{a_1, a_2, \cdots, a_m\}$ such that

$$Range(f) = \{a_1, a_2, a_3, \cdots, a_m\}.$$

This class of functions is denoted by $\mathcal{S}$. For such an element f, its integral with respect to the measure μ is given by

$$L(f) \equiv \int_\Omega f \, d\mu = \sum_{i=1}^{m} a_i \mu(E_i)$$

where

$$E_i \equiv \{x \in \Omega : f(x) = a_i\}$$

are disjoint $\mathcal{B}_\Omega$ measurable sets with $\mu(E_i) < \infty$. It is clear that for $f \in \mathcal{S}$ and $\alpha \in R$ we have

$$L(\alpha f) = \alpha L(f)$$

and for $f_1, f_2 \in \mathcal{S}$ we have

$$L(f_1 + f_2) = L(f_1) + L(f_2).$$

Thus Lebesgue integration is certainly a linear operation on $\mathcal{S}$. For $f \in M$, let $\{f_n\} \in \mathcal{S}$ be such that

$$\lim_{n\to\infty} f_n = f \;\; \mu \;\; a.e.$$

That is, f is the almost every where limit of a sequence of simple functions $\{f_n\}$. Then the Lebesgue integral of f is defined by

$$L(f) \equiv \lim_{n\to\infty} L(f_n)$$

possibly taking values from the extended real line.

Next we consider Lebesgue-Stieltjes integral denoted by,

$$L_s(f) \equiv \int_\Omega f(x)d\alpha(x),$$

where α is a real valued function on Ω having bounded total variation. For any such α we can define a signed measure (measures that can assume positive as well as negative values) as follows

$$\mu_\alpha(\Gamma) \equiv \int_\Gamma d\alpha(x), \Gamma \in \mathcal{B}_\Omega.$$

This measure can be decomposed into two positive measures μ_α^+ and μ_α^- giving

$$\mu_\alpha = \mu_\alpha^+ - \mu_\alpha^-, \qquad \| \mu_\alpha \| \equiv \mu_\alpha^+ + \mu_\alpha^-,$$

where $\| \mu_\alpha \|$ denotes the total variation norm. This is called the Jordan decomposition of μ_α. Since α is of bounded variation, these are finite (positive) measures of bounded total variation. Now we can define Lebesgue integral of $f \in M$ with respect to these measures giving

$$L^+(f) = \int_\Omega f(x)\mu_\alpha^+(dx), \; L^-(f) = \int_\Omega f(x)\mu_\alpha^-(dx).$$

The Lebesgue-Stieltjes integral is then given by

$$L_s(f) = L^+(f) - L^-(f)$$

which may take values in the extended real number system.

A.5 Modes of Convergence

There are many different notions of convergence in measure theory and functional analysis. Here we present only those that have been used in this book. For detailed study the reader is referred to any of the standard books on real analysis and measure theory, for example, Halmos [41], Munroe [69], Hewit and Stromberg [44], Berberian [19], Royden [80]. Let (Ω, B_Ω, μ) be a measure space and $M(\Omega, B_\Omega, \mu) \equiv M$ the linear space of μ measurable real valued functions defined on Ω.

Uniform Convergence: A sequence $\{f_n\} \in M$ is said to converge to $f \in M$ uniformly if for every $\varepsilon > 0$, there exists an integer n_ε such that whenever $n > n_\varepsilon$,

$$|f_n(\omega) - f(\omega)| < \varepsilon \; \forall \; \omega \in \Omega.$$

Recall that for any set $E \subset \Omega$ we have used E' to denote its complement, that is, $E' \equiv \Omega \setminus E$.

Almost Uniform Convergence: A sequence $\{f_n\} \in M$ is said to converge to $f \in M$ almost uniformly, denoted by $a.u$, if, for every $\varepsilon > 0$, there exists a set $E_\varepsilon \in B_\Omega$ with $\mu(E'_\varepsilon) < \varepsilon$, such that

$$f_n(\omega) \longrightarrow f(\omega) \text{ uniformly on } E_\varepsilon.$$

Clearly uniform convergence implies almost uniform convergence.

Almost Everywhere Convergence: A sequence $\{f_n\} \in M$ is said to converge to f μ-almost everywhere, indicated by $a.e$, if,

$$\mu\{\omega \in \Omega : \lim_{n\to\infty} f_n(\omega) \neq f(\omega)\} = 0.$$

In other words, the set on which f_n fails to converge to f is a negligible set or a set of μ-measure zero. If μ is a probability measure, $(\mu(\Omega) = 1)$, this mode of convergence is known as *almost sure convergence.*

The space of measurable functions M is closed with respect to convergence almost every where. In other words, the a.e. limit of a sequence of measurable functions is also a measurable function. This is presented in the following result.

Theorem A.5.1. Let (Ω, B_Ω, μ) be a measure space and $M \equiv M(\Omega, B_\Omega, \mu)$ denote the class of measurable functions. Then, if $\{f_n\} \in M$ and $f_n \longrightarrow f$ $\mu - a.e$, we have $f \in M$.

Convergence in Measure: A sequence $\{f_n\} \in M$ is said to converge to $f \in M$ in measure, indicated by $f_n \longrightarrow f$ $meas.$, if, for every $\varepsilon > 0$,

$$\lim_{n\to\infty} \mu\{\omega \in \Omega : |f_n(\omega) - f(\omega)| > \varepsilon\} = 0.$$

Again if μ is a probability measure, this mode of convergence is known as *Convergence in probability.*

Convergence in the Mean-p: A sequence $\{f_n\} \in M$ is said to converge to $f \in M$ in the mean of order p, if,

$$\lim_{n\to\infty} \| f_n - f \|_p \equiv \lim_{n\to\infty} \left(\int_\Omega |f_n(\omega) - f(\omega)|^p \mu(d\omega) \right)^{1/p} = 0.$$

It is clear that convergence in the mean-p applies only to the subspace $L_p(\Omega, B_\Omega, \mu) \subset M$. The family of vector spaces $\{L_p(\Omega, B_\Omega, \mu), 1 \leq p \leq \infty\}$

is called the Lebesgue spaces. These constitute a very large class of Banach spaces as seen in Chapter 1. For $1 \leq p < \infty$, $L_p(\Omega, B_\Omega, \mu)$ denotes the class of functions whose $p-th$ power is Lebesgue integrable with respect to the measure μ, that is

$$\int_\Omega |f(x)|^p \mu(dx) < \infty.$$

For $p = \infty$, we have the class of essentially bounded measurable functions. By this one means that there exists a finite positive number β which is exceeded by the absolute value of the function f only on a set of μ-measure zero. That is,

$$\mu\{x \in \Omega : |f(x)| > \beta\} = 0.$$

The norm of the function f is defined by

$$\| f \|_\infty \equiv ess - sup\{|f(x)|, x \in \Omega\}$$

and it is given by the smallest number β for which

$$\mu\{x \in \Omega : |f(x)| > \beta\} = 0.$$

There are many subtle relationships between all these modes of convergence. For details the reader may consult any book on measure and integration. The most popular books are those of Halmos, Munroe, Royden, Berbarian, Hewittand Stromberg as mentioned above.

Consider the space M and define

$$d(f,g) \equiv \mu\{\omega \in \Omega : f(\omega) \neq g(\omega)\}$$

for $f, g \in M$. This is a very important metric space used in control theory.

Theorem A.5.3. The vector space M furnished with the metric d, denoted by (M, d), is a complete metric space.

Proof. (Outline) Let $\{f_n\}$ be a Cauchy sequence with respect to the metric d. Then $\{f_n\}$ is a Cauchy sequence in μ measure. Hence there exists a subsequence which is a Cauchy sequence *a.e* and therefore there exists an f to which this subsequence converges a.e. Hence f is measurable by Theorem A.5.1. The limit is independent of the choice of the subsequence. Hence the sequence $\{f_n\}$ itself converges to f. Thus (M, d) is complete. □

Another function space used frequently in this book is the vector space $C(I, R^k)$ of continuous functions defined on the interval I and taking values from R^k.

Theorem A.5.4. Furnished with the norm topology $\| x \| \equiv \sup\{\| x(t) \|_{R^k}, t \in I\}$, the vector space $C(I, R^k)$ is a Banach space.

Proof. Let $\{x_n\} \in C(I, R^k)$ be a Cauchy sequence. That is $\lim_{n,m\to\infty} \| x_n - x_m \| = 0$. Then for each $s \in I$, $\{x_n(s)\}$ is Cauchy sequence in R^k. Since R^k is a Banach space there exists a function x defined on I and taking values from R^k such that $\lim_{n\to\infty} x_n(s) = x(s)$ for each $s \in I$. To complete the proof, we must show that $x \in C(I, R^k)$. For any $\varepsilon > 0$, and for any $t \in I$, it suffices to show that there exists a $\delta > 0$ such that $\| x(t+h) - x(t) \| < \varepsilon$ whenever $|h| < \delta$. Since for any $s \in I$, $x_n(s) \to x(s)$, there exists an integer $n_\varepsilon \in N$ such that

$$\| x_n(t+h) - x(t+h) \| < \varepsilon/3, \quad \| x_n(t) - x(t) \| < \varepsilon/3$$

for all $n > n_\varepsilon$. Choose any integer $n_0 > n_\varepsilon$ and note that

$$\begin{aligned} \| x(t+h) - x(t) \| &\leq \| x(t+h) - x_{n_0}(t+h) \| + \| x_{n_0}(t+h) - x_{n_0}(t) \| \\ &\quad + \| x_{n_0}(t) - x(t) \| \\ &< (2/3)\varepsilon + \| x_{n_0}(t+h) - x_{n_0}(t) \| . \end{aligned}$$

Since $x_{n_0} \in C(I, R^k)$, for the given ε there exists a $\delta > 0$ such that

$$\| x_{n_0}(t+h) - x_{n_0}(t) \| < \varepsilon/3 \ \text{ for } \ |h| < \delta.$$

Thus it follows from the above inequalities that for any given $\varepsilon > 0$ there exists a $\delta > 0$ such that

$$\| x(t+h) - x(t) \| < \varepsilon \ \text{ for } \ |h| < \delta.$$

Since the choice of $t \in I$ is arbitrary, this proves that $x \in C(I, R^k)$. □

A.6 Frequently Used Results from Measure Theory

In this section we present some results from measure theory which have been frequently used in this book. These are monotone convergence theorem, Fatou's Lemma and the celebrated Lebesgue dominated convergence theorem, and Fubini's theorem. First we state the monotone convergence theorem.

Theorem A.6.1 (Monotone Convergence Theorem (MCT)) Let $\{f_n\} \in L_1(\Omega, B_\Omega, \mu)$ be a nondecreasing sequence of nonnegative functions and let f_0 be a function such that

$$\lim_{n\to\infty} f_n = f_0 \ \ a.e.$$

Then $f_0 \in L_1(\Omega, B_\Omega, \mu)$ if, and only if, $\lim_{n\to\infty} \int_\Omega f_n d\mu < \infty$, and if this is the case then

$$\lim_{n\to\infty} \int_\Omega f_n d\mu = \int_\Omega f_0 d\mu.$$

Proof. Since $\{f_n\}$ is monotone nondecreasing sequence, $f_n(x) \le f_0(x)$ μ *a.e.* Thus if $f_0 \in L_1(\Omega, B_\Omega, \mu)$, then $\lim \int_\Omega f_n \, d\mu < \infty$. So this is necessary for integrability of f_0. Clearly by monotonicity of $\{f_n\}$ we have $\int_\Omega f_n \, d\mu \le \int_\Omega f_0 \, d\mu$ for all $n \in N$ and hence

$$\lim_{n\to\infty} \int_\Omega f_n \, d\mu \le \int_\Omega f_0 \, d\mu. \tag{D1}$$

Let $\mathcal{S}(\Omega) \equiv \mathcal{S}$ denote the class of simple functions, that is, functions assuming only a finite number of values and let $\mathcal{S}^+ \subset \mathcal{S}$ denote the class of nonnegative simple functions. For each $n \in N$, let $\{S_{n,k}\} \subset \mathcal{S}^+, k \in N$, be a nondecreasing sequence of integrable simple functions such that for each $n \in N$

$$\lim_{k\to\infty} S_{n,k} = f_n \ \mu \ a.e.$$

Define the sequence of functions $\{R_{n,k}\}$ by

$$R_{n,k} \equiv \sup_{1\le i\le n} S_{i,k}.$$

Clearly this is also a sequence of nonnegative integrable simple functions and it is monotone nondecreasing in both the indices $\{n, k\}$. Since $\{f_n\}$ is a nondecreasing sequence, we have, for each $n \in N$ and each $k \ge n$,

$$S_{n,k} \le R_{n,k} \le R_{k,k} \le \sup_{1\le i\le k} f_i = f_k. \tag{D2}$$

Letting $k \to \infty$ in the above inequality we obtain,

$$f_n \le \lim_{k\to\infty} R_{k,k} \le f_0 \ \mu \ a.e.$$

Now letting $n \to \infty$, it follows from the above expression that

$$\lim_{k\to\infty} R_{k,k} = f_0 \ \mu \ a.e. \tag{D3}$$

For simple functions, it follows from the definition of Lebesgue integral that

$$\lim_{k\to\infty} \int_\Omega R_{k,k} \, d\mu = \int_\Omega \lim_{k\to\infty} R_{k,k} \, d\mu = \int_\Omega f_0 \, d\mu. \tag{D4}$$

From (D2) it is clear that for all $k \in N$,

$$\int_\Omega R_{k,k}\, d\mu \le \int_\Omega f_k\, d\mu. \tag{D5}$$

and hence, letting $k \to \infty$, it follows from from (D4) and (D5) that

$$\int_\Omega f_0\, d\mu \le \lim_{k\to\infty} \int_\Omega f_k\, d\mu. \tag{D6}$$

This shows that for integrability of f_0 it is sufficient that $\lim_{k\to\infty} \int_\Omega f_k\, d\mu < \infty$. Combining (D1) and (D6) we arrive at the following identity

$$\lim_{k\to\infty} \int_\Omega f_k\, d\mu = \int_\Omega f_0\, d\mu.$$

This completes the proof. □

This theorem is due to Lebesgue. Later, this result was extended by B.Levi lifting the nonnegativity hypothesis. We present this in the following theorem without proof. First let us define

$$f^- \equiv -\{f \wedge 0\} \equiv -inf\{f, 0\}.$$

Theorem A.6.2 (B.Levi's Theorem (GMCT)) Let $\{f_n\}$ be a nondecreasing sequence of extended real valued measurable functions such that $\int_\Omega f_k^- d\mu < \infty$ for some $k \in N$. Then

$$\lim_{n\to\infty} \int_\Omega f_n d\mu = \int_\Omega \lim_{n\to\infty} f_n d\mu.$$

Fatou's lemma and Lebesgue dominated convergence theorem are frequently used in this book. On the basis of the preceding results we can give simple proof of these as follows.

Theorem A.6.3 (Fatou's Lemma) Let $\{f_n\}$ be a sequence of nonnegative μ-integrable functions on the measure space (Ω, B_Ω, μ) satisfying $\liminf f_n = f_0$ μ *a.e.* Then

$$\int_\Omega \liminf_{n\to\infty} f_n d\mu \le \liminf_{n\to\infty} \int_\Omega f_n d\mu.$$

Note that the expression on the right hand side may assume the value $+\infty$.

Proof. The proof is based on the monotone convergence theorem A.6.1. Define the sequence $g_n \equiv \inf_{k\ge n} f_k$. Since f_n is integrable, this is a nondecreasing sequence of nonnegative integrable functions satisfying $0 \le g_n \le f_n$. Clearly

$$\int_\Omega g_n d\mu \le \int_\Omega f_n d\mu$$

and hence

$$\liminf_{n\to\infty} \int_\Omega g_n d\mu \le \liminf_{n\to\infty} \int_\Omega f_n d\mu.$$

By the monotone convergence theorem and the fact that $\lim g_n = \liminf f_n$, we have

$$\lim_{n\to\infty} \int_\Omega g_n d\mu = \int_\Omega \lim g_n d\mu \equiv \int_\Omega \liminf_{n\to\infty} f_n d\mu.$$

It follows from these two expressions that

$$\int_\Omega \liminf_{n\to\infty} f_n d\mu = \lim_{n\to\infty} \int_\Omega g_n d\mu = \liminf_{n\to\infty} \int_\Omega g_n d\mu \le \liminf_{n\to\infty} \int_\Omega f_n d\mu.$$

Thus

$$\int_\Omega f_0 d\mu \le \liminf_{n\to\infty} \int_\Omega f_n d\mu.$$

This completes the proof. □

Remark. In view of this result,it is clear that if $\liminf_{n\to\infty} \int_\Omega f_n d\mu < \infty$ then $f_0 \in L_1(\Omega, \mathcal{B}_\Omega, \mu)$.

Remark. Fatou's Lemma also applies to any sequence of integrable functions which are bounded from below by an integrable function. For example, consider the sequence of integrable function f_n and suppose $f_n \ge g$ μ $a.e$ where $g \in L_1(\Omega, \mathcal{B}_\Omega, \mu)$. Then

$$\int_\Omega \liminf f_n d\mu \le \liminf \int_\Omega f_n d\mu.$$

The reader is encouraged to verify this.

Now we present the Lebesgue dominated convergence theorem. For abbreviation we denote this by LDCT.

Theorem A.6.4 (LDCT-1) Let $\{f_n\} \in L_1(\Omega, B_\Omega, \mu)$ and suppose

$$(i) : \lim_{n\to\infty} f_n(\omega) = f(\omega) \quad \mu - a.e$$

$$(ii) : \exists \text{ a } g \in L_1^+(\Omega, B_\Omega, \mu) \text{ such that } |f_n(\omega)| \le g(\omega) \quad \mu - a.e.$$

Then $f \in L_1(\Omega, B_\Omega, \mu)$ and

$$\lim_{n\to\infty} \int_\Omega f_n \, d\mu = \int_\Omega \lim_{n\to\infty} f_n \, d\mu = \int_\Omega f d\mu.$$

Proof. The proof is based on Fatou's lemma. By hypothesis, the sequence f_n is dominated by the integrable function g and hence the two sequences

$f_n + g$ and $g - f_n$ are nonnegative integrable functions. Applying Fatou's Lemma to the first sequence, we obtain

$$\int_\Omega \liminf f_n d\mu \leq \liminf \int_\Omega f_n d\mu.$$

Applying it to the second sequence we obtain

$$\int_\Omega \liminf (g - f_n) d\mu \leq \liminf \int_\Omega (g - f_n) d\mu.$$

This means

$$\limsup \int_\Omega f_n d\mu \leq \int_\Omega \limsup f_n d\mu.$$

Since $f_n \to f$ μ-a.e, it follows from the first inequality that

$$\int_\Omega f d\mu = \int_\Omega \liminf f_n d\mu \leq \liminf \int_\Omega f_n d\mu.$$

Similarly it follows from the second inequality that

$$\limsup \int_\Omega f_n d\mu \leq \int_\Omega \limsup f_n d\mu = \int_\omega f d\mu.$$

From the last two inequalities we conclude that

$$\lim_{n\to\infty} \int_\Omega f_n d\mu = \int_\Omega \lim_{n\to\infty} f_n d\mu = \int_\Omega f d\mu.$$

This completes the proof.

Two Examples

(E1): Consider $\Omega = R$ and define the sequence of real valued (positive) functions given by

$$f_n(x) \equiv (n/\sqrt{2}\pi) exp\{-(1/2)n^2x^2\}, x \in R, n \in N.$$

This is a Gaussian density function with zero mean and variance $(1/n^2)$. It is easy to see that $\int_R f_n(x)dx = 1$ for all $n \in N$, and that $f_n \longrightarrow 0$ $a.e$, but clearly $\lim_{n\to\infty} \int_R f_n(x)dx \neq 0$. What is lacking here is the condition (ii) of the LDCT-1. There is no integrable function that dominates the sequence $\{f_n\}$. The sequence $\{f_n\}$ given by

$$f_n(x) \equiv (1/\sqrt{2}\pi) exp - (1/2)(x - n)^2$$

has similar behavior. Its integral is one for all $n \in N$ and it converges to zero for each $x \in R$, but the $\lim_{n\to\infty} \int f_n(x) = 1 \neq 0$. The reason is same as discussed above.

(E2): Consider the interval $J \equiv [0, \pi]$ and the sequence of functions $f_n(t) \equiv \sin nt, t \in J$. Note that

$$\lim_{n\to\infty} \int_J f_n(t)dt = 0.$$

Here $|f_n| \le 1$ and so dominated by an integrable function but $\lim_{n\to\infty} f_n(t) \neq 0$ *a.e.* Here condition (i) is missing.
However, note that the Gaussian sequence converges, in generalized sense, to the Dirac measure with the point $\{0\}$ being the support. Indeed, for any bounded continuous function $\varphi \in C_b(R)$ we have

$$\int_R \varphi(x) f_n(x)dx \longrightarrow \varphi(0).$$

In other words $f_n \longrightarrow \delta_0$ in the sense of (Schwarz) distribution.

Let Q be an open connected set and let C_0^∞ denote the class of all C^∞ functions with compact supports in Q. The dual of this space, that is $(C_0^\infty)^*$, denoted by $\mathcal{D}$, is called the space of distributions in the sense of Schwarz. For example if $f \in L_1(Q)$, then it has distributional derivatives of all orders, even though it is not differentiable in the usual sense. We let L_f denote the distribution generated by f in the sense that its action on a test function $\varphi \in C_0^\infty(Q)$ is given by

$$L_f(\varphi) \equiv \int_Q f(x)\varphi(x)dx.$$

Distribution theory is extremely powerful and it is extensively used in the study of partial differential equations. Since here we hardly use distribution theory, we do not want to go into details any further. We end this discussion with another simple example.

An Example. Consider the characteristic function

$$f_n(x) \equiv (n/2)\chi_{[s-1/n, s+1/n]}(x), x \in R.$$

Clearly this is not differentiable in the usual sense, but it has generalized derivatives of all orders. For any $\varphi \in C_0^\infty(R)$, we have

$$L_{f_n}(\varphi) \longrightarrow \varphi(s)$$
$$L_{Df_n}(\varphi) \longrightarrow -D\varphi(s)$$
$$L_{D^k f_n}(\varphi) \longrightarrow (-1)^k D^k \varphi(s).$$

In other words $f_n \to \delta_s, Df_n \to D\delta_s, D^k f_n \to D^k \delta_s$, all in the sense of distribution. The reader can easily verify these limits by integration by parts using the basic principles of calculus. This ends our discussion of distribution theory.

Lebesgue dominated convergence theorem also holds for $L_p(\Omega, B_\Omega, \mu)$ spaces. This is stated in the following theorem.

Theorem A.6.5 (LDCT-2) Let $\{f_n\} \in L_p(\Omega, B_\Omega, \mu)$ for some $p \in [1, \infty)$ and suppose

$$(i) : \lim_{n\to\infty} f_n(\omega) = f(\omega) \quad \mu - a.e.$$

$$(ii) : \exists \text{ a } g \in L_p^+(\Omega, B_\Omega, \mu) \text{ such that } |f_n(\omega)| \leq g(\omega) \quad \mu - a.e..$$

Then $f \in L_p(\Omega, B_\Omega, \mu)$ and

$$\lim_{n\to\infty} \int_\Omega |f_n - f|^p d\mu = 0.$$

Fubini's Theorem

Some times we have to deal with multiple integrals. For example, let us consider double integrals. Multiple integrals are similar. Clearly the process of integration simplifies considerably if the double integrals can be evaluated iteratively in the following sense. First, one integrates with respect to one variable keeping the other variable fixed and then complete the integration by integrating with respect to the remaining variable. This process is admissible if the end result is the same irrespective of the order of integration. This is the basic content of Fubini's theorem. Let $(\Omega_i, \Sigma_i, \mu_i)$ $i = 1, 2$ be two measure spaces. Let $\Sigma = \Sigma_1 \times \Sigma_2$ denote the product σ-field on the product space$\Omega \equiv \Omega_1 \times \Omega_2$ and $\mu \equiv \mu_1 \times \mu_2$ denote the product measure on the σ-field Σ.

Theorem A.6.6 (Fubini's Theorem) Let f be a measurable function with respect to the product σ-field and suppose the following integral exists

$$\int_{\Omega_1 \times \Omega_2} f \, d\mu.$$

If the functions f_1 and f_2 given by

$$f_1(x) \equiv \int_{\Omega_2} f(x, y)\mu_2(dy)$$

$$f_2(y) \equiv \int_{\Omega_1} f(x, y)\mu_1(dx),$$

exist μ_1-a.e and μ_2-a.e, respectively, then f_1 and f_2 are integrable with respect to the measures μ_1 and μ_2 respectively and the following identity holds

$$\int_\Omega f\, d\mu = \int_{\Omega_1} f_1(x)\mu_1(dx) = \int_{\Omega_2} f_2(y)\mu_2(dy). \qquad (*)$$

Proof. First we prove that the theorem is true for characteristic functions. Let $\Gamma \in \Sigma$ and denote by χ_Γ the characteristic function of the set Γ and set $f = \chi_\Gamma$. Clearly the x and y-sections of Γ, denoted by Γ_x and Γ^y, are given by

$$\Gamma_x \equiv \{y \in \Omega_2 : (x, y) \in \Gamma\},\ \Gamma^y \equiv \{x \in \Omega_1 : (x, y) \in \Gamma\}.$$

We must verify that $\Gamma_x \in \Sigma_2$ and $\Gamma^y \in \Sigma_1$. The second being identical, it suffices to consider the first only. Define

$$\mathcal{F} \equiv \{K \in \Sigma : K_x \in \Sigma_2\ \forall\ x \in \Omega_1\}.$$

We show that $\mathcal{F}$ contains all measurable rectangles. For $A \in \Sigma_1$ and $B \in \Sigma_2$, $(A \times B)_x = B$ for $x \in A$ and empty set for $x \notin A$. Since both $\{B, \emptyset\} \in \Sigma_2$ we have $(A \times B)_x \in \Sigma_2$. Thus it suffices to verify that $\mathcal{F}$ is a sigma algebra. Consider the sequence $\Gamma_n \in \mathcal{F}$. We show that the set $D \equiv \bigcup_n \Gamma_n$ is in $\mathcal{F}$. Clearly $D_x = (\bigcup_n \Gamma_n)_x = \bigcup_n (\Gamma_n)_x \in \Sigma_2$. Similarly for $\Gamma \in \mathcal{F}$ we have $(\Gamma')_x = (\Gamma_x)' \in \Sigma_2$ implying that $\Gamma' \in \mathcal{F}$. This shows that $\mathcal{F}$ is a sigma algebra. Then

$$f_1(x) \equiv \int_{\Omega_2} \chi_\Gamma(x, y)\mu_2(dy) = \int_{\Gamma_x} \mu_2(dy) = \mu_2(\Gamma_x), x \in \Omega_1$$
$$f_2(y) \equiv \int_{\Omega_1} \chi_\Gamma(x, y)\mu_1(dx) = \int_{\Gamma^y} \mu_1(dx) = \mu_1(\Gamma^y), y \in \Omega_2.$$

For a Σ measurable rectangle $\Gamma = A \times B$, we have seen that

$$\Gamma_x = \begin{cases} B \text{ if } x \in A \\ \emptyset \text{ otherwise.} \end{cases}$$

Thus $\mu_2(\Gamma_x) = \mu_2(B)c_A(x)$ and similarly $\mu_1(\Gamma^y) = \mu_1(A)c_B(y)$ where C_K denotes the characteristic of the set K. Then we have

$$f_1(x) = \mu_2(\Gamma_x) = \mu_2(B)c_A(x), x \in \Omega_1, f_2(y) = \mu_1(\Gamma^y)$$
$$= \mu_1(A)c_B(y), y \in \Omega_2.$$

Clearly f_i is a Σ_i measurable function for $i = 1, 2$. Now integrating f_1 and f_2 with respect to μ_1 and μ_2 respectively we have

$$\int_{\Omega_1} f_1(x)\mu_1(dx) = \mu_2(B)\mu_1(A), \quad \int_{\Omega_2} f_2(y)\mu_2(dy) = \mu_1(A)\mu_2(B).$$

Integrating $f \equiv \chi_\Gamma$ with respect to μ we have

$$\int_\Omega f(x,y)\mu(dx \times dy) = \int_\Omega \chi_\Gamma(x,y)\mu(dx \times dy) = \mu(\Gamma) = \mu(A \times B)$$
$$= \mu_1(A)\mu_2(B).$$

Thus we see that for characteristic functions χ_Γ with $\Gamma \in \Sigma$, a measurable rectangle, all the three integrals coincide. In the case of an arbitrary $\Gamma \in \Sigma$, the measurability of the functions $f_i, i = 1,2$ follow from the fact that $\Sigma \equiv \Sigma_1 \times \Sigma_2$ is closed under all finite union of disjoint measurable rectangles. Hence the function f_i, is Σ_i measureable and μ_i integrable for $i = 1,2$. Thus for any $\Gamma \in \Sigma$ all the three integrals coincide. Next we consider simple functions, that is, functions having a finite range, which we denote by $\mathcal{S}(\Omega)$. For $f \in \mathcal{S}$, let the range of f be given by

$$Range(f) = \{\alpha_1, \alpha_2, \cdots, \alpha_m\}$$

with $\alpha_i \in \bar{R}$, $i = 1,2,\cdots,m$. Define

$$\Gamma_i \equiv \{(x,y) \in \Omega : f(x,y) = \alpha_i\}, i = 1,2\cdots,m.$$

Since f is Σ measurable, $\Gamma_i \in \Sigma$ for all $i = 1,2,\cdots,m$. Thus it follows from the above result, that all the three integrals coincide for $f \equiv \chi_{\Gamma_i}$. Since integration is a linear operation and $f = \sum_{i=1}^m \alpha_i \chi_{\Gamma_i}$, it follows that for every $f \in \mathcal{S}$ all the three integrals coincide. Thus the statements of the theorem hold for all simple functions. Next we consider any nonnegative extended real valued Σ measurable function f. For such functions we can choose a sequence of nonnegative, nondecreasing, simple function $f_n \in \mathcal{S}$ such that

$$\lim_{n\to\infty} f_n(x,y) = f(x,y) \;\; \forall \;\; (x,y) \in \Omega.$$

Since f_n is a simple function it follows from the previous result that

$$\int_\Omega f_n \; \mu(dx \times dy) = \int_{\Omega_1}\Big(\int_{\Omega_2} f_n(x,y)\mu_2(dy)\Big)\mu_1(dx)$$
$$= \int_{\Omega_2}\Big(\int_{\Omega_1} f_n(x,y)\mu_1(dx)\Big)\mu_2(dy). \qquad (**)$$

Thus if $f \in L_1(\Omega,\Sigma,\mu)$, it follows from Lebesgue monotone convergence theorem, that

$$\lim_{n\to\infty}\int_\Omega f_n(x,y)\;\mu(dx \times dy) = \int_\Omega f(x,y)\;\mu(dx \times dy).$$

Define

$$\tilde{f}_{1,n}(x) \equiv \int_{\Omega_2} f_n(x,y)\mu_2(dy)$$
$$\tilde{f}_{2,n}(y) \equiv \int_{\Omega_1} f_n(x,y)\mu_1(dx)$$

and note that $\tilde{f}_{1,n} \in L_1(\Omega_1, \Sigma_1, \mu_1)$, $\tilde{f}_{2,n} \in L_1(\Omega_2, \Sigma_2, \mu_2)$ and since the measures $\{\mu, \mu_1, \mu_2\}$ are nonnegative these functions are also monotone, nonnegative and nondecreasing. Hence, again by the monotone convergence theorem we obtain

$$\lim_{n\to\infty} \tilde{f}_{1,n}(x) = \int_{\Omega_2} f(x,y)\mu_2(dy) \equiv \tilde{f}_1(x) \quad \mu_1 \ a.e$$
$$\lim_{n\to\infty} \tilde{f}_{2,n}(y) = \int_{\Omega_1} f(x,y)\mu_1(dx) \equiv \tilde{f}_2(y) \quad \mu_2 \ a.e.$$

Being the a.e limit of a sequence of Σ_1 measurable functions, $\tilde{f}_1$ is Σ_1 measurable and hence μ_1 integrable. Similarly $\tilde{f}_2$ is Σ_2 measurable and μ_2 integrable. Letting $n \to \infty$ in the expression (**) it follows from these facts that all the three integrals coincide and so the statements (*) of the theorem hold for nonnegative extended real valued functions. Next let $f \in L_1(\Omega, \Sigma, \mu)$ be arbitrary and define $f^+ \equiv \sup\{f, 0\}$ and $f^- \equiv -\inf\{f, 0\}$ and note that these are nonnegative functions and that $f = f^+ - f^-$. Since the assertions of the theorem hold for both f^+ and f^- and integration is a linear operation, the theorem holds for the sum f. This completes the proof. □

A.7 Frequently Used Results from Analysis

In this section we present without proof some important results from functional analysis which have been frequently used in this book.

Weak and Weak* Convergence: Let X be a Banach space with the first and the second duals denoted by X^* and X^{**} respectively. A sequence $x_n^* \in X^*$ is said to converge weakly to x^*, denoted by

$$x_n^* \xrightarrow{w} x^*,$$

if, for every $x^{**} \in X^{**}$,

$$x^{**}(x_n^*) \longrightarrow x^{**}(x^*)$$

and it is said to converge in the weak star topology to x^*, denoted by

$$x_n^* \xrightarrow{w^*} x^*,$$

if, for every $x \in X$,

$$x_n^*(x) \longrightarrow x^*(x).$$

Since every element of X induces a continuous linear functional on X^* through the relation $\hat{x}(x^*) \equiv x^*(x)$, we have the canonical embedding $X \hookrightarrow X^{**}$. Hence the weak star topology is weaker than the weak topology.

Definition A.7.1. A Banach space X is said to be reflexive, if $X^{**} = X$.

For $1 < p < \infty$, the L_p spaces are reflexive Banach spaces. Indeed, for $(1/p) + (1/q) = 1$, $(L_p)^* = L_q$ and $(L_q)^* = L_p$. Hence $(L_p)^{**} = L_p$ and so these spaces are reflexive.

It is well known that a closed bounded subset of a finite dimensional space is compact. Though this is false in infinite dimensional spaces, there is a similar result with respect to weak topologies. This is presented in the following theorem.

Theorem A.7.2. A closed bounded subset of a reflexive Banach space is weakly compact.

It is an amazing fact that the closed convex hull of a compact set inherits the compactness of its parent set. This is stated in the following theorem.

Theorem A.7.3. The closed convex hull of a weakly compact set is weakly compact.

Under certain (geometric) properties, the weak and strong closures (topological) are equivalent. This is the content of Mazur's theorem.

Theorem A.7.4. (Mazur's Theorem.) A convex subset of a Banach space is closed if and only if it is weakly closed.

As a result of this theorem we have the following corollary which is known as Banach-Sacks-Mazur Theorem.

Corollary A.7.5. For any weakly convergent sequence, there exists a proper convex combination of the given sequence, that converges strongly to the same limit.

In a finite dimensional space the closed unit ball is compact. An analogous result in infinite dimensional setting is Alaoglu's theorem.

Theorem A.7.6. (Alaoglu's Theorem) The unit ball $B_1(X^*)$ of the dual X^* of the Banach space X is weak star compact.

Remark. As seen above $B_1(X^*)$ is compact in the weak star topology (ie X topology of X^*) but it is not necessarily weakly compact (ie X^{**} topology of X^*).

Definition A.7.7. Let C be a convex subset of a Banach space X. A point $e \in C$ is said to be an extreme point of C, if it is not a point in the interior of any non-degenerate line segment in C. In other words for $\alpha \in [0,1]$ and $x_1, x_2 \in C$, if $e = (1-\alpha)x_1 + \alpha x_2$ then $\alpha = 0$ or 1.

For any set B, let coB denote the convex hull of B and $clcoB \equiv \overline{coB}$ its closed convex hull.

Theorem A.7.8. (Krein-Millman Theorem) For any weakly compact convex set K of a Banach space X, $K = \overline{co}^w Ext(K)$ where $Ext(K)$ denotes the extreme points of K. Similarly, for a weak star compact convex set K of the dual X^* of a Banach space X, $K = \overline{co}^{w^*} Ext(K)$.

Theorem A.7.9. (Kuratowski-Ryll Nardzewski)[Ahmed-Teo [4], Hu-Papageorgiou [48]] Let $(E, \mathcal{E})$ be a measurable space, F a Polish space (complete separable metric space) and $c(F)$ the class of nonempty closed subsets of F, and $G : E \longrightarrow c(F)$ a measurable multi function. Then G has measurable selections, that is, there exist measurable functions $g : E \longrightarrow F$ such that $g(e) \in G(e)$ for all $e \in E$. .

Theorem A.7.10. (Cesari Lower Closure Theorem [12, 24]) Let $(t,x) \longrightarrow F(t,x)$ be a multi function (Borel) measurable in t and upper semi continuous in x defined on $I \times R^n$ with closed convex values from $\mathcal{P}(R^n)$. Let $\{x_k\} \in C(I, R^n)$ and $\{y_k\} \in L_1(I, R^n)$ be two sequences of functions satisfying $y_k(t) \in F(t, x_k(t))$ for $t \in I$, and that $x_k(t) \longrightarrow x_0(t)$ for each $t \in I$ and $y_k \xrightarrow{w} y_0$ in $L_1(I, R^n)$. Then for almost all $t \in I$, $y_0(t) \in F(t, x_0(t))$.

Let $\mathcal{M}(I, R^n)$ denote the space of countably additive bounded vector measures. Furnished with the total variation norm this is a Banach space. An important result giving the necessary and sufficient conditions for weak compactness of a subset of this space is given in the following theorem.

Theorem A.7.11. (Bartle-Dunford-Schwartz, [30]) A set $\Gamma \subset \mathcal{M}(I, R^n)$ is relatively weakly compact if and only if (i): Γ is bounded (ii): there exists a countably additive bounded positive measure ν such that $\lim_{\nu(\sigma) \to 0} |\mu|(\sigma) = 0$ uniformly with respect to $\mu \in \Gamma$. If in addition Γ is weakly closed, then it is also weakly compact.

Remark. This theorem is in fact a special case of a much more general theorem (Bartle-Dunford-Schwartz, [30]) that holds for general Banach spaces E

under some additional assumptions: both E and E^* satisfy Radon-Nikodym property and that for each $\sigma \in \mathcal{B}_I$ the set $\{\mu(\sigma), \mu \in \Gamma\}$ is a weakly conditionally compact subset of E. For our purpose the theorem as presented is sufficient.

Another very important result having strong similarity with the previous result is the celebrated Dunford-Pettis theorem. This result has been frequently used in the proof of existence of optimal controls in Chapter 8.

Theorem A.7.12. (Dunford-Pettis Theorem IV.8.9) [33] Consider the Banach space $L_1(I, R^n)$. A subset B of $L_1(I, R^n)$ is relatively weakly sequentially compact if (1): it is bounded in norm and (ii) it is uniformly integrable in the sense that

$$\lim_{\lambda(\sigma)\to 0} \int_\sigma f(t)dt = 0 \text{ uniformly with respect to } f \in B,$$

where λ denotes the Lebesgue measure. Further, if B is also weakly closed, then it is weakly sequentially compact.

Another very interesting result that has been used in the study of time optimal control problems [43] and relaxation problems arising from non convexity of the contingent set [139] is concerned with the properties of integrals of measurable set valued functions. Let $k(R^n)$ denote the class of nonempty compact subsets of R^n and $V : I \longrightarrow k(R^n)$ a measurable set valued function. By the integral of such a function one means the integral of all its measurable selections, that is,

$$\int_I V(t)dt \equiv \left\{ \int_I v(t)dt : v \text{ measurable } v(t) \in V(t), t \in I \right\}.$$

Theorem A.7.13. (Aumann, [43] Theorem 8.4) For any measurable and integrable set valued function $V(\cdot)$ with values in $k(R^n)$,

$$\int_I V(t)dt = \int_I co\ V(t)dt,$$

and that both these sets are convex and compact.

In separable reflexive Banach spaces E, a similar result holds

$$cl \int_I V(t)dt = \int_I clco\ V(t)dt,$$

where $V : I \longrightarrow cb(E)$, with $cb(E)$ denoting the class of closed bounded (so weakly compact) subsets of E. This later result, due to Datko (see [139]), has been used in the study of optimal control of nonlinear systems in infinite dimensional spaces.

A.8 Bibliographical Notes

Most of the results presented in sections A1 to A6 can be found in standard books on measure theory and real analysis. The classical books on measure theory are Halmos [41], Munroe [69], Hewit and Stromberg [44], Berberian [19], Royden [80]. The materials in section A7 can be found in books on funtional analysis. The most celebrated books on functional analysis are Dunford and Schwartz [33], Yosida [98]. An excellent book is Larsen [57] containing most widely used results from analysis. For nonlinear analysis, the book of Zeidler is the most outstanding one. For multi functions and their analysis the books by Hu and Papageorgiou [48], and Aubin and Frankowska [12] are very popular. See also Hermes and LaSalle [43] and Ahmed and Teo[4]. For vector measures, the book by Diestel and Uhl Jr. [30] is the most outstanding one. For measures on metric spaces the book by Parthasarathy [74] is classical.

Bibliography

Books

[1] Adams R.A. (1975) *Sobolev Spaces*, Academic Press, New York

[2] Ahmed N.U. (1988) *Elements of Finite Dimensional Systems and Control Theory, Pitman Monographs and Syrveys in Pure and applied Mathematics*, Longman Scientific and Technical, U.K, Copublished by John Wiley & sons, New York, Vol 37

[3] Ahmed N.U. (1998) *Linear and Nonlinear Filtering for Scientists and Engineers*, World Scientific, Singapore, New Jersey, London, Hong Kong

[4] Ahmed N.U. and Teo K.L. (1981) *Optimal Control of Distributed Parameter Systems*, North Holland, New York, Oxford

[5] Ahmed N.U. (1991) *Semigroup Theory with applications to Systems and Control*, Pitman Research Notes in Mathematics, Longman Scientific & Technical, Harlow, Essex,UK

[6] Ahmed N.U. (1988) *Optimization and Identification of Systems Governed by evolution Equations on Banach Space*, Pitman Research Notes in Mathematics, Longman Scientific & Technical, Harlow, Essex,UK

[7] Ahmed N.U. and Skowronski J.M. (1994) *Boundary Stabilization of Nonlinear Flexible Systems in Mechanics and Control*, Plenum Press, New York, 213-221

[8] Ames W.F. (1968) *Nonlinear Ordinary Differential Equations in Transport Processes*, Academic Press, New York, London

[9] Aplevich J.D. (2000) *The Essentials of Linear State-Space Systems*, John Wiley & sons, New York, London

[10] Aris R. (1961) *The Optimal Design of Chemical Reactors*, Academic Press, New York, London

[11] Aubin J.P., Cellina A. (1984) Differential Inclusions, Springer-Verlag, Grundlehren der math. Wiss 264.

[12] Aubin J.P., Frankowska H. (1990) *Set-Valued Analysis*, Birhauser, Boston, Basel, Berlin

[13] Balakrishnan A.V. (1976) *Applied Functional Analysis*, Springer-Verlag, New York, Heidelberg, Berlin

[14] Barbu V. (1976) *Nonlinear Semigroups and Differential Equations in Branch Spaces*, Noordhoff International Publishing, Leyden

[15] Barbu V and Da Prato G. (1983) *Hamilton Jacobi Equations in Hilbert Spaces*, Pitman Research Notes in Mathematics,Pitman Advanced Publishing Program, Boston London Melbourne

[16] Basar Tamar & Olsder Geert (1999) (1st edn. 1982, 2nd edn. 1995) *Dynamic Noncooperative Game Theory* 2nd Edn. Revised, SIAM, Philadelphia

[17] Bellman R., Cooke K.L. (1963) *Differential-Difference Equations*, Academic Press, New York, London

[18] Bellman R. (1957) *Dynamic Programming*, Princeton University Press, Princeton, New Jersey

[19] Berberian S.K. (1962) *Measure and Integration*, The Macmilian Company, New York

[20] Berkovitz L.D. (1974) *Optimal Control Theory*, Springer-Verlag, New York, Heidelberg, Berlin

[21] Boltyanskii V.G. (1971) *Mathematical Methods of Optimal Control*, Holt, Rinehart and Winston, New York, Chicago, San Francisco, Altanta

[22] Bolza O.(1904) *Calculus of Variations*,Ch elsea Publishing Company, New York, N.Y.

[23] Brogan W.L. (1985) *Modern Control Theory*, Prentice Hall, Englewood Cliffs, New Jersey

[24] Cesari Lamberto (1983) *Optimization Theory and Applications:Problems with Ordinary Differential equations*, Springer-Verlag, New York,Heidelberg,Berlin

[25] Chuong N.M, Nirenberg L. and Tutschke W. (2002) *Abstract and Applied Analysis*, World Scientific, Singapore, New Jersey, London, Hong Kong

[26] Clark F.H. (1983) *Optimization and Nonsmooth Analysis*, John Wiley & Sons, New York, Chichester, Brisbane, Toronto, Singapore

[27] Deimling K. (1980) *Nonlinear Functional Analysis*, Springer-Verlag, New York, Heidelberg, Berlin

[28] Deimling K. (1980) *Differential Inclusions*, Springer-Verlag, New York, Heidelberg, Berlin

[29] Denkowski Z., Migorski S. and Papageorgiou N.S. (2003), *An Introduction to Nonlinear Analysis: Theory and Applications I,II*, Kluwer Academic/Plenum Publishers, Boston, Dordrecht, London, New York

[30] Diestel J. and Uhl, Jr. J.J. (1977) *Vector Measures*, American Mathematical Society, Providence, Rhode Island

[31] Dorato P., Abdallah C. and Cerrone V. (1995) *Linear Quadratic Control: An Introduction*, Prentice Hall, Englewood Cliffs, Nerw Jersey

[32] Driver R.D. (1977) *Ordinary and Delay Differential Equations*, Springer-Verlag, New York, Heidelberg, Berlin

[33] Dunford N. and Schwartz J.T. (1964) *Linear Operators part I : General Theory*, Interscience Punlishers, INC., New York, London

[34] Elsgolc L.E. (1962) *Calculus of Variations*, Pergmon Press Ltd, London, Paris, Frankfurt: Addison-Wesley Publishing Company Inc. Reading, Massachusetts, USA

[35] Fattorini H.O. (1999) *Infinite Dimensional Optimization and Control Theory*, Cambridge University Press, Cambridge, U.K

[36] Fleming W.H. and Rishel R.W. (1979) *Deterministic and Stochastic Optimal Control*, Springer-Verlag, New York, Heidelberg, Berlin

[37] Friedman Avner (1975) *Stochastic Differential Equations and Applications (Vol. 1,2)*, Academic Press, New York, San Franciso, London

[38] Gamkrelidze R.V. (1978) *Principals of Optimal Control Theory*, Plenum Press, New York, London

[39] Gihman I.I. and Skorohod A.V. (1979) *Controlled Stochastic Processes*, Springer-Verlag, New York, Heidelberg, Berlin

[40] Hale J. (1971) *Functional Differential Equations*, Springer-Verlag, New York, Heidelberg, Berlin

[41] Halmos P.R. (1950) *Measure Theory*, D. Van Nostrand Company, INC., Princeton, Toronto, London, New York

[42] Hasminskii R.Z. (1980) *Stochastic Stability of Differential Equation*, Sijthoff & Noordhoff, Alphen, Rockville

[43] Hermes H., LaSalle J.P. (1969) *Functional Analysis and Time Optimal Control*, Academic Press, New York, London

[44] Hewitt E., Stromberg K. (1965) *Abstract Analysis*, Springer-Verlag New York, Inc.

[45] Hida T. (1980) *Brownian Motion*, Springer-Verlag, New York, Heidelberg, Berlin

[46] Edited by Hinrichsen D., Martensson B. (1990) *Control of Uncertain Systems*, Birkhauser, Boston, Basel, Berlin

[47] Ho Y.C. and Mitter S.K. (1962) *Directions in Large-Scale System*, Plenum Press, New York, Lodon

[48] Hu S. and Papageorgiou N.S. (1997) *Handbook of Multivalued analysis*, Kluwer Academic Publishers, Dordrecht, Boston, Lodon

[49] Isidori A. (1999) *Nonlinear Control Systems II*, Springer, Berlin

[50] Kailath T. (1980) *Linear Systems*, Prentice-Hall, Englewood Cliffs, NJ

[51] Kalman, R.E., Falb, P.L. and Arbib, M.A. (1969) *Topics in Mathematical System Theory*, McGraw-Hill, New York

[52] Kisielewicz M. (1991) *Differential Inclusions and Optimal Control*, PWN-Polish Scientific Publishers, Warsaw; Kluwer Academic Publishers, Dordrecht, Boston, London

[53] Kolmanovskii V.B. and Nosov V.R. (1986) *Stability of Functional Differential Equations*, Academic Press, New York, London

[54] Krylov N.V. (1980) *Controlled Diffusion Processes*, Springer-Verlag, New York, Heidelberg, Berlin

[55] Lagnese J.E. (1989) *Boundary Stabilization of Thin Plates*, SIAM, Philadephia

[56] Lakshmikantham V., Bainov D.D. and Simeonov P.S. (1989) *Theory of Impulsive Differential Equations*, World Scientific, Singapore

[57] Larsen R. (1973) *Functional Analysis*, Marcel dekker, INC., New York, Inc.

[58] LaSalle J.P. and Lefschetz S. (1961) *Stability by Liapunov's Direct Method with Applications*, Academic Press, New York

[59] Lasdon L.S. (1970) *Optimization Theory for Large Systems*, The MacMillan Limited, London

[60] Lasiecka I. and Triggiani R. (1999),*Deterministic Control Theory for Infinite Dimensional Systems, Vol 1,2*, Encyclopedia of Mathematics, Cambridge University Press, Cambridge

[61] Leo-Garcia A. (2000) *Communication Networks*, McGraw Hill, Boston, New York, Sydney, Toronto

[62] Lewis L.J., Reynolds D.K. Bergseth F.R. and Alexxandro F.J.,Jr. (1969) *Linear Systems Analysis*, McGraw-Hill Book Company, New York, San Francisco, Toronto, London

[63] Lions J.L., (1971), *Optimal Control of Sytems Governed by Partial Differential Equations*,Springer-Verlag Berlin, Heidelberg, New York

[64] Liptser R.S. and Shiryayev A.N. (1977) *Statistics of Random Processes I: General Theory*, Springer-Verlag, New York, Heidelberg, Berlin

[65] Luenberger D.G. (1969) *Optimization By Vector Space Methods*, John Wiley & Sons, Inc., New York, London, Sydney, Toronto

[66] Martin R. and Teo K.L. (1994) *Optimal Control of Drug Administration in Cancer Chemotherapy*, World Scientific, Singapore, New Jersey, London, Hong Kong

[67] Martynyuk A.A. (2002) *Qualitative Methods in Nonlinear Dynamics*, Marcel Derker, Inc., New York, Basel

[68] Mosca E. (1995) *The Optimal, Predictive, and Adaptive Control*, Prentice Hall, Englewood Cliffs, New Jersey

[69] Munroe M.E. (1953) *Introduction to Measure and Integration*, Addison-Wesley Publishing Company, INC., Reading, Massachusetts

[70] Nemytskii V.V., Stepanov V.V. (196) *Qualitative Theory of Differential Equations*, Princeton University Press, Princeton New Jersey

[71] Neustadt L.W. (1976), *Optimization: A Theory of Necessary Conditions*, Princeton University Press, Princeton, New Jersey

[72] Ogata K. (2001) *Modern Control Engineering (4th Edition)*, Prentice Hall, Upper Saddle River, New Jersey

[73] Oguztoreli M.N. (1966) *Time-Lag Control Systems*, Academic Press, New York, London

[74] Parthasarathy K.R. (1967) *Probability Measures on Metric spaces*, Academic Press, New York, London

[75] Pavel N.H. (1987) *Nonlinear Evolution Operatiors and Semigroups*, Springer-Verlag, New York, Heidelberg, Berlin, London, Paris, Tokyo

[76] Petrov I.P. (1968) *Variational Methods in Optimum Control Theory*, Academic Press, New York, London

[77] Pontryagin L.S., Boltyanskii V.G., Gamkrelidze R.V. and Mishchenko E.F. (1962) *Mathematical Theory of Optimal Process*, Interscience Publishers, New York, Lodon, Sydney

[78] Prato G.D. and Zabczyk J. (1996) *Ergodicity for Infinite Dimensional Sysyems*, Cambridge University, Cambridge

[79] Prato G.D. and Zabczyk J. (1992) *Stochastic Equations in Infinite Dimensions*, Cambridge University, Cambridge

[80] Royden H.L., (1963) *Real Analysis*, The Macmillan Company, New York, Collier-Macmillan ltd, London

[81] Rozovskii B.L. (1983) *Stochastic Evoluation Systems*, Kluwer Academic Publishers, Dordrecht, Boston, London

[82] Sell G.R. and You Y. (2002) *Dynamics of Evolutionary Equations*, Springer, New York, Berlin, Heidelberg, London

[83] Shinners S.M. (1978) *Modern Control System Theory and Application*, Addison-wesley Publishing Company, Reading, Menlo Park, London, Sydney

[84] Skorohod A.V. (1965) *Studies in the Theory of Random Processes*, Addison-wesley Publishing Company, Reading, Massachusetts

[85] Skowronski J.M. (1991) *Control of Nonlinear Mechanical Systems*, Plenum Press, New York, London

[86] Stallings W. (2004) *Computer networking with Internet protocols and technology*, Prentice Hall, Upper Saddle River, New Jersey

[87] Stroock D.W. and Varadhan S.R.S. (1979) *Multidimensional Diffusion Processes*, Springer-Verlag, New York, Heidelberg, Berlin

[88] Tanabe H. (1979) *Equations of Evolution*, Pitman, London, San Francisco, Melbourne

[89] Teo K.L., Goh C.J. and Wong K.H. (1991) *A Unified Computational Approach to Optimal Control Problems*, Longman Scientific & Technical, New York

[90] Thompson J.M.T. and Stewart H.B. (1986) *Nonlinear Dynamics and Chaos*, John Wiley and Sons, Chichester, New York, Brisbane, Toronto, singapore

[91] Udwadia F.E., Webber H.I. and Leitmann G. (2004) *Dynamical Systems and Control*, Chapman & Hall/CRC, Boca Raton, London, New York, Washington,D.C.

[92] Vainberg M.M. (1973) *Variational Method and Method of Mnotone Operators in the Theory of Nonlinear Equations*, (Eng. Trans. by A. Libin), John Wiley & sons, New York, London

[93] Voit J. (2001) *The statistical Mechanics of Financial Markets*, Springer, Berlin Heidelberg New York Barcelona Hong Kong London Milan Paris Singapore Tokyo

[94] Vrabie I.I. (1987) *Compactness Methods for Nonlinear Evolutions*, Longman Scientific & Technical, New York

[95] Warga J.(1972), *Optimal Control of Differential and Functional Equations*, Academic Press, New York San Francisco London

[96] Wonham W.M. (1970), *Random Differential Equations in Control Theory*, in Probabilistic Methods in Applied Mathematics, Vol II, (Ed. A.T.Bharucha Reid), Academic Press, New York London

[97] Yong J. and Zhou X.Y. (1999) *Stochastic Controls*, Springer-Verlag, New York Heidelberg Berlin Barcelona Hong Kong London Milan Paris Singapore Tokyo

[98] Yosida K. (1968) *Functional Analysis*, Springer-Verlag New York, Inc.

[99] Zabczyk J. (1992), *Mathematical Control Theory;An Introduction*, Birkhauser

[100] Zeidler E. (1985) *Nonlinear Functional Analysis and its Applications (II/B)*, Springer-Verlag, New York, Heidelberg, Berlin

Papers

[101] Ahmed N.U. (1986) *Existence of Optimal Controls for a Class of Systems Governed by Differential Inclusions on a Banach Space*, Journal of Optimization Theory and Applications(JOTA), 50(2), 213-237

[102] Ahmed N.U. and Li Cheng (2004) *Stochastic Models for Traffic Dynamics and Access Control Mechanism in Computer Network*, Electronic Modeling, T.26, No.4, 53-71

[103] Ahmed N.U., Dabbous T.E. and Wong H.W. (1987) *Gradient Method for Computing Optimal Controls for Stochastic Differential Equations*, Journal of Stochastic Analysis and Applications, Vol. 5, No.2, 121-150

[104] Ahmed N.U. and Mouadeb A.R. (1991) *Optimal Regulators for Linear Systems with No Control Cost*, Dynamics and Control, 1, 341-355

[105] Ahmed N.U. (2000) *Optimal Impulse Control for Impulsive Systems in Banach Spaces*, International Journal of Differential Equations and Applications, 1,1, 37-52.

[106] Ahmed N.U. (2001) *Some Remarks on the Dynamics of Impulsive Systems in Banach Spaces*, Dynamics of Continuous , Discrete and Impulsive Systems, Series A, 8, 261-274.

[107] Ahmed N.U. (2002) *Necessary conditions of Optimality for Impulsive Systems on Banach Spaces*, Nonlinear Analysis,, Nonlinear Analysis: Series A, 51, 409-424.

[108] Ahmed N.U. and Yan Hong (2004) *Access Control for MPEG Video Applications Using Neural Network and Simulated Annealing*, Mathematical Problems in Engineering, 3, 291-304.

[109] Ahmed N.U. and Li Cheng (2004) *Optimum Feedback Strategy for Access Control Mechanism Method Modelled as Stochastic Differential Equation in Computer Network*, MPE (Mathematical Problems in Engineering), 3, 263-276

[110] Ahmed N.U., Yan Hong and Barbosa L.Orozco (2004) *Performance Analysis of the Token Bucket Control Mechanism Subject to Stochastc Traffic*, DCDIS, 11, 363-391.

[111] Ahmed N.U., Wang Qun and Barbosa L. Orozco (2002) *Systems Approach to Modeling the Token Bucket Algorithm in Computer Networks*, MPE (Mathematical Problems in Engineering), 8(3), 265-279

[112] Ahmed N.U. (2001) *Systems Governed by Impulsive Differential Inclusions on Hilbert Space*, Nonlinear Analysis: Theory, Methods and Applications, 45, 693-706

[113] Ahmed N.U. (2000) *Vector Measures for Optimal Control of Impulsive Systems in Banach Spaces*, Nonlnear Functional Analysis and Applications (An International Mathematical Journal for Theory and Applications, Korea), 5, 2, 95-106

[114] Ahmed N.U. (2001) *State Dependent Vector Measures as Feedback Controls for Impulsive Systems in Banach Spaces*, Dynamics of Continuous, Discrete and Impulsive Systems,8, 251-161

[115] Ahmed N.U. and Rahim M.A. (2001) *Deterministic and Stochastic Dynamic Models for Demography*, Journal of Dynamic Systems and Applications, 10, 325-358

[116] Ahmed N.U. (2004) *Stability of Torsional and Vertical Motion of Suspension Bridges Subject to Stochastic Wind Forces*, Stability and Control: Theory, Methods and Applications, Vol. 22, 145-162

[117] Ahmed N.U. (2001) *Optimal Control of Infinite Dimensional Stochastic Systems via Generalized Solutions of HJB Equations*, Journal of Differential Inclusions, Control and Optimization, 21, 97-126

[118] Ahmed N.U. (2002) *Necessary Conditions of Optimality for Impulsive Systems on Banach Spaces*, Nonlinear Analysis: Series A, 51, 409-424

[119] Ahmed N.U. (2001) *Differential Inclusions on Banach Spaces with Nonlocal State Constraints*, Nonlnear Functional Analysis and Applications (An International Mathematical Journal for Theory and Applications, Korea), 6,3, 395-409

[120] Ahmed N.U. and Charalambos C.D. (2002) *Filtering for Linear Systems Driven by Fractional Brownian Motion*, SIAM Journal on Contr. and Optim. Vol.41, No.1, 313-330

[121] Ahmed N.U. and Ding X. (2001) *Controlled McKean-Vlasov Equations*, Communications in Applied Analysis, 5,2,183-206

[122] Ahmed N.U. (2001) *Measure Solutions Impulsive Evolutions Differential Inclusions and Optimal Control*, Nonlinear Analysis 47,13-23

[123] Ahmed N.U. (2001) *Stone-Cech Compactification with Applications to Evolution Equations on Banach Spaces*, Publicationes Mathematicae, Debrecen, Hungary, 59,3-4, 289-301

[124] Ahmed N.U. (2002) *Nonstandard Impulsive Evolution Equations in Banach Spaces*, J. of Nonlinear Functional Analysis, 7(3), 437-453

[125] Ahmed N.U. and Teo K.L. (2002) *Dynamic Models for Computer Communication Networks and Their Mathematical Analysis*, DCDIS (Dynamics of Continuous, Discrete and Impulsive Systems), series B, 9, 507-524

[126] Ahmed N.U. (2001) *Distributed Parameter Systems*, Encyclopedia of Physical Science and Technology, 3rd Edition, Volume 4, 561-587

[127] Ahmed N.U. (2002) *Impulsive Perturbation of C_0-Semigroups and Evolution Inclusions*, Nonlinear Functional Analysis and Applications, 7(4), 555-580

[128] Ahmed N.U. (2002) *Impulsive Perturbation of C_0 -Semigroups and Stochastic Evolution Inclusions*, Discussiones Mathematicae, Differential Inclusions, Control and Optimization, 22,125-149

[129] Ahmed N.U. (2003) *Existence of Optimal Controls for a General Class of Impulsive Systems on Banach Spaces*, SIAM Journal on Contrl. and Optim. Vol.42, No.2, 669-685

[130] Ahmed N.U. (2002) *Optimal Control of Impulsive Stochastic Evolution Inclusions*, Differential Inclusions, Control and Optimization, 22, 155-184

[131] Ahmed N.U. (2003) *Generalized Solutions of HJB Equations Applied to Stochastic Control on Hilbert Space*, Nonlinear Analysis, 54 495-523

[132] Ahmed N.U., Teo K.L. and Hou S.H. (2003) *Nonlinear Impulsive Systems on Infinite Dimensional Spaces*, Nonlinear Analysis, 54, 907-925

[133] Ahmed N.U. and Xiang X. (2003) *Optimal Relaxed Controls for Differential Inclusions on Banach Spaces*, Dynamic Systems and Applications, 12, 235-250

[134] Ahmed N.U. (2004) *Controllability of Evolution Equations and Inclusions Driven by Vector Measures*, Discussiones Mathethicae Differential Inclusions, Control and Optimization, 24, 49-72.

[135] Ahmed N.U. (2004) *Measure Solutions for Evolution Equations with Discontinous Vector Fields*, Nonlinear Funct. Anal. & Appl., Vol. 9, No.3, 467-484

[136] Ahmed N.U. (2004) *Optimal Control for Evolution Equations with Discontinuous Vector Fields*, DCDIS 11 (Series A), 105-118

[137] Ahmed N.U. (1974) *Existence of Optimal Controls for a Class of Hereditary Systems with Lagging Control*, Information and Control, Vol.26, No.2, 178-185

[138] Ahmed N.U. (1996) *Optimal Relaxed Controls for Infinite-Dimensional Stochastic Systems of Zakai Type*, SIAM J. Control and Optimization, Vol.34, No.5, 1592-1615

[139] Ahmed N.U. (1983) *Properties of Relaxed Trajectories for a Class of Nonlinear Evolution Equations on a Banach Space*, SIAM J. Control and Optimization, Vol.21, No.6, 953-967

[140] Ahmed N.U. and He Y.J. (2004) *Dynamic Model for Population Distribution and Optimum Immigration and Job Creation Policies*, Canadian Studies in Population,31(2),261-295

[141] Ahmed N.U. (2004) *Optimal Control of Systems Governed by Impusive Differential Inclusions*, Proceedings of Dynamic Systems & Applications, Vol.4, 1-11

[142] Ahmed N.U. and Bo Li (to appear) *Modelling and Optimization of Computer Network Traffic Controllers*, Mathematical Problems in Engineering

[143] Ahmed N.U. and Harbi H. (1998) *Mathematical Analysis of Dynamic Models of Suspension Bridge*, SIAM, Journal of Applied Mathematics, 58(3), 853-874

[144] Ahmed N.U. and Harbi H. (1998) *Torsional and Longitudinal Vibration of Suspension Bridge Subject to Aerodynamic Forces*, Mathematical Problems in Engineering, 3(1), 1-29

[145] Ahmed N.U. and Harbi H. (1998) *Stability of Suspension Bridge II: Aerodynamic vs Structural Damping*, In Proc. of SPIE, Smart Structures and Materials, Smart Systems for Bridges, Structures, and Highways, San Diego, California, 276-289

[146] Ahmed N.U. (2000) *A General Mathematical Framework for Stochastic Analysis of Suspension Bridge*, Nonlinear Analysis, Series B, 1, 451-483

[147] Ahmed N.U. and Song Hui (2005) *Real-time Feedback Control of Computer Networks Based on Predicted State Estimation*, Mathematical Problems in Engineering, 1, 7-32

[148] Ahmed N.U. and Li Cheng (2005) *Suboptimal Feedback Control of TCP Flows in Computer Network Using Random Early Discard (RED) Mechanism*, Mathematical Problems in Engineering, to appear

[149] Ahmed N.U. and Li P. (1991) *Quadratic Regulator Theory and Linear Filtering under System constraints*, Journal of Mathematical Control and Information, Vol.8, 93-107

[150] Ahmed N.U. and Mouadeb A. N. (1991) *Optimal Regulators for Linear Systems with no Control Cost*, Journal of Dynamics and Control, Vol. 1, 341-355

[151] Ahmed N.U. (2005) *Optimal Relaxed Controls for Systems Governed by Impulsive Differential Inclusions*, Nonlinear Funct. Anal. & Appl., 10(3),427-460

[152] Ahmed N.U., Dabbous T.E., Wong H.W. (1987) *Gradient Method for Computing Optimal Controls for Stochastic Differential Equations*, Journal of Stochastic Analysis and Applications, 5(2), 121-150

[153] Ahmed N.U. and Schenk K.F. (1978) *Optimal Availability of Maintainable Systems*, IEEE Trans. on Reliability, R-27,1, 41-45.

[154] Ahmed N.U. (1989) *Identification of Operators in Systems Governed by Evolution Equations on Banach Space*, Lect. Notes in Control and Information Sciences (Proc. of IFIP WG 7.2, Working Conference, Santiago de Compostela, Spain, July 6-9, 1987), Springer-Verlag, Berlin, Heidelberg, New York, London, Paris, Tokyo, 114, 73-83

[155] Ahmed, N.U., Dabbous, T.E. and Lee, Y.E. (1988) *Dynamic routing for computer queuing networks*, Int. J. Systems SCL, 19(6).

[156] Ahmed, N.U., Li, P. (1991) *Quadratic Regulator Theory and Linear Filtering Under System Constraints*, IMA Journal of Mathematical Control and Information,8(8),93-107

[157] Aikat Jay, Kaur Jasleen, Smith F.Donelson and Jeffay Kevin (2003) *Variability in TCP Roundtrip Times* IMC03, October 2729, Miami Beach, Florida, USA

[158] Akhmet, M.U.(2003) *On the General Problem of Stability for Impulsive Differential Equations*, J. Math. Anal. Appl. 288, 182-196

[159] Benchohra M., Henderson J., Ntouyas S. K. and Ouahab A. (2003) *Existence Results for Impulsive Semilinear Damped Differential Inclusions*, Electronic Journal of Qualitative Theory of Differential Equations, 11, 1-19

[160] Biswas S.K. and Ahmed N.U. (1986) *Stablization of a Class of Hybrid Systems Arising in Flexible Spacecraft*, Journal of Optimization Theory and Applications, 50(1), 83-108

[161] Biswas S.K. and Ahmed N.U. (2001) *Optimal Control of Flow-induced Vibration of Pipelines*, Dynamics and Control, 11(2), 187-201.

[162] Braden B. et al. (1998) *Recommendations on queue management and congestion avoidance in the Internet*, RFC 2309

[163] Butto, M., Cavallero, E. and Tonietti, A. (1991) *Effectiveness of the leaky bucket policing mechanism in ATM networks*, IEEE Journal on Selected Areas in Communications, 9(3)

[164] Chen Y.C. and Ahmed N.U. (1983) *An Application of Optimal Control Theory in Fisheries Management*, INT. J. Systems Sci., 14(4), 453-462

[165] Cnodder Stefaan De, Pauwels Kenny and Elloumi Omar (2000) *A Rate Based RED Mechanism*, The 10th International Workshop on Network and Operating System Support for Digital Audio and Video, June 26-28, Chapel Hill, North Carolina, USA

[166] Crovella M. and Bestavros A. (1996) *Self-similarity in World Wide Web traffic: evidence and possible causes*, IEEE/ACM Transactions on Networking, 5(6), 835-846

[167] Dai W. and Heyde C. C. (1996) *Itos formula with respect to fractional Brownian motion and its application*, Journal of Appl. Math. and Stoch. An., Vol.9, 439-448.

[168] Dal Maso G. and Rampazzo F. (1991),*On systems of Ordinary Differential Equat with Measures as Controls*,Differential and Integral Equations,4(4),739-765.

[169] Duncan T.E., Maslowski B. and Duncan B.P. (2005-) *Fractional Brownian Motion and Linear Stochastic Equations in Hilbert Space*, Stochastics Dynamics, to appear

[170] Feng W., Kandlur D., Saha D. and Shin K. (1999) *A self-configuring RED gateway*, Proceedings of INFOCOM '99, Eighteenth Annual Joint Conference of the IEEE Computer and Communications Societies, Vol. 3, 1320-1328.

[171] Floyd Sally and Paxson Vern (2001) *Difficulties in Simulating the Internet*, IEEE/ACM Transactions on Networking, February

[172] Floyd S. and Jacobson V. (1993) *Random early detection gateways for congetsion avoidance*, IEEE/ACM Transactions on Networking, 1: 397-413

[173] Heinanen, J. and Guerin, R. (1999) *A two rate three color marker*, RFC 2698

[174] Huang C.S. Wang S. and Teo K.L. (2004) *On Application of an Alternating Direction Method to Hamilton-Jacobi-Bellman Equations*, J. of Comput. Appl. Math., Vol. 166, 153-166

[175] Jennings L.S., Teo K.L. and Goh C.J., (1997) *MISER3.2 Optimal Control Software: Theory and User Manual*, Department of Mathematics, University of Western Australia, Australia

[176] Kalman RE, Ho Y, Narendra KS. (1963) *Controllability of linear dynamical systems*, Contributions to Differential Equations, 1(2), 189-213.

[177] Kaul S.K. and Liu X. (2000) *Generalized Variation of Parameters and Stability of Impulsive Systems*, Nonlinear Analysis, 40, 295-307

[178] Kaul S.K. and Liu X. (1999) *Vector Lyapunov Functions for Impulsive Differential Systems with Variable Times*, DCDIS,6,25-38

[179] Kholodnyi V.A (2005)*Universal Contingent Claims in a General Market Environment and Multiplicative Measures: Examples and Applications*, Nonlinear Analysis,62,1437-1452.

[180] Leland W., Taqqu M., Willinger W. and Wilson D. (1994) *On the Self-Similar Nature of Ethernet Traffic (Extended Version)*, IEEE/ACM Transactions on Networking, 2(1), 1-15, February

[181] Lim S.S. and Ahmed N.U. (1992) *Modeling and Control of Flexible Space Stations*, Dynamics and Control, 2, 5-33

[182] Liu X. and Shen J. (2000) *Razumikhin-type Theorems on Boundedness for Impulsive Functional Differential Equations*, Dynamic Systems & Appl., 9, 389-404

[183] Liu X. and Vatsala A.S. (1998) *Variation of Parameters in Terms of Lyapunov-like Functions and Stability of Perturbed Systems*, Nonlinear Studies, 5,47-58

[184] Liu B., Liu X. and Liao X. (2003) *Stability and Robustness of Quasilinear Impulsive hybrid Systems* , Journal of Mathematical analysis and Applications, (To appear)

[185] Liu Y., Ito S., Lee H.W.J. and Teo K.L. (2001) *Semi-Infinite Programming Approch to Continously Constrained LQ Optimal Control*

Problems, Journal of Optimization Theory and Applications, Vol. 108, 617-632

[186] Liu Y., Teo K.L., Jennings L.S. and Wang S. (1998) *On a Class of Optimal Control Problems with State Jumps*, Journal of Optimization Theory and Applications, Vol. 98, 65-82

[187] Mandelbrot B.B. & Van Ness J.W. (1968) *Fractional Brownian Motions, Fractional Noises and Applications*, SIAM Review, Vol.10, No.4,422-437

[188] Maso G.D. and Rampazzo F. (1991) *On Systems of Ordinary Differential Equations with Measures as Controls*, Differential and Integral Equations, 4(4), 739-765

[189] Misra Vishal, Gong Wei Bo and Towsley Don (2000) *Fluid-based analysis of a network of AQM routers supporting TCP flows with an application to RED*, ACM SIGCOMM Computer, Vol. 30, Issue 4, 151-160, October

[190] Nagai H. & Peng S. (2002) *Risk-sensitive Dynamic Portfolio Optimization with Partial Information on Infinite Time Horizon*, The Annals of Applied Probobability, Vol.12, No.1, 1-23

[191] Paxson V. and Floyd S. (1994) *Wide-area traffic: the failure of Poisson modeling*, IEEE/ACM Transactions on Networking 3, 226-244

[192] Rabitz H. and Shi S. (1991) *Optimal Control of Molecular Motion: Making Molecules Dance*, Advances in Molecular Vibrations and Collision Dynamics, 1A, 187-214

[193] Rehbock V., Teo K.L., Jennings L.S., and Lee H.W. (1999) *A Survey of the Control Parameterization Enhancing Transform for Constrained Optimal Control Problems*, in "Progress in Optimization: Contributions from Australia", edited by Eberhard, R. Hill, D. Ralph and B.M. Glover

[194] Sellami T. and Ahmed N.U. (1989) *An algorithm for computing time optimal controls with examples from robotics and ecology*, Math. Engng Ind., Vol 2, No 2, 113-125

[195] Shenker, S. and Wroclawski, J. (1997) *General Characterization Parameters for Integrated Service Network Elements*, RFC 2215

[196] Silva G.N. and Vinter R.B. (1996) *Measure Driven Differential Inclusions*, Journal of Mathematical Analysis and Applications, Vol.202, 727-746

[197] Subrahmanyam M.B. (1986) *A Computational Method for the Solution of Time-Optimal Control Problems by Newton's Method*, International Journal of Control, 44, 1233-1243

[198] Teo K.L., Ahmed N.U. and Fisher M.E. (1989) *Optimal Feedback Control for Linear Stochastic Systems Driven by Counting Processes*, Journal of Engineering Optimaztion, 36, 479-486

[199] Teo K.L., Jennings L.S., Lee H.W. and Rehbock V., (1999) *The Control Parameterization Enhancing Transform for Constrained Optimal Control Problems*, Journal of Australian Mathematicl Society, Series B Vol. 40, 314-335

[200] Wang W. Jennings L.S. and Teo K.L. (2003) *Numerical Solution of Hamilton-Jacobi-Bellman Equations by an Upwind Finite Volume Method*, Journal of Global Optimization, Vol. 27, 177-192

[201] Xiang X. Ahmed N.U. (1993) *Properties of Relaxed Trajectories of Evolution Equations and Optimal Control*, SIAM J. Control and Optimization, Vol.31, No.5, 1135-1142

[202] Ye Tao, Kalyanaraman Shivkumar (2002) *Adaptive Tuning of RED Using On-line Simulation*, GLOBECOM 2002 - IEEE Global Telecommunications Conference, No. 1, 2222-2226, November

Index